Progress in the Chemistry of Organic Natural Products

More information about this series at http://www.springer.com/series/10169

A. Douglas Kinghorn • Heinz Falk •
Simon Gibbons • Jun'ichi Kobayashi
Editors

Progress in the Chemistry of Organic Natural Products

Volume 105

With contributions by

H. Ashihara · A. Crozier · K. Mizuno · T. Yokota

C. Lavaud · G. Massiot

W. Robien

Springer

Editors
A. Douglas Kinghorn
Division of Medicinal Chemistry & Pharmacognosy, College of Pharmacy
The Ohio State University
Columbus, Ohio
USA

Heinz Falk
Institute of Organic Chemistry
Johannes Kepler University Linz
Linz, Austria

Simon Gibbons
Research Department of Pharmaceutical and Biological Chemistry
UCL School of Pharmacy
London, United Kingdom

Jun'ichi Kobayashi
Graduate School of Pharmaceutical Science
Hokkaido University
Sapporo, Japan

ISSN 2191-7043 ISSN 2192-4309 (electronic)
Progress in the Chemistry of Organic Natural Products
ISBN 978-3-319-49711-2 ISBN 978-3-319-49712-9 (eBook)
DOI 10.1007/978-3-319-49712-9

Library of Congress Control Number: 2017933078

Printed on acid-free paper

This Springer imprint is published by Springer Nature
The registered company is Springer International Publishing AG
The registered company address is: Gewerbestrasse 11, 6330 Cham, Switzerland

Contents

Xanthine Alkaloids: Occurrence, Biosynthesis, and Function in Plants

Hiroshi Ashihara, Kouichi Mizuno, Takao Yokota, and Alan Crozier

Contents

H. Ashihara (✉)
Department of Biology, Ochanomizu University, Tokyo 112-8610, Japan
e-mail: ashihara.hiroshi@ocha.ac.jp

K. Mizuno
Faculty of Bioresource of Science, Akita Prefectural University, Akita 010-0195, Japan
e-mail: koumno@akita-pu.ac.jp

T. Yokota
Department of Biosciences, Teikyo University, Utsunomiya 320-8551, Japan
e-mail: yokota@nasu.bio.teikyo-u.ac.jp

A. Crozier
Department of Nutrition, University of California, Davis, CA 951616, USA
e-mail: alan.crozier44@gmail.com

A.D. Kinghorn, H. Falk, S. Gibbons, J. Kobayashi (eds.), *Progress in the Chemistry of Organic Natural Products 105*, DOI 10.1007/978-3-319-49712-9_1

1 Introduction

Xanthine alkaloids, also known as purine alkaloids, consist of methylxanthines and methyluric acids and their structures are based on the xanthine (**1**) and uric acid (**9**) skeletons (Fig. 1). Caffeine (1,3,7-trimethylxanthine) (**8**) and theobromine (3,7-dimethylxanthine) (**5**) occur in plant species such as coffee (*Coffea arabica*), tea (*Camellia sinensis*), maté (*Ilex paraguariensis*), cacao (*Theobroma cacao*), and guaraná (*Paullinia cupana*), which are used for popular non-alcoholic beverages [1]. The isolation of **8** from coffee seeds was first reported independently in 1820 by the German researchers, Runge [2] and von Giese [3]. Caffeine (**8**) was found as "thein" in tea leaves by Oudry [4] in 1827. Subsequently, it was also detected in maté by Stenhouse [5] in 1843 and in kola nuts (*Cola acuminata*) by Daniell [6] in 1865. Theobromine (**5**) was discovered in cacao seeds by Woskresensky [7] in 1842. Paraxanthine (1,7-dimethylxanthine) (**7**) was isolated from human urine by Salomon [8] in 1883, but it was not detected in coffee seeds until 1980 by Chou and

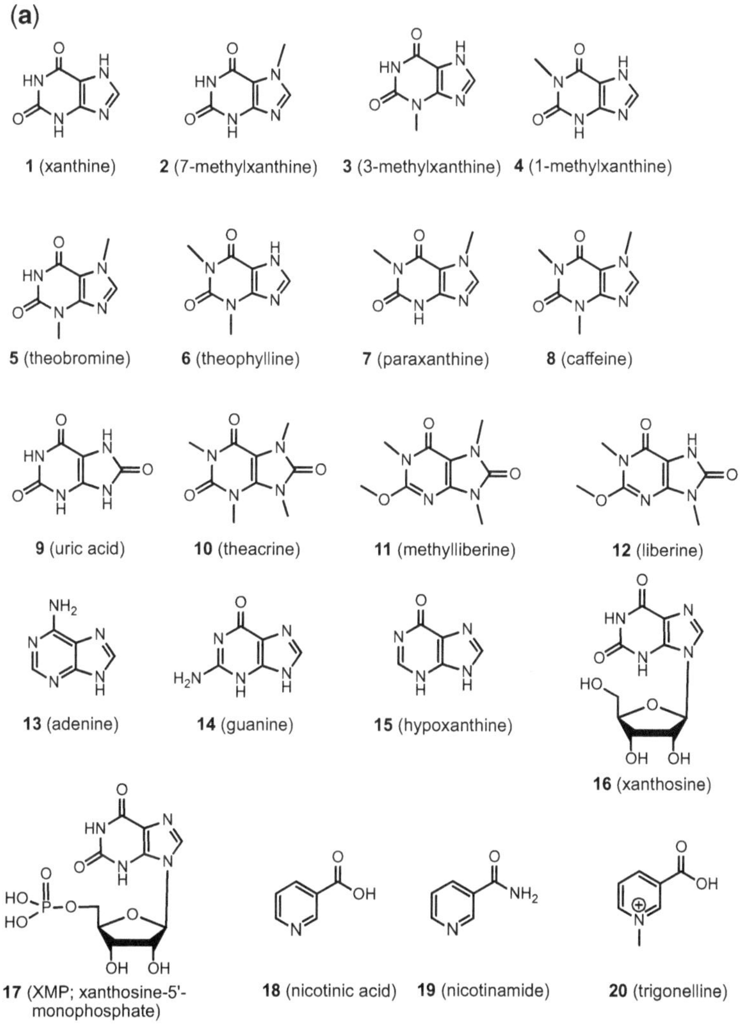

Fig. 1 (**a**) Structures of **1–20**. (**b**) Structures of **21–36**. (**c**) Structures of **37–51**. (**d**) Structures of **52–61**

Waller [9]. The complete chemical synthesis of **8** was reported by Fischer and Ach [10] in 1895.

Studies on caffeine biosynthesis were initiated in the 1960s, while highly purified caffeine synthase was isolated by Kato et al. [11] in 1999 and the gene encoding this enzyme was cloned in 2000 [12]. The major caffeine biosynthesis pathway, of xanthosine (**16**) → 7-methylxanthosine (**23**) → 7-methylxanthine (**2**) → theobromine (**5**) → caffeine (**8**), is now confirmed at both the biochemical and molecular level.

Initially, caffeine (**8**) was considered to be a waste end-product in plants, although it was also hypothesized that this purine alkaloid plays a role in chemical

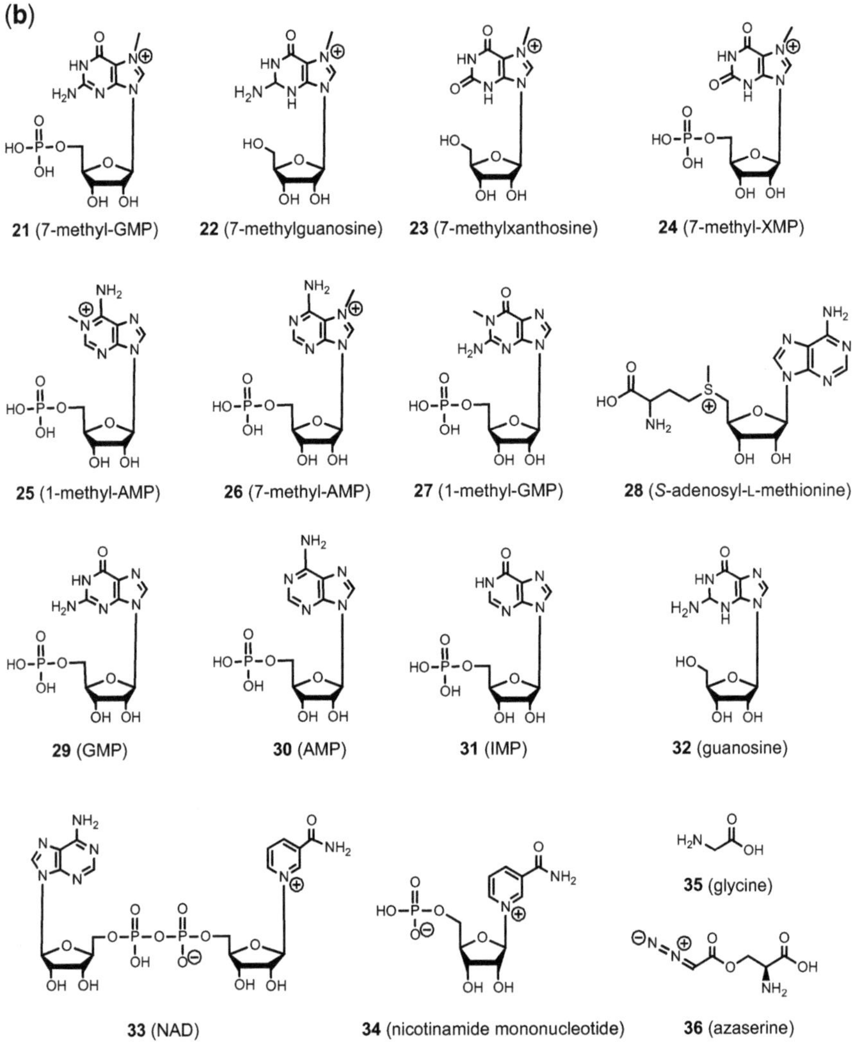

Fig. 1 (continued)

defenses and had an allelopathic function [13]. Recent research using transgenic plants has provided convincing evidence for the chemical defense theory [14].

A comprehensive review of xanthine alkaloid metabolism was published over 15 years ago by Ashihara and Crozier [13]. The present contribution will focus on findings published since 2000, while background information, such as general purine metabolism and various hypotheses on caffeine biosynthesis, which were covered in the initial review [13], will not be described here in detail. The physical and chemical properties of xanthine alkaloids, which are not a topic of this chapter, have been summarized by Tarka and Hurst [15].

(c)

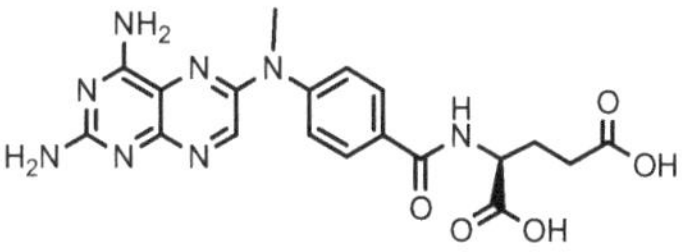

37 (aminopterin)

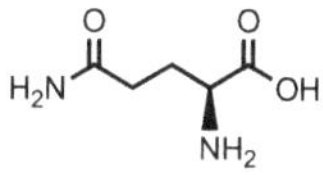

38 (glutamine)

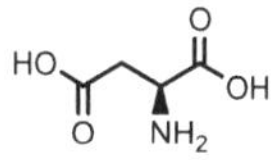

39 (aspartic acid)

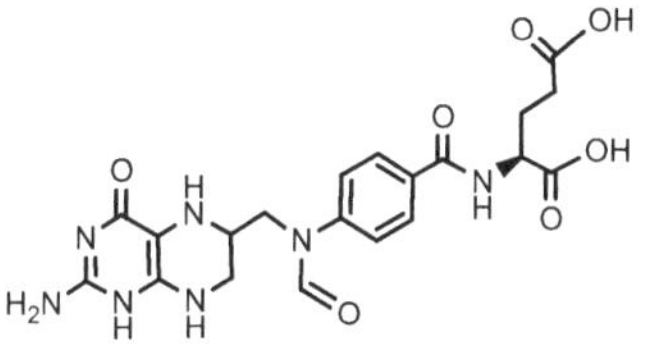

40 (10-formyltetrahydrofolate)

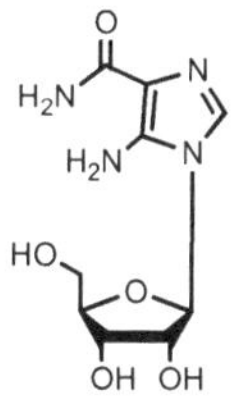

41 (5-aminoimidazole-4-carboxamide riboside)

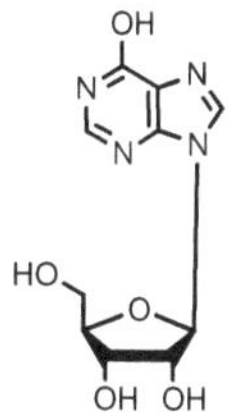

42 (inosine)

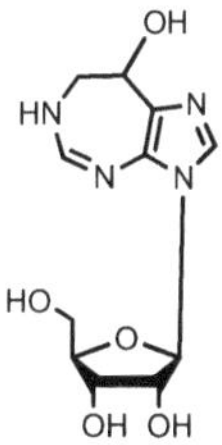

43 (coformycin)

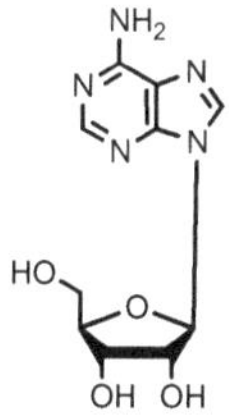

44 (adenosine)

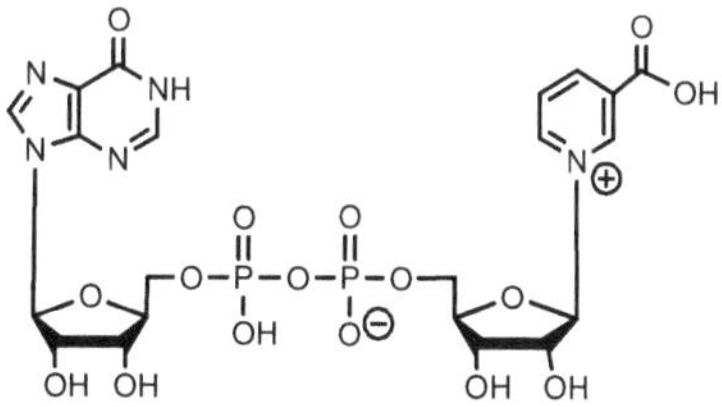

45 (nicotinic acid hypoxanthine dinucleotide)

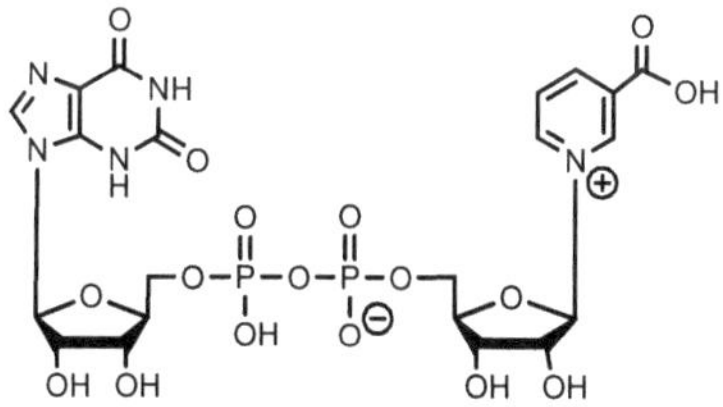

46 (nicotinic acid xanthine dinucleotide)

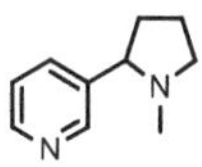

47 (nicotine)

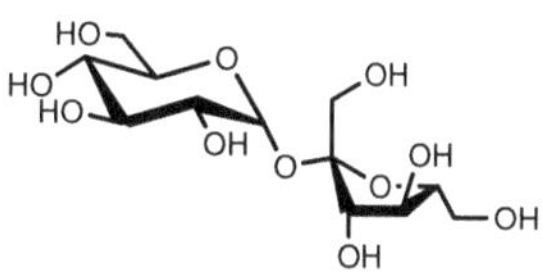

48 (sucrose)

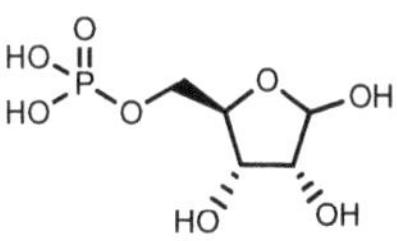

49 (ribose-5-phosphate)

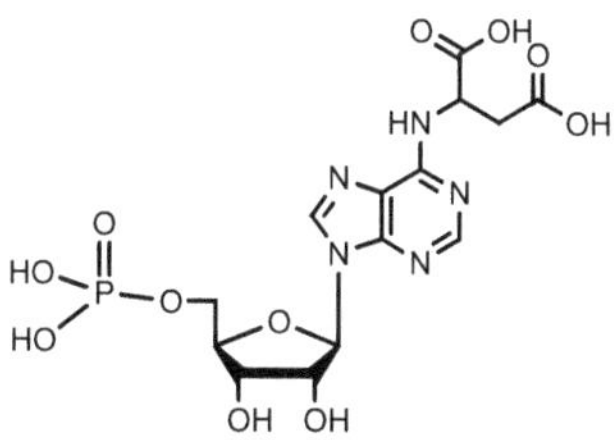

50 (adenylosuccinate)

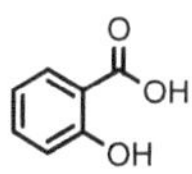

51 (salicylic acid)

Fig. 1 (continued)

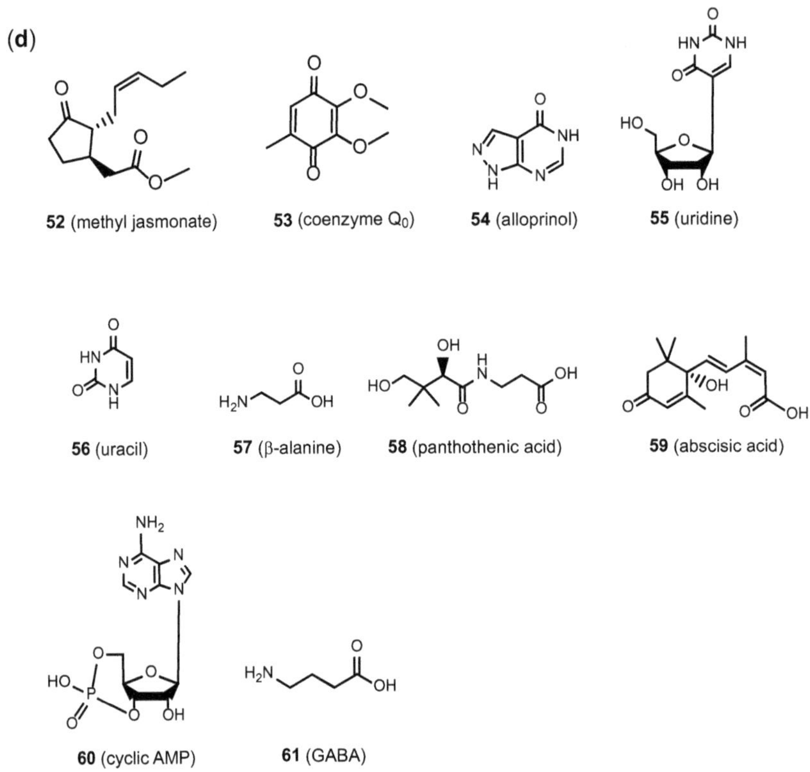

Fig. 1 (continued)

2 Occurrence of Xanthine Alkaloids in the Plant Kingdom

In our earlier review, it was noted that caffeine (**8**) had been detected in at least 80 species in 13 orders of the plant kingdom [13]. This was based mainly on a review by Kihlman [16], which quoted Willaman and Schubert [17] and O'Connell [18], and more recent information published by Stewart [19] and Kretschmar and Baumann [20]. Phylogenetic relationships of orders, which include xanthine alkaloid-accumulating plants, are shown in Fig. 2, and the order of the plant species is according to Smith et al. [21] for ferns and APG III for Angiospermae [22]. Typical species and their common names and the major xanthine alkaloids present are listed in Table 1.

2.1 Ericales (Tea and Related Species)

The occurrence of caffeine (**8**) in young leaves of various *Camellia* plants (Plate 1) was reported by Nagata and Sakai [23] and sizable amounts of **8** occur in species of the section Thea, namely, *C. sinensis* var. *sinensis* (2.8% dry weight (d.w.)),

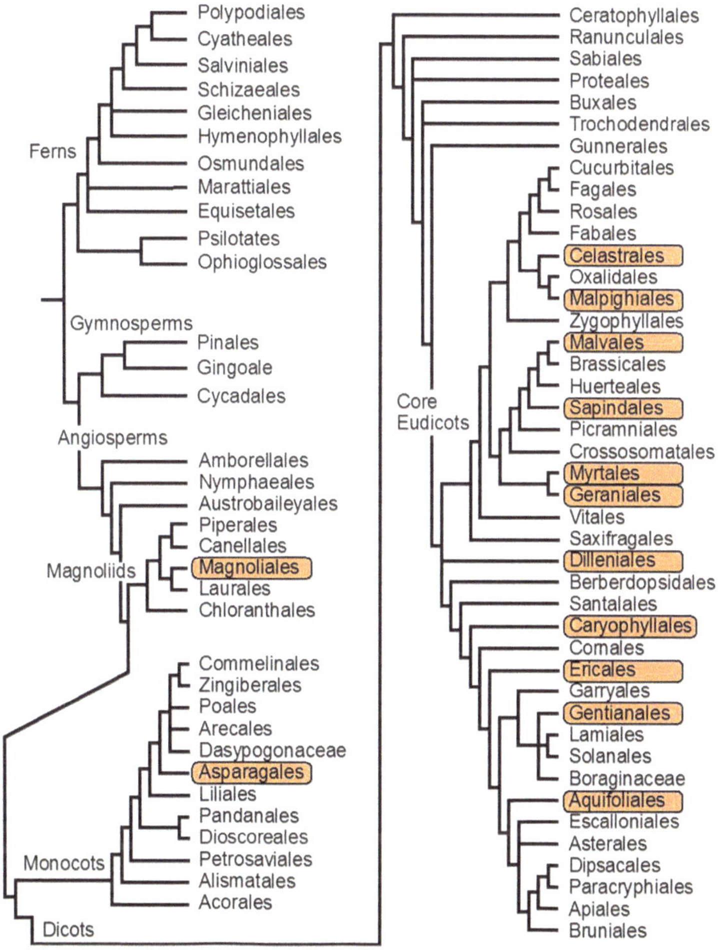

Fig. 2 Phylogenetic relationship of xanthine alkaloid-accumulating plant orders. The order of the plant species is according to Smith et al. [21] for ferns and APG III for Angiospermae [22]. The colored boxes show the orders that include a species accumulating caffeine and related xanthine alkaloids (see Table 1)

C. sinensis var. *assamica* (2.4%), and *C. taliensis* (2.5%). Caffeine (**8**) was not detected in species belonging to the section Paracamellia, *C. sasanqua* and *C. oleifera*, although small amounts (<0.02%) occur in *C. kissi*. Caffeine does not occur in *Camellia* species belonging to other sections, such as *C. japonica*. The hybrids *C. sinensis* x *C. japonica* and *C. sasanqua* x *C. sinensis*, contain 0.1–2.8% of **8**, indicating that the ability to synthesize the purine alkaloid is a dominant characteristic [24]. Nagata and Sakai [25] noted that theobromine (**5**), but not caffeine, accumulated in the young leaves of *C. irrawadiensis* (<0.8%). Ye et al. [26] reported that theobromine (**5**) is the predominant xanthine alkaloid in *C. ptilophylla* (5.0–6.8%). A small amount of theobromine is found invariably in young leaves of *C. sinensis*, where it is an intermediate in caffeine biosynthesis.

Table 1 Occurrence of xanthine alkaloids in the plant kingdom

Order	Typical species	Common name	Methylxanthine[a]
Magnoliales	*Annona cherimola*	Cherimoya	Cf
Asparagales	*Scilla maritima*	Red squill	Cf
Celastrales	*Maytenus* sp.		Cf
Malpighiales	*Banisteriopsis caapi*	Caapi	Cf
Malvales	*Cola acuminata*	Kola nut	Cf > Tb
	Herrania purpurea	Monkey cocoa	Tc
	Theobroma cacao	Cacao (cocoa)	Tb > Cf
Sapindales	*Citrus paradisi*	Grapefruit	Cf > Tp > Tb > Px
	Paullinia cupana	Guaraná	Cf > Tb > Tp
Myrtales	*Combretum jacquinii*	Red bushwillow	Cf
Geraniales	*Erodium cicutarium*	Pinweed	Cf
Dilleniales	*Davilla rugosa*	Fire vine	Cf
Caryophallales	*Cereus jamacaru*	Cactus	Cf
Ericales	*Camellia sinensis*	Tea	Cf
Gentinales	*Coffea arabica*	Coffee (arabica)	Cf
	Coffea canephora	Coffee (robusta)	Cf
Aquifoliales	*Ilex paraguariensis*	Maté	Cf
	Villaresia mucronata	Chilean citronella tree	Cf

Adapted from Kihlman [16] and Ashihara and Crozier [13]
[a]Tb theobromine (**5**), Tp theophylline (**6**), Px paraxanthine (**7**), Cf caffeine (**8**), Tc theacrine (**10**)

Plate 1 Tea (*Camellia sinensis*); photograph by H. Ashihara

In contrast, theophylline (**6**) and paraxanthine (**7**) do not accumulate in detectable quantities in tea leaves. Johnson [27] reported that leaves of *C. sinensis* contain small amounts of theacrine (1,3,7,9-tetramethyluric acid (**10**)) while Ye et al. found sizable amounts of **10** in *C. assamica* cv. *kucha* (kucha) [28]. The content of **10** in young leaves of kucha was 2.8% of d.w. and comprised 75% of the total xanthine alkaloids [29].

Recently, Deng, using HPLC with photodiode array detection, screened the purine alkaloid content of the leaves of 28 plants in the Theaceae and Pentaphyllaceae families grown at the Koishikawa Botanical Gardens of the University of Tokyo [30]. Detectable amounts of caffeine (**8**) were found in 17 samples, but substantial quantities were found in only five species: namely, *C. sinensis* var. *assamica* (28 μmol/g fresh weight [f.w.]) > *C. sinensis* var. *sinensis* (20 μmol/g) > *C. sinensis* f. *rosea* (12 μmol/g) > *C. chekiangoleosa* (10 μmol/g) > *C. japonica* var. *macrocarpa masam* (3.5 μmol/g). The quantities detected in other species were very low (0.1–0.6 μmol/g). Small amounts of theophylline (**6**) were detected in five species: *C. japonica* (2.9 μmol/g) > *C. sinensis* var. *sinensis* (1.8 μmol/g) > *Eurya osimensis* (0.6 μmol/g) > *Schima mertensiana* (0.2 μmol/g) > *E. japonica* (0.1 μmol/g). Theobromine (**5**) was found only in *C. sinensis* var. *sinensis* (0.4 μmol/g). These estimates need to be confirmed by HPLC-MS and further insights into purine alkaloid biosynthesis are required, such as the occurrence of theobromine synthase genes in *C. japonica* [31] and theophylline synthesis from 8-^{14}C-labeled adenine (**13**) in *E. japonica* and *S. mertensiana* [32].

Ashihara and Kubota [33] reported that more than 99% of the total caffeine (**8**) in tea seedlings is found in the leaves, and only a small amount occurred in other parts of the plant (Table 2). The concentration of **8** was higher in mature than in younger

Table 2 Distribution of caffeine (**8**) and theobromine (**5**) in 4-month-old seedlings of *Camellia sinensis*

	Caffeine (**8**)		Theobromine (**5**)	
Sample	(μg/organ)	(μg/g f.w.)	(μg/organ)	(μg/g f.w.)
Leaf 1[a]	1.19	24.0	0.50	10.1
2	3.73	19.9	1.11	6.0
3	4.34	19.5	0.91	4.1
4	4.94	33.5	1.21	8.2
5	2.88	39.8	tr	–
6	1.83	26.3	tr	–
7	2.60	50.9	nd	–
8	1.41	52.8	nd	–
9	0.38	86.4	nd	–
Stem 1[a]	0.04	4.9	nd	–
2	0.07	3.4	nd	–
3	0.04	0.5	nd	–
Cotyledons	0.02	0.0	nd	–
Roots	0.03	0.1	nd	–

Adapted from Ashihara and Kubota [33]

[a]Numbered from shoot apex to the base of the seedling. tr = trace amounts; nd = not detected

leaves, while theobromine was detected only in younger leaves. In tea seeds, nearly all of **8** was located in the seed coat at a concentration of ~5 μmol/g f.w.

In addition to these xanthine alkaloids, mono-methylxanthines, which are intermediates of the biosynthesis and catabolism of **8**, have been detected in *C. sinensis* leaves [13]. Young tea leaves contain 7-methylxanthosine (**21**) (179 nmol/g f.w.), 7-methylxanthine (**2**) (67 nmol/g), 3-methylxanthine (**3**) (5 nmol/g) and 1-methylxanthine (**4**) (2 nmol/g). In mature leaves, **21** and **2** levels dropped to 23 and 16 nmol/g, respectively, while the content of **3** and **4** content increased to 38 and 14 nmol/g, respectively [13].

The anatomical localization of caffeine (**8**) within young leaves of *Camellia sinensis* has been investigated using immunohistochemical methods and confocal scanning laser microscopy [34]. The results demonstrated that caffeine is localized within vascular bundles, mainly in the precursor phloem. Palisade and spongy parenchyma also contain caffeine but in much lower concentrations, undetectable by immune-labeling and confocal scanning microscopy analysis. However, this could be misleading as the area of palisade and spongy parenchyma was much larger than that of the vascular tissues. Thus, more caffeine might be distributed within these tissues as opposed to the much smaller vascular material where it is concentrated, and as a consequence gives a higher fluorescence signal.

2.2 Gentianales (Coffee and Related Species)

Two *Coffea* species, *Coffea arabica* (Plate 2) and *Coffea canephora*, used extensively to make the beverage coffee, are known, respectively, as Arabica coffee and Robusta coffee. Arabica coffee is the major coffee produced in Brazil, Colombia, Mexico, Ethiopia, and Guatemala while Robusta coffee is produced mainly in Vietnam, Indonesia, and the Ivory Coast [35]. Caffeine (**8**) contents in different

Plate 2 Coffee (*Coffea arabica*); photograph by H. Ashihara

varieties of *C. arabica* and *C. canephora* and other *Coffea* species have been surveyed by Mazzafera and Carvalho [36] and Anthony et al. [37]. Although small differences were found between the data obtained by the two groups, the content of **8** in *C. canephora* (2–3% d.w.) is generally higher than in *C. arabica* (1–2%). Silvarolla et al. [38] discovered three low-caffeine *C. arabica* plants, designated as AC1, AC2, and AC3, the seeds of which contained only 0.08% of the purine alkaloid.

Examples of levels of **8** in different *Coffea* species are summarized in Table 3. Substantial amounts were detected in the seeds of *C. liberica* var. *liberica* (~2% d. w.) and *C. racemosa* (~1%) with lower amounts in other *Coffea* species. In addition to the methylxanthines, mature leaves of *C. liberica*, *C. dewevrei*, and *C. abeokutae*

Table 3 Caffeine (**8**) content in *Coffea* species

Species and cultivar or variety	Source	Caffeine content/%[a]
C. arabica cv. *typica*	Tanzania	1.67
C. arabica cv. *purpurescens*	Tanzania	0.76
C. arabica cv. *laurina*	Tanzania	0.79
C. arabica cv. *erecta*	Tanzania	1.49
C. arabica cv. *murta*	Tanzania	1.75
C. arabica cv. Bourbon amarelo	Brazil	1.54
C. arabica cv. Bourbon vermelho	Brazil	1.26
C. arabica cv. Mokka	Tanzania	0.87
C. arabica cv. Maragogype	Kew	1.39
C. arabica cv. SanRamon	Tanzania	1.44
C. arabica cv. Caturra	Tanzania	1.59
C. arabica cv. Mundo novo	Brazil	1.57
C. arabica cv. Catuai vermelho	Brazil	1.53
C. arabica cv. AC1	Ethiopia	0.08
C. canephora cv. Ugandae	Tanzania	3.02
C. canephora cv. Robusta	Brazil	2.48
C. canephora cv. Kouillou	Tanzania	2.22
C. canephora cv. Caféier de la Nana	Cote-d'Ivoire	1.98
C. eugenioides	Brazil	0.55
C. farafanganensis	Madagascar	0.09
C. homollei	Madagascar	0.06
C. liberica var. *dewevrei*	Cote-d'Ivoire	1.97
C. perrieri	Madagascar	0.00
C. pseudozanguebariae	Kew	0.02
C. racemosa	Brazil	1.16
C. salvatrix	Brazil	0.28
C. sessiliflora	Kew	0.46
C. kapakata	Brazil	0.72

Based on the data of Anthony et al. [37] and Mazzafera and Carvalho [36]
[a]Values are expressed as % of d.w.

contained the methyluric acids, theacrine (**10**), liberine (*O*(2)1,9-trimethyluric acid) (**12**) and methylliberine (*O*(2)1,7,9-tetramethyluric acid) (**11**) [39, 40].

Caffeine (**8**) is distributed mainly in leaves and cotyledons of coffee seedlings but small amounts are also found in the stems and roots (Table 4) [41]. Young expanding leaves of *C. arabica* also contain theobromine (**5**) [42, 43]. The theobromine content is slightly higher than caffeine in the young small-sized developing coffee leaves [43], but the caffeine content becomes higher than that of theobromine during leaf development [42, 43]. Keller et al. [44] reported that caffeine in the pericarp fell from 2 to 0.2% d.w., while it remained constant at 1.3% during the development of seeds in *C. arabica* fruits. Koshiro et al. [45] monitored changes in caffeine content in the pericarp and seeds of *C. arabica* and *C. canephora* during ripening of fruits (Table 4). In ripened fruits, 80% and 89% of total caffeine of *C. arabica* and *C. canephora*, respectively, was found in the seeds and the remainder was in the pericarp. Caffeine (**8**) accumulated in the seeds gradually, but a similar build up not was found in the pericarp.

Table 4 Distribution of caffeine (**8**) in seedlings and fruits of coffee plants

Sample	Caffeine (**8**) (μg/organ)	(μg/g f.w.)
(1) *C. arabica* seedlings		
Leaves (upper)	6.6	55.8
Leaves (middle)	7.7	47.2
Leaves (lower)	6.8	42.6
Cotyledons	30.0	103.6
Stems (upper)	0.16	6.2
Stems (lower)	0.11	0.9
Roots	0.19	1.1
(2) *C. arabica* fruits		
Seeds (green)	11.7	22.9
Seeds (pink)	15.2	26.2
Seeds (red)	18.0	26.5
Pericarp (green)	1.8	7.3
Pericarp (pink)	4.4	9.5
Pericarp (red)	4.6	8.2
(3) *C. canephora* fruits		
Seeds (green)	22.6	54.7
Seeds (pink)	21.3	59.0
Seeds (red)	28.3	49.8
Pericarp (green)	4.2	16.7
Pericarp (pink)	8.3	19.9
Pericarp (red)	3.4	6.1

Adapted from Zheng and Ashihara [41] and Koshiro et al. [45]

2.3 Aquifoliales (Maté and Related Species)

Since the young leaves of maté (*Ilex paraguariensis*) (Plate 3) are used to produce the beverage maté tea, xanthine alkaloids in the tea products have been analyzed. For example, Clifford and Ramirez-Martinez [46] reported that the concentrations of caffeine (**8**) and theobromine (**5**) in dried maté leaves were 0.9–1.7% and 0.5–0.9% d.w. The presence of **5** in various *Ilex* species was reported by Bohinc and coworkers [47]. They detected **5** in the following *Ilex* species: *I. aquifolium*, *I. crenatu*, *I. caroliunu*, *I. cassine*, *I. perudo*, and *I. umbiguu*. Filip et al. [48] reported that **5**, but not **8**, was present in *I. argentina*. Recently, Edwards and Bennett examined the diversity of methylxanthine levels in *I. cassine* and *I. vomitoria* [49]. The caffeine (**8**) and theobromine (**5**) contents were 0.12% and 0.22% in *I. cassine* and 0.56% and 0.11% in *I. vomitoria*.

The distribution of xanthine alkaloids within maté plants was reported by Mazzafera [50]. Caffeine (**8**) was found in the developing and old leaves, bark, wood, and in immature, and mature fruits. Smaller amounts of theobromine (**5**) and theophylline (**6**) were present, but **5** was not detected in mature fruits, and **6** did not accumulate in the fruits, bark, and wood. The concentrations of xanthine alkaloids in young leaves and immature fruits were higher than that occurring in old leaves and mature fruits (Fig. 3). Young and old leaves collected from branches containing fruits showed significantly lower amounts of theobromine and caffeine than those from the branches not bearing fruits.

Plate 3 Maté (*Ilex paraguariensis*); photograph from Wikimedia Commons, L. Kibisz

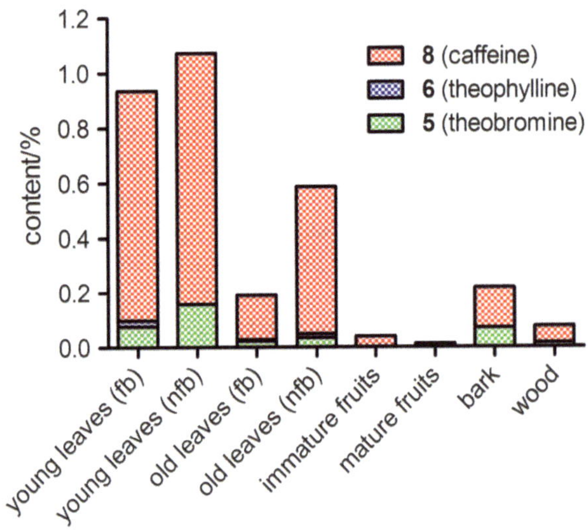

Fig. 3 Occurrence of xanthine alkaloids in different organs of maté. Xanthine alkaloid contents are shown as % of dry weight. Leaves were collected from branches bearing fruits (fb) and from branches not bearing fruits (nfb). Based on the data of Mazzafera [50]

2.4 Malvales (Cacao, Cola, and Related Species)

Seeds of cacao (*Theobroma cacao*) (Plate 4) are used in the manufacture of cocoa and chocolate products [51]. Three major cacao varieties, Forastero, Criollo, and Trinitario, are grown commercially [52]. Theobromine (**5**) is the dominant purine alkaloid in seeds of cacao. Timbie et al. [53] analyzed seeds of different cacao varieties and the total xanthine alkaloid content ranged from 2.4 to 5.0% of defatted cocoa with an average of 3.7% for ten samples. Theobromine (**5**) consisted of 52–99% of total xanthine alkaloids averaging 87%. Hammerstone et al. [54] examined the xanthine alkaloid content of several genotypes representing the three horticultural races of *T. cacao*. Theobromine (**5**) and caffeine (**8**) concentrations were varied as follows: Criollo type (1.5–3.3% theobromine, 0.3–1.3% of defatted seeds), Forastero type (0.9–1.2% theobromine, 0.08–1.2% of defatted seeds), and Trinitario type (0.9–1.2% theobromine, 0.3–0.6% of defatted seeds).

Theacrine (**10**) is a major xanthine alkaloid in seeds of many *Theobroma* species except for cacao: *T. speciosum* (0.12% of defatted seed), *T. microcarpum* (0.02%), *T. angustifolium* (0.06%), *T. grandiforum* (cupuassu, 0.26%), *T. subincanaum* (0.28%), and *T. mammosum* (0.06%). Immature seeds of *T. bicolor* (jaguar tree) accumulate **5** (0.26%), but this is replaced by **10** in mature seeds (0.35%) [54]. *T. bicolor* has been one of the most useful species, aside from *T. cacao*, as since prehistoric times, it has been used as a food [55].

Herrania species, belonging to the Sterculioideae, a subfamily of the Malvaceae, are morphologically similar to *Theobroma* plants. Hammerstone et al. [54] reported that nine species of *Herrania* contained theacrine (0.23–2.0%). The species with a high concentration of theacrine (**10**) were *H. purpurea* (2.0%),

Plate 4 Cacao (*Theobroma cacao*); photograph by H. Ashihara

Table 5 Changes in methylxanthine concentrations during the development of cacao leaves[a]

Methylxanthine	Stage I	Stage II	Stage III	Stage IV
1-Methylxanthine (**4**)	nd	25	8	3
3-Methylxanthine (**3**)	26	23	2	2
7-Methylxanthosine (**23**)	9	nd	nd	nd
7-Methylxanthine (**1**)	114	15	9	5
Theophylline (**6**)	2	1	6	12
Theobromine (**5**)	4530	112	33	77
Paraxanthine (**7**)	nd	nd	nd	nd
Caffeine (**8**)	754	122	36	10

Adapted from Koyama et al. [56]
[a]Data expressed as nmol/g fresh weight; nd = not detected

H. cuatrecasana (1.9%), *H. balaensis* (1.8%), and *H. columbia* (1.7%). Trace amounts of caffeine were found in some species, but no theobromine was detected in any *Herrania* species.

Changes in the levels of theobromine (**5**) and caffeine (**8**) during the development of leaves of cacao have been reported by Hammerstone et al. [54] and Koyama et al. [56]. The results of Hammerstone et al. [54] indicated that the content of **5** was highest in the young leaves and it decreased gradually with age. In contrast, the content of **8** remained almost constant, but both of the purine alkaloids disappeared in the oldest leaves. Koyama et al. [56] surveyed several methylxanthines, as well as **5** and **8**, in leaves of four different developmental stages: young small red leaves (stage I), developing pale green intermediate size leaves (stage II), fully developed green leaves (stage III) from flush shoots, and aged leaves (stage IV) from 1-year-old shoots (Table 5). A high concentration of xanthine alkaloids was found in young leaves (stage I), in which the major alkaloid was theobromine (4.5 μmol/g f.w.) and the next was caffeine (0.75 μmol/g). In addition to these major alkaloids, small amounts of 1-methylxanthine (**4**), 3-methylxanthine (**3**), 7-methylxanthine (**2**), 7-methylxanthosine (**23**), and theophylline (**6**) were also present although paraxanthine (**7**) did not accumulate in detectable quantities. The concentrations and the absolute amounts of these purine alkaloids decreased markedly with leaf development. Almost 75% of purine alkaloids present in the stage I leaves disappeared during leaf development. The theobromine (**5**) and caffeine (**8**) concentrations in young leaves (stage I) shown in Table 5 are, respectively, equivalent to 0.037 and 0.007% d.w. Consistent with this observation, Gurney et al. [57] reported that small amounts of **5** (0.051%) and trace amounts of **8** were present only in very young red flush leaves of cacao plants, and that levels declined by ~90% when the leaves became green. Thus, small amounts of theobromine and caffeine appear in young cacao leaves, but their levels fall substantially with leaf development. The trends are, therefore, different from those occurring in tea and coffee leaves.

In contrast to the leaves, the seeds of cacao accumulate xanthine alkaloids [52, 54, 57–61]. Zheng et al. [61] compared the xanthine alkaloid contents in cacao fruits at different developmental stages: stage A (young small size fruits, ~2 g f.w.), stage B (medium size fruits, ~100 g f.w.), and stage C (~500 g f.w.). As samples, whole fruits (stage A), pericarp and seed tissue (mainly endosperm) (stage B), and pericarp, placenta, embryo (mainly cotyledons), and seed coat (stage C) were used (Table 6). In stage A fruits, theobromine (**5**) was the major purine alkaloid, followed by caffeine (**8**). In stage B fruits, **5** was the predominant xanthine alkaloid in the pericarp but **8** comprised 74% of the xanthine alkaloids in the seeds. In stage C fruits, the highest concentration of purine alkaloids was found in the seeds (cotyledons and an embryonic axis), in which the major component was **5** (22 μmol/g fresh weight) and the next was **8** (4.9 μmol/g). Compared with the seeds, the total alkaloid content of the pericarp was very low (0.05 μmol/g); the major components were 3-methylxanthine (**3**) (44%), **8** (38%) and **5** (15%). On a dry weight basis, the theobromine content in

Table 6 Xanthine alkaloid concentration in cacao fruits at different stages of growth[a]

	Stage						
	A	B			C		
Sample	Whole fruits	Pericarp	Seed	Pericarp	Seed coat	Cotyledon	Placenta
1-Methylxanthine (**4**)	nd	nd	nd	nd	4	21	nd
3-Methylxanthine (**3**)	1	nd	nd	23	5	25	nd
7-Methylxanthosine (**23**)	nd	nd	nd	nd	nd	12	nd
7-Methylxanthine (**2**)	1	2	nd	1	6	23	nd
Theophylline (**6**)	2	2	18	nd	2	nd	11
Paraxanthine (**7**)	1	nd	nd	nd	3	2	nd
Theobromine (**5**)	671	206	10	8	779	21,900	238
Caffeine (**8**)	89	80	82	20	161	4860	87

Adapted from Zheng et al. [61]
[a]Data are expressed as nmol/g fresh weight; nd = not detected

Plate 5 Kola nut (*Cola acuminata*); photograph from Wikimedia Commons, D. Culbert

seeds of stage C fruits was equivalent to 0.62%. This value is comparable to the 0.43–0.89% content of **5** of maturing cacao seeds 125 days after pollination reported by Bucheli et al. [60]. As well as these major alkaloids, small amounts of 1-, 3-, and 7-methylxanthines (**4**, **3**, **2**), 7-methylxanthosine (**23**), theophylline (**6**), and paraxanthine (**7**) were detected (Table 6).

Cola is a tropical African genus that belongs to the Sterculiaceae. The genus comprises about 140 species and the most commonly consumed are *Cola acuminata* (Plate 5), *C. nitida*, and *C. anomala*. Belliardo et al. [62] reported that caffeine (**8**) occurred in seeds of *Cola* spp. at 1.4–2.2% d.w. Recently, a more detailed investigation was performed by Niemenak et al. [63]. The respective mean contents of **8** and theobromine (**5**) were 1.1% and 0.04% in *C. acuminata*, 1.4% and 0.05% in *C. nitida*, and 0.7% and 0.06% in *C. anomala*.

2.5 Sapindales (Guaraná, Citrus, and Related Plants)

Sapindales, *Paullinia* (Sapindaceae) and *Citrus* (Rutaceae) species, contain xanthine alkaloids. Seeds of guaraná (*Paullinia cupana* var. *sorbilis*), which are utilized commercially in South America, contain high concentrations (2.5–6.5% d.w.) of caffeine (**8**) with smaller quantities of theobromine (**5**) and theophylline (**6**)

Plate 6 Guaraná (*Paullinia cupana*); photograph courtesy of Flávia Camila Schimpl

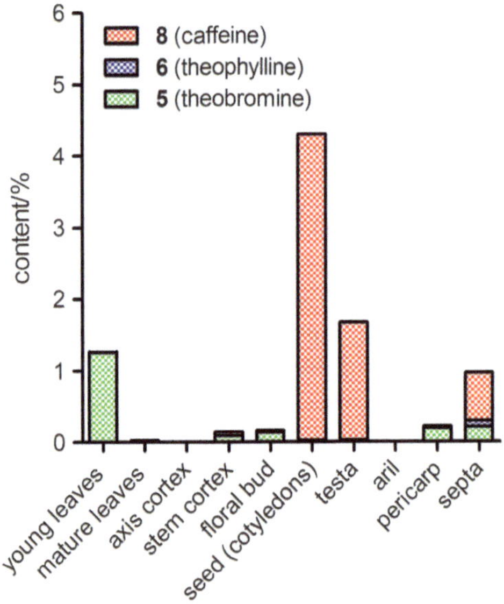

Fig. 4 Occurrence of xanthine alkaloids in different organs of guaraná. Xanthine alkaloid contents are shown as % of dry weight. Based on the data of Weckerle et al. [65]

[64]. Weckerle et al. [65] screened 34 species of *Paullinia* and related genera and found only three species, *P. cupana* (Plate 6), *P. yoco*, and *P. pachycarpa* to contain xanthine alkaloids. In *P. cupana* seeds (cotyledons), **8** (4.3% d.w.) was the major alkaloid and very small amounts of **5** (0.015%) and **6** (0.007%) were also detected (Fig. 4). Caffeine (**8**) also occurs in the testa and septa while, in contrast, **5** was detected mainly in the leaves, stem, and pericarp (Fig. 4) [65, 66].

In *P. yoco*, caffeine is found principally in the stem cortex (0.45% d.w.), stem wood (0.28%), and floral buds (0.19%). Smaller amounts of theobromine (**5**) (0.03–0.05%) accumulate in the mature leaves, axis cortex, and floral buds, while theophylline (**6**) has not been detected in any part of this plant. *P. pachycarpa* accumulated only small amounts of **5** (<0.03%) in the leaves and stems, and no xanthine alkaloids were detected in the fruits [65].

Stewart [19] found low a concentration of **8** in the flower buds of eight cultivars of citrus and in the leaves of Valencia orange (*Citrus sinensis*). The caffeine content of flower buds varied from 19–50 μg/g d.w., and the maximum value was obtained from Lisbon lemon flowers. The content in leaves was 6 μg/g d.w.

Detailed studies of the distribution of xanthine alkaloids in citrus flowers were reported by Kretschmar and Baumann [20]. They detected substantial quantities of caffeine and theophylline (**6**) and trace amounts of theobromine and paraxanthine (**7**) in the flowers of *Citrus paradisi* (grapefruit) (Plate 7), *C. maxima* (shaddock), *C. limon* (lemon), and *C. trifoliata* (syn. *Poncirus trifoliata*, trifoliate orange). Caffeine (**8**) was undetectable in the small buds but it appeared at anthesis. Xanthine alkaloids, mainly **8**, were located in the stamens. For example, a high level of xanthine alkaloids (96.5%) was found in the stamens of *C. limon* flowers and the remainder occurred in the petals (2.5%) and pistils (1.0%). Concentrations of **8** and **5** expressed on a fresh weight basis in the stamens of the following *Citrus* species were as follows: *C. paradisi* (3.2 and 0.2 μmol/g), *C. maxima* (1.1 and 0.3 μmol/g), *C. trifoliata* (0.8 and 0.1 μmol/g), and *C. limon* (1.4 and 0.1 μmol/g). In some samples, stamens were separated into the filament and anthers and this revealed that caffeine and theophylline were located mainly in the anthers, at high concentrations (10.1 and 10.7 μmol/g f.w.) in *C. paradisi* and *C. maxima*, respectively. These values corresponded to 0.7–0.8% d.w. Xanthine alkaloids in pollen were analyzed in *C. medica* (citron). Caffeine (6.9 μmol/g fresh weight) and

Plate 7 Grapefruit (*Citrus paradisi*); photograph from Wikimedia Commons, F. and K. Starr

theophylline (1.9 μmol/g fresh weight) were found in the microspores and small amounts (<0.6 $\mu mol/cm^{-3}$) of xanthine alkaloids, mainly caffeine, were also detected in the nectar of *Citrus* flowers.

2.6 Other Species

In addition to the species involved in the five orders of plants mentioned above, older literature [16] has reported the occurrence of caffeine in plants involved in eight other orders.

The species and family names are as follows:

Magnoliales:	*Annona cherimola* (Annonaceae)
Asparagales:	*Scilla maritima* (syn. *Drimia maritima*) (Asparagaceae)
Celastrales:	*Maytenus* sp. (Celastraceae)
Malpighiales:	*Banisteriopsis caapi* (Malpighiaceae), *Turnera ulmifolia* (Passifloraceae), *Populus alba* (Salicaceae)
Myrtales:	*Combretum jacquinii*, *C. loeflingii* (Combretaceae)
Geraniales:	*Erodium cicutarium* (Geraniaceae)
Dilleniales:	*Davilla rugosa* (Dilleniaceae)
Caryophallales:	*Cereus jamacaru*, *Harrisia adscendens*, *Leocereus bahiensis*, *Pilocereus gounellei*, *Trichocereus* sp. (Cactaceae)

The occurrence of xanthine alkaloids has been investigated mainly in plants of economic significance [67] and medicinal plants grown on the American continent [16]. Therefore, their actual occurrence in the plant kingdom may well be more diverse. On the other hand, most of the identifications and quantifications reported in the older literature were based on paper chromatography and colorimetry-based techniques, and have not been confirmed with more modern analytical methodology.

Recently, two phytochemical studies using HPLC coupled with tandem mass spectrometry confirmed the occurrence of caffeine. Pierattini et al. [68] reported the presence of endogenous caffeine (**8**) (<0.4 μg/g f.w.), theobromine (<1.5 μg/g f. w.), and theophylline (**6**) (~60 μg/g f.w.) in the roots of *Populus alba* plantlets (Order Malpighiales). These xanthine alkaloids were also detected in the leaves and stems at lower concentrations. Small amounts of **8** (86 ng/g d.w.) and theobromine (**5**) (70 ng/g d.w.) have been detected in a cactus, *Epiphyllum oxypetalum* (order Caryophallales) (Gema Peireira-Caro and Hiroshi Ashihara, unpublished results).

Strictly speaking, methylxanthine-synthesizing and accumulating plants are not the same. For example, a discrepancy on the ability of biosynthesis and accumulation was observed with another *N*-methyl compound, trigonelline (**20**). Although accumulation of trigonelline has been recorded in a limited number of species, many plants possess the ability to produce trigonelline if nicotinic acid (**18**) is supplied [69]. Therefore, a different approach, such as the determination of the occurrence of gene expression and enzyme activity may be required to clarify whether or not the capacity to synthesize xanthine alkaloids is present in a particular species. This type of confirmation is necessary for chemotaxonomic analysis of caffeine (**8**) and related alkaloids.

Caffeine (**8**), used in beverages, pharmaceutical, and personal care products, is a recognized contaminant of aquatic ecosystems [68, 70]. Thus, small amounts of **8** might be found in plants grown in the wild that do not synthesize xanthine alkaloids. To determine the occurrence of biosynthesis is also important to confirm the origin of the xanthine alkaloids.

3 Biosynthesis of Xanthine Alkaloids

3.1 A Brief History of the Elucidation of the Biosynthesis Pathways

In 1950, James Bonner speculated in his famous book, "Plant Biochemistry", that caffeine (**8**) was synthesized by a xanthine (**1**) → heteroxanthine (7-methylxanthine) (**2**) → theobromine (**5**) → caffeine (**8**) pathway [71]. The proposed pathway was based on the following: (i) the occurrence of heteroxanthine in *Beta vulgaris* reported in 1904 [72], and (ii) the conversion of **5** to **8** that was demonstrated in *Theobroma cacao* seedlings by Weevers in 1930 [73].

In the 1960s, Anderson and Gibbs, and Inoue and co-workers, demonstrated the incorporation of various precursors into caffeine (**8**) in leaves of coffee [74] and tea [75–77]. Up to 1960, the carbon precursors utilized in the biosynthesis of the purine ring had been identified in birds, mammals, bacteria, and yeast, but there was no information on the biosynthesis of purines in plants. In addition to general purine compounds, such as ATP and GTP, it was known that some higher plants synthesize methylated purines, including **8**, but no attention had been given to the relationship between methyl group and purine ring formation. Anderson and Gibbs [74] used ^{14}C-labeled CO_2, glycine, serine, HCOONa, HCHO, CH_2OH, and methionine and demonstrated that **8** is synthesized from the same carbon precursors utilized for purines while the methyl group is derived from a different route. Inoue et al. [75] reported incorporation of root-absorbed $(^{15}NH_4)_2SO_4$ into **8** in leaves in tea seedlings. They also demonstrated the biosynthesis of **8** from $(^{15}NH_4)_2SO_4$ and ^{14}C-glycine (**35**) by excised shoots [76, 77]. In addition, they established that the methyl groups of **8** are derived from [methyl-^{14}C]methionine [78]. Caffeine biosynthesis from several purine precursors was investigated by Russian scientists using tea leaves, and similar conclusions were drawn [79, 80].

In the 1970s, extensive research on the biosynthesis of **8** was carried out using intact plant tissues [81–87] and callus cultures [88–90]. These studies confirmed that the purine ring of caffeine was synthesized from the same precursors that are utilized for general purine metabolism and several caffeine biosynthesis pathways from purine compounds were postulated. Ogutuga and Northcote [89] proposed two routes: one derived from nucleic acids, i.e. methylated nucleic acids → 7-methyl-GMP (**21**) → 7-methylguanosine (**22**) → 7-methylxanthosine (**23**) → 7-methylxanthine (**2**) → theobromine (**5**) → caffeine (**8**); with another leading from xanthine (**1**) in the purine pool, i.e. xanthine (**1**) → 3-methylxanthine (**3**) → theophylline (**6**) → caffeine (**8**). In contrast, Suzuki and Takahashi [91] proposed other pathways: (i) nucleic acids (tRNA) → 1-methyl-AMP (**25**) → 1-methylxanthine (**4**) → theophylline (**6**) → caffeine, and (ii) purine nucleotides 7-methyl-AMP (**26**) or 1-methyl-GMP (**27**) → 7-methylxanthine (**2**) → theobromine (**5**) → caffeine (**8**). However, several pieces of evidence seriously questioned the validity of these proposals [87, 92].

In contrast to these studies with plant tissues, cell-free systems proved more advantageous in obtaining direct proof for the involvement of *S*-adenosyl-L-methionine (SAM) (**28**) and the sequence of methylation of the purine ring of **8**. At that time, one of the difficulties in extracting active enzymes from tea and coffee leaves was that they contained high concentrations of phenolic compounds. This was overcome with the use of insoluble polyvinylpyrrolidone (PVP) [93]. Suzuki and Takahashi [94] prepared cell-free extracts from tea leaves using insoluble PVP to remove phenolics and demonstrated two methyltransferase activities that catalyze the transfer of methyl groups from SAM (**28**) to 7-methylxanthine (**2**) and theobromine (**5**), resulting, respectively, in the production of theobromine and caffeine. These findings suggested that **8** is synthesized from **2** via **5** but not from xanthine (**1**), xanthosine (**16**), XMP (**17**), or hypoxanthine (**15**). In contrast to tea extracts, the extraction of active enzyme activity from coffee was difficult. Indeed, Suzuki and Takahashi reported that their attempts to extract enzyme activity capable of synthesizing **2** from coffee leaves were unsuccessful [94].

Active cell-free extracts from coffee tissues were first obtained by Roberts and Waller [95] in 1979. This was achieved with the use of insoluble PVP, and rigorous exclusion of oxygen in the presence of solid CO_2 and liquid N_2, and rapid gel filtration to remove residual phenolic materials. The resultant enzyme preparation was not only able to convert 7-methylxanthosine (**23**) to theobromine (**5**) but also **5** to caffeine (**8**).

In the 1980s, *N*-methyltransferase activity for caffeine biosynthesis was demonstrated in tissue and cell cultures of coffee by Baumann et al. [96] and Waller et al. [97]. Their enzyme system catalyzed the conversion of 7-methylxanthosine (**23**) to caffeine (**8**). The 7-methylxanthosine synthase activity that catalyzed the first methylation of the xanthine (**1**) ring of **8** was demonstrated initially in tea extracts by Negishi et al. [98] in 1985. Later, *N*-methylnucleosidase activity, which converts 7-methylxanthosine (**23**) to 7-methylxanthine (**2**), was isolated from tea extracts by the same investigators [99]. Thus, a xanthosine (**16**) → 7-methylxanthosine (**23**) → 7-methylxanthine (**2**) → theobromine (**5**) → caffeine (**8**) pathway was postulated as the major route of caffeine biosynthesis.

The initial methyl acceptor for the main caffeine (**8**) biosynthesis pathway in tea and coffee plants is xanthosine (**16**), and there are several pathways leading to **16**. It can be derived from XMP (**17**), produced by de novo purine biosynthesis, but a contribution from cellular pools of AMP (**30**) and GMP (**29**), which are derived from both de novo and some purine salvage pathways, was implied from studies on general purine metabolism in plants [100, 101]. Suzuki and Takahashi [86] reported that adenine (**13**) is the most efficient precursor for caffeine synthesis in the tea shoot. Ashihara and Kubota [33] demonstrated that [8-^{14}C]adenine was converted to **30** by the salvage enzyme, adenine phosphoribosyltransferase, and hence to xanthosine (**16**) by an AMP (**30**) → IMP (**31**) → XMP (**17**) → xanthosine (**16**) pathway. Subsequently, there were several reports that purine precursors, such as adenosine (**44**), hypoxanthine (**15**), and inosine (**42**), were also utilized as precursors for caffeine biosynthesis. It was proposed that these compounds entered the caffeine (**8**) biosynthesis pathway after their conversion to **30** and/or **29** [102–104]. Until the early 1990s, three routes: (i) the de novo route, (ii) the AMP (**30**) route, and (v) the GMP (**29**) route, had been proposed as the source of **16** [105]. In addition to these three routes, a new pathway that utilized adenosine released from the SAM (**28**) cycle (iii) was proposed by Koshiishi et al. [106] in 2001. Thus, four routes have been proposed for the formation of the purine skeleton of caffeine (Fig. 5) [1].

Alternative routes of caffeine biosynthesis were put forward in the 1990s. In 1993, Nazario and Lovatt [107] argued that theobromine (**5**) was not the immediate precursor of caffeine (**8**). They hypothesized that **5** might be synthesized from

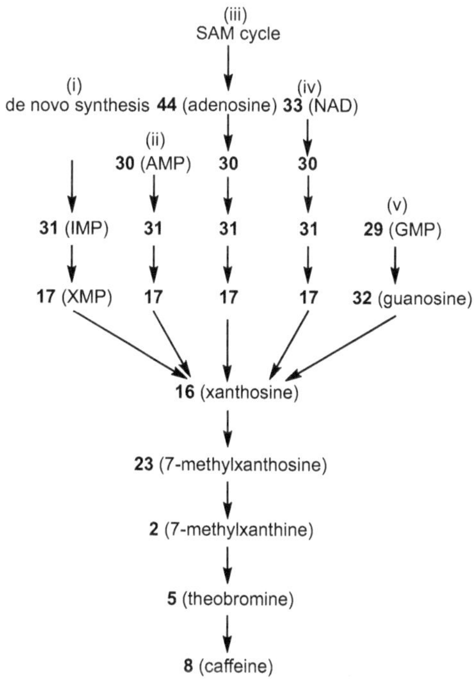

Fig. 5 Outline of five possible pathways of caffeine biosynthesis: (i) De novo route, (ii) AMP route, (iii) SAM route, (iv) NAD route, and (v) GMP route

adenine nucleotides without a xanthine species serving as an intermediate, whereas **8** was likely to be derived from a methylated xanthine. However, they could not identify the steps by which xanthine (**1**) was methylated. In 1996, Ashihara et al. [108] provided strong evidence against the proposal by Nazario and Lovatt [107] and confirmed the operation of an AMP (**30**) → IMP (**31**) → XMP (**17**) (or GMP (**29**) → guanosine (**32**)) → xanthosine (**16**) → 7-methylxanthosine (**23**) → 7-methylxanthine (**2**) → theobromine (**5**) → caffeine (**8**) pathway in young coffee leaves, using pulse-chase experiments with ^{14}C-precursors (routes ii and v in Fig. 5).

Schulthess et al. [109, 110] reported the results of a study on the initial methylation step of caffeine biosynthesis. They proposed that caffeine biosynthesis starts with XMP (**17**), and not xanthosine (**16**), with the metabolically channeled formation of 7-methyl-XMP (**24**) from **17**. However, this pathway is not supported by more recent research [111].

Attempts to purify the *N*-methyltransferases involved in caffeine (**8**) biosynthesis were made by several research groups. However, it was found to be very difficult to purify these enzymes because they are extremely labile. For example, Gillies et al. [112] reported that the half-life of the enzyme activity in coffee cell-free extracts is ~90 min at 4°C. Partially purified enzymes were reported from laboratories in Brazil [113], Japan [114, 115], Switzerland [116] and the UK [112]. Finally, in 1999, Kato et al. [11] successfully isolated a highly purified *N*-methyltransferase that catalyzed the last two steps of caffeine biosynthesis. The enzyme referred to as caffeine synthase was purified to apparent homogeneity and detailed properties, including the *N*-amino-terminal sequence were reported.

In 2000, using the *N*-terminal sequence of highly purified caffeine synthase [11], a gene encoding caffeine synthase was cloned from tea leaves by Kato et al. [117]. The substrate specificity of the recombinant enzyme was very similar to that of the native enzyme purified from young tea leaves [11]. The recombinant enzyme catalyzed N-3- and N-1-methylation but not N-7-methylation. Shortly before the publication of the Kato et al. paper [117], a Hawaiian group claimed to have cloned the 7*N*-methyltransferase gene involved in caffeine biosynthesis from coffee leaves [118, 119]. However, two Japanese groups have subsequently revealed that the reported gene did not encode any *N*-methyltransferase involved in caffeine (**8**) biosynthesis, and that the sequence very closely resembled that of a lipase gene [111, 120].

In 2001–2003, genes encoding *N*-methyltransferases involved in caffeine biosynthesis were cloned from coffee plants by two independent Japanese groups [111, 121–124]. Currently, three different coffee genes encoding *N*-methyltransferases associated with caffeine biosynthesis have been cloned, namely, 7-methylxanthine synthase, theobromine synthase and caffeine synthase [125]. These results support the operation of the four-step caffeine biosynthesis pathway leading from xanthosine (**16**) as originally proposed by Suzuki and Takahashi [94], Negishi et al. [126] and Ashihara et al. [108]. There is now widespread agreement that a xanthosine (**16**) → 7-methylxanthosine (**23**) → 7-methylxanthine (**2**) → theobromine (**5**) → caffeine (**8**) pathway is the main route of caffeine biosynthesis in tea and coffee plants.

3.2 Caffeine Biosynthesis Pathways from Xanthosine

Xanthine alkaloids are formed from purine nucleotides in plants [13]. As discussed in Sect. 3.1, there have been a number of proposals on the pathways involved in caffeine biosynthesis. However, current data from a series of studies on in situ metabolism of labeled precursors and characterization of enzymes and genes have established that the main caffeine (**8**) biosynthesis pathway is a four-step sequence consisting of three methylations and one nucleosidase reaction starting with xanthosine (**16**) serving as the initial substrate. The conversion of **16** to **8** involves three N-methylation steps (steps 1, 3, and 4 in Fig. 6) and the removal of a ribose moiety (step 2 in Fig. 6). The main pathway is via theobromine (**5**). However, a route with paraxanthine (**7**) as an intermediate is also feasible. 7-Methylxanthine (**2**), **5**, and **7** may exist as different tautomers. Three tautomers for **2** are shown in Fig. 6, but the tautomer with the hydroxy group at C-6 is not considered here, because there seems to be no potential general base/acid catalyst for the methyl transfer in the X-ray structure [127]. Different tautomers can undergo N-3 and N-1 methylation.

To date, the information on caffeine biosynthesis has been obtained mainly from coffee and tea plants. The available evidence suggests that the pathway is essentially the same in both species, but some minor differences are found in other species. Therefore, this chapter will consider data from tea (*Camellia sinensis*), coffee (*Coffea arabica* and *C. canephora*), and other species separately.

There is no consistency in the nomenclature of *N*-methyltransferases involved in caffeine biosynthesis in the literature. Kato et al. [11] used product names of each enzyme: 7-methylxanthosine synthase, theobromine synthase, and caffeine synthase. On the other hand, Ogawa et al. [123] used substrate names: xanthosine methyltransferase, monomethylxanthine methyltransferase, and dimethylxanthine methyltransferase. Since the former system is registered with the IUBMB enzyme nomenclature, the terms 7-methylxanthosine synthase (EC 2.1.1.158), theobromine synthase (EC 2.1.1.159), and caffeine synthase (EC 2.1.1.160) are used herein. However, when comparing the various *N*-methyltransferases, substrate names, such as XMP-*N*-methyltransferase and xanthosine *N*-methyltransferase are also used, because it is more common in the nomenclature for methyltransferases.

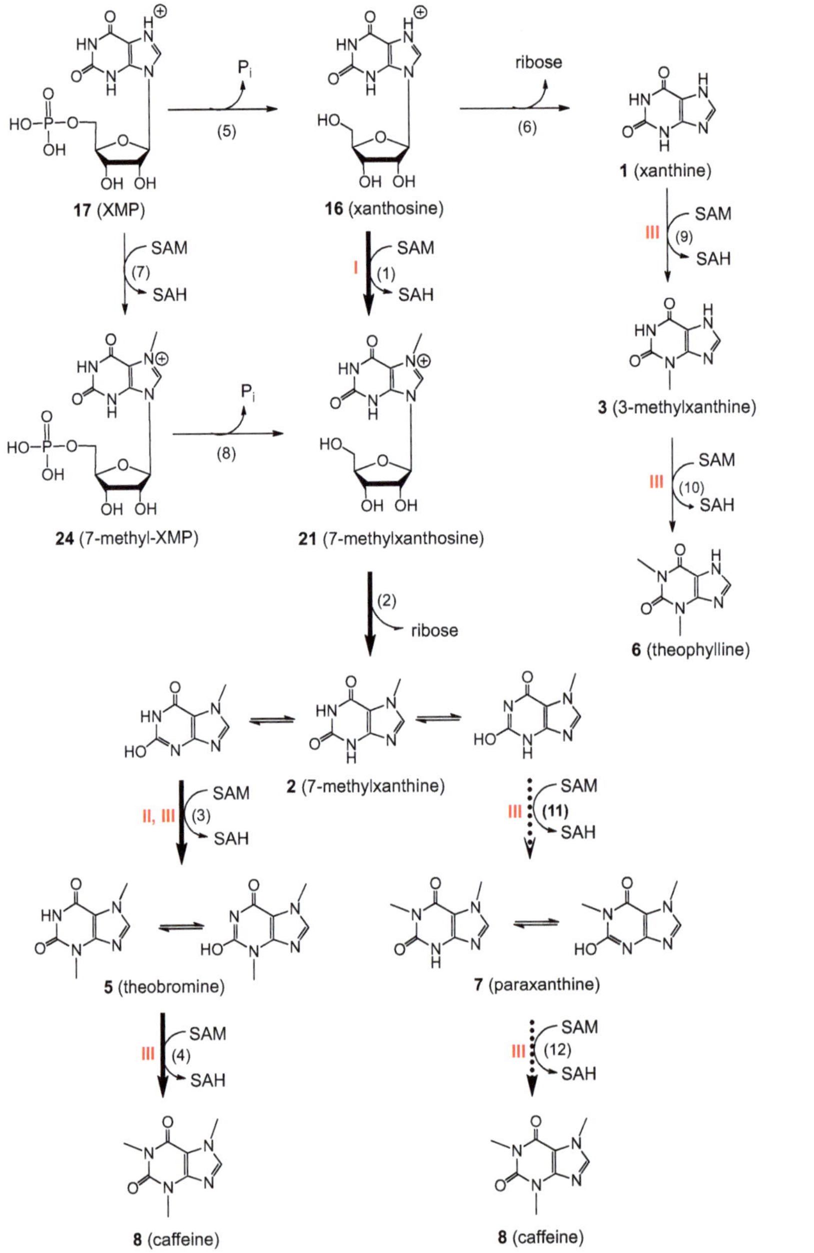

Fig. 6 The biosynthesis pathways of caffeine from xanthosine. The major pathway consisting of four steps are shown in solid arrows (steps 1–4). Three types of *N*-methyltransferases, 7-methylxanthosine synthase (EC 2.1.1.158), theobromine synthase (EC 2.1.1.159), and caffeine synthase (EC 2.1.1.160), are shown as **I, II, III**. Conversion of 7-methylxanthosine to 7-methylxanthine (step 2) is catalyzed by *N*-methylnucleosidase (EC 3.2.2.25). Minor pathways, shown with dotted arrows, may occur because of the broad substrate specificities of caffeine synthase (**III**). The route of 7-methylxanthosine formation from XMP via 7-methyl-XMP (steps 7–8) was proposed by Schulthess et al. [109], but these conversions are not catalyzed by any reported recombinant *N*-methyltransferases

3.2.1 Formation of 7-Methylxanthine

Feeding Experiments

In tea plants, the formation of monomethylxanthine in the main caffeine biosynthesis pathway is initiated by xanthosine (**16**) derived from purine nucleotides. This reaction was first demonstrated in ^{14}C-feeding experiments using excised tea shoots by Negishi et al. [126]. Approximately 50% of [2-^{14}C]xanthosine (**16**) taken up by the tea shoots was incorporated into caffeine within 24 h. Compound **16** disappeared within 1 h and the radioactivity appeared transiently in theobromine. 7-Methylxanthosine (**23**) and 7-methylxanthine (**2**) also appeared in small amounts at the earlier incubation periods.

Negishi et al. [98] also found 7-methylxanthosine synthase activity that catalyzed the conversion of xanthosine (**16**) to 7-methylxanthosine (**23**) in cell-free extracts of tea shoots. When XMP (**17**) is included in the incubation medium instead of xanthosine (**16**), small amounts of 7-methylxanthosine (**23**) (12% of the amount obtained with **16**) were detected but 7-methyl-XMP (**24**) was not. This suggests that **17** was not methylated directly. The apparent activity may be due to **16**, which was produced from **17** by nucleotidase contamination of the cell-free extracts. To confirm this speculation, Kato and Ashihara partially purified tea *N*-methyltransferase by ion-exchange chromatography after ammonium sulfate fractionation [13]. The major peak of xanthosine *N*-methyltransferase did not methylate **17**. Therefore, **17** is unlikely to be the direct methyl acceptor in tea plants as suggested by Negishi et al. [98].

Enzyme Studies

7-Methylxanthine synthase (EC 2.1.1.158) catalyzes the following reaction: xanthosine (**16**) + SAM (**28**) → 7-methylxanthosine (**23**) + *S*-adenosyl-L-homocysteine (SAH) (**73**). The enzyme activity has been detected in vitro in crude and partially purified extracts from tea [98, 128] and coffee [109, 116].

In the earlier paper by Suzuki and Takahashi [94], which demonstrated the existence of *N*-methyltransferases involved in caffeine biosynthesis in cell-free tea leaf extracts, they indicated that it was difficult to detect the methylation of xanthosine. However, this is probably due to the fact that 7-methylxanthine synthase activity is low compared with other *N*-methyltransferase activities and is extremely unstable in vitro. 7-Methylxanthosine synthase activity was detected only in young leaves and its activity is ~6% of theobromine synthase activity [128]. Kato et al. [115] partially purified tea leaf *N*-methyltransferases involved in caffeine (**8**) biosynthesis by anion-exchange and gel-filtration column chromatography. When xanthosine (**16**) was used as a substrate only the N-7 position was methylated. In this study, 7-methylxanthosine synthase could not be separated from the other *N*-methyltransferases, but subsequent extensive purification of

N-methyltransferases revealed that this enzyme, which is involved in the first methylation step, is distinct from the enzymes that catalyze second and third methylation steps of biosynthesis of **8** [11]. Neither highly purified native 7-methylxanthine synthase nor a cloned gene encoding this enzyme has been isolated from tea plants.

Using tea leaf extracts, Negishi et al. [99] partially purified *N*-methyl nucleosidase activity (7-methylxanthosine nucleosidase, EC 3.2.2.25). The substrate specificity of this enzyme was broad, and high activity was found with several methyl purine nucleosides including 7-methylxanthosine (**23**). The estimated molecular weight was 55,000 and the pH optimum was 8.0–8.5. This enzyme is distinct from adenosine nucleosidase (EC 3.2.2.7) that has a pH optimum of 4.5. *N*-Methylnucleosidase was readily separated from adenosine nucleotidase by DEAE-cellulose chromatography. The results obtained with the native tea leaf enzymes suggest that the conversion of xanthosine (**16**) to 7-methylxanthine (**2**) via **23** is catalyzed by two enzymes, 7-methylxanthosine synthase and *N*-methylnucleosidase.

The genes encoding 7-methylxanthosine synthase have been isolated from *Coffea arabica* and the properties of the recombinant enzymes were reported [111, 124]. In contrast to tea, 7-methylxanthosine nucleosidase has not yet been isolated from coffee plants. There is also some debate on the synthesis of 7-methylxanthine (**2**) in coffee and the substrate of the 7*N*-methyltransferase. Schulthess et al. [109] reported that in addition to xanthosine (**16**), XMP (**17**) could act as a substrate for 7*N*-methyltransferase. As shown in Table 7, **17** was converted to 7-methyl-XMP (**24**), 7-methyxanthosine (**23**) and 7-methylxanthine (**2**) in crude enzyme extracts of young coffee leaves, with SAM (**28**) acting as the methyl donor. Conversion to **24** was enhanced at higher concentrations of **17** in the

Table 7 Substrate specificity and products of native and recombinant 7-methylxanthosine synthase from coffee plants[a]

Sample	Substrate	Product(s)	Relative rate/%	Ref.
C. arabica (native, crude)	Xanthosine (**16**)	7mXR (**23**)	29	[109]
		7mX (**2**)	71	
	XMP (**17**)	7mXMP (**24**)	18	
		7mXR(**23**)	57	
		7mX (**2**)	23	
C. arabica (recombinant)	Xanthosine (**16**)	7mXR (**23**)	100	[111]
	XMP (**17**)	None	–	
C. arabica (recombinant)	Xanthosine (**16**)	7mXR (**23**)	100	[124]
	XMP (**17**)	None	–	
C. canephora (recombinant)	Xanthosine (**16**)	7mX (**2**)	100	[129]
	XMP (**17**)	None	–	

[a]7mX = 7-methylxanthine (**2**), 7mXR = 7-methylxanthosine (**23**)

presence of Na_2MoO_4, a nucleotidase inhibitor. On the basis of these data and in situ feeding experiments, which showed a failure to detect ^{14}C-labeled **23** in coffee leaves, a XMP (**17**) → 7-methyl-XMP (**24**) → 7-methylxanthosine (**23**) → 7-methyxanthine (**2**) pathway was proposed.

Recombinant enzyme proteins of 7-methylxanthosine synthase have been prepared using the coffee gene sequences, and the biochemical properties characterized. The recombinant enzyme uses xanthosine (**16**) as a substrate, but not XMP (**17**) nor xanthine (**1**) [111]. The K_M for **16** is 74–78 μ*M* [111, 124] and for SAM (**28**) is 13 μ*M* [124]. These findings do not support the hypothesis proposed by Schulthess et al. [109] that caffeine biosynthesis starts with the metabolically channeled formation of 7-methyl-XMP (**24**) (Table 7).

There has also been debate about the product of 7-methylxanthosine synthase. Mizuno et al. [111] and Uefuji et al. [124] reported that 7-methylxanthosine (**23**) is the product of 7-methylxanthosine synthase from *Coffea arabica*. In contrast, McCarthy and McCarthy [129], who used a recombinant 7-methylxanthosine synthase from *Coffea canephora*, proposed that the methyl transfer and nucleoside cleavage were coupled and catalyzed by a single enzyme, 7-methylxanthosine synthase/nucleosidase (Table 7). Uefuji et al. noted that crude recombinant enzyme, but not the purified enzyme, catalyzed the conversion of xanthosine (**16**) to 7-methylxanthine (**2**), possibly as the result of the presence of nucleoside phosphorylase(s) derived from *E. coli* contamination of the bacterial extracts. Recently, the conclusions of McCarthy and McCarthy [129] were revised by the group after they obtained data showing that the main product of the 7-methylxanthosine synthase is **23**, and not **2** as originally reported. However, small amounts of **2** were detected, suggesting that 7-methylxanthosine synthase may have some weak secondary nucleosidase activity or that **23** is unstable and slowly degrades to **2** (Amandine Lallemand and Andrew A. McCarthy, personal communication). Thus, *N*-methylnucleosidase appears to be involved in caffeine (**8**) biosynthesis in coffee as well as in tea [99].

Genes and Proteins

A single gene encoding 7-methylxanthosine synthase has been identified. It encodes a polypeptide consisting of 372 amino acids with an apparent molecular mass of 41.8 kDa and is expressed almost uniformly in all parts of coffee plants, including the leaves, floral buds, and the immature but not mature beans [111, 124]. In contrast, at least four genes encoding theobromine synthase have been isolated [121, 124]. The number of amino acids in the putative polypeptides are 378 for CTS1 and MXMT1 (42.7 kDa) and 384 for CTS2 and MXMT2 (43.4 kDa). They differ by insertion or deletion of blocks of several residues in the *C*-terminal region. Their catalytic properties as judged from kinetic parameters, such as K_M values, are apparently different. They are expressed in young leaves, floral buds and immature but not mature beans. At least two genes were also identified for caffeine synthase [121, 124], *CCS1* and *DXMT*, each encoding a 43-kDa polypeptide consisting of 384 amino acids. Patterns of gene expression of *CCS1* and *DXMT* were different, with *DXMT* being expressed exclusively in immature beans, while *CCS1* expression occurred in all organs. The isoforms of theobromine synthase and caffeine synthase with different properties may be present and theobromine (**5**) and caffeine (**8**) may be synthesized through multiple pathways that appear in different organs and at different stages of development.

3.2.2 Conversion of 7-Methylxanthine to Caffeine via Theobromine

Feeding Experiments

Negishi et al. [126] reported that [2-^{14}C]7-methylxanthine (**2**) is metabolized to theobromine and then further converted to caffeine in young tea shoots. Ashihara et al. [130] found that [2-^{14}C]theobromine (**5**) is efficiently incorporated into caffeine (**8**) in young and mature tea leaf segments. These feeding experiments support the operation of a 7-methylxanthine (**2**) → theobromine (**5**) → caffeine (**8**) pathway (Fig. 5) in tea.

Enzyme Studies

The second and third methylation steps in the caffeine biosynthesis pathway (steps 3 and 4 in Fig. 6) are catalyzed by SAM (**28**)-dependent *N*-methyltransferase(s), namely, theobromine synthase (EC 2.1.1.159) and/or caffeine synthase (EC 2.1.1.160), but the participating enzymes are distinct from the 7-methylxanthosine synthase (EC 2.1.1.158) involved in the first step of caffeine (**8**) biosynthesis (step 1 in Fig. 6) [11]. As described below, caffeine synthase is a dual-functional enzyme that catalyzes the two-step reaction 7-methylxanthine (**2**) → theobromine (**5**) → caffeine (**8**) (steps 3 and 4 in Fig. 5), while theobromine synthase catalyzes only the conversion of **2** to **5** (step 3 in Fig. 6). Up to now, only a caffeine synthase has been found in tea, but both theobromine and caffeine synthases occur in coffee (see below).

Highly purified native caffeine synthase (EC 2.1.1.160) was obtained from young tea leaves by Kato et al. [11]. The enzyme was purified 520-fold to apparent homogeneity by fractionation with ammonium sulfate and hydroxyapatite, anion-exchange chromatography, affinity chromatography with adenosine agarose, and gel-filtration chromatography. The enzyme displayed a sharp pH optimum at 8.5. The final preparation exhibited 3*N*- and 1*N*-methyltransferase activity and had a rather broad substrate specificity, showing high activity towards paraxanthine (1,7-dimethylxanthine) (**7**), 7-methylxanthine (**2**), and theobromine (**5**), and low activity with 3-methylxanthine (**3**) and 1-methylxanthine (**4**) (Table 8). Xanthosine

(**16**) and XMP (**17**) were also methylated by the enzyme. The 20-amino acid *N*-terminal sequence was determined. Highly purified enzymes have not yet obtained from other plants including coffee.

Genes and Proteins

Rapid amplification of complementary DNA ends (RACE) was used with degenerate gene-specific primers based on the amino-terminal sequence of caffeine synthase to obtain a 1.31-kb sequence of cDNA. The 5′ untranslated sequence of the cDNA fragment was isolated by 5′-RACE. The isolated cDNA termed *TCS1* [117] was expressed in *Escherichia coli* and the lysates of the bacterial cells incubated with a variety of xanthine substrates in the presence of SAM. The substrate specificity of the recombinant enzyme was similar to that of the native enzyme purified from young tea leaves [117]. The recombinant and native enzymes principally catalyze 3-N- and N-1-methylation of the purine ring of mono- and dimethylxanthines (Table 8).

In coffee, plural genes encoding methylxanthine *N*-methyltransferases have been cloned. The recombinant enzymes obtained from these genes show different substrate specificities. The recombinant coffee caffeine synthase (EC 2.1.1.160), namely, CCS1 and CaDXMT1, can utilize paraxanthine (**7**), theobromine (**5**), and 7-methylxanthine (**2**) as substrates (Table 8). Compound **7** is the most active substrate of tea and coffee caffeine synthase (Table 8), but only limited amounts of **7** appear to be synthesized in these tissues. The synthesis of **7** from **2** by caffeine synthase (step 11 in Fig. 6) has not been reported. Therefore, in planta, caffeine synthase is involved principally in the conversion of **2** to caffeine (**8**) via **5** (steps 3 and 4 in Fig. 6).

Recently, recombinant caffeine synthase (PcCS) was obtained from guaraná (*Paullinia cupana* var. *sorbilis*) by a Brazilian group (Table 8) [64]. Like the tea

Table 8 Comparison of substrate specificity of caffeine synthase of tea (*Camellia sinensis*), coffee (*Coffea arabica*) and guaraná (*Paullinia cupana*)[a]

	Tea		Coffee		Guarana
Substrate	Native	TCS1	CCS1	CaDXMT1	PcCS
7-Methylxanthine (**2**)	100.0	100.0	100.0	100	100
3-Methylxanthine (**3**)	17.6	1.0	3.4	nd	nd
1-Methylxanthine (**4**)	4.2	12.3	2.2	nd	nd
Theobromine (**5**)	26.8	18.5	104.0	380	189
Theophylline (**6**)	<0.1	<0.1	nd	nd	nd
Paraxanthine (**7**)	210.0	230.0	417.0	10,000	nd
Xanthine (**1**)	<0.1	–	–	–	–
Xanthosine (**16**)	nd	nd	–	–	–
XMP (**17**)	nd	–	–	–	–
Reference	[11]	[12]	[121]	[124]	[64]

[a]Caffeine synthase activity is expressed as a percentage of the activity on 7-methylxanthine (**2**). – = not determinded; nd = not detected

and coffee enzymes, this is a dual-functional enzyme able to catalyze the conversion of 7-methylxanthine (**2**) to theobromine (**5**) and also that of **5** to caffeine (**8**) (steps 3 and 4 in Fig. 6). However, in contrast to tea and coffee caffeine synthases, this enzyme does not use paraxanthine (**7**) as a substrate (Table 8). This unique substrate affinity of guaraná caffeine synthase is probably explained by the distinct amino acid residues found in the active site of the predicted protein [64].

In coffee, in addition to the dual-functional caffeine synthase, genes encoding theobromine synthase (EC 2.1.1.159) have been cloned; the recombinant enzyme is specific for the conversion of 7-methylxanthine (**2**) to theobromine (**8**) (step 3 in Fig. 6) [122, 123]. The enzyme obtained from coffee genes catalyzes N-3- but not N-1-methylation. Furthermore, the activity of N-3-methylation of **2** was much greater than the N-3-methylation activity of paraxanthine (**7**) (Table 9).

As shown in Sect. 2, some purine alkaloid-producing species including *Camellia irrawadiensis*, *Camellia ptilophylla* and *Theobroma cacao* accumulate theobromine (**5**) rather than caffeine (**8**). Yoneyama et al. [131] cloned genes encoding theobromine synthase (*ICS1*, *PCS1*, and *BTS1*) from these three species. The recombinant enzymes had only 3*N*-methyltransferase activity, mainly converting 7-methylxanthine (**2**) to **5** (step 3 in Fig. 5). Limited activity that catalyzed N-3-methylation of paraxanthine (**7**) to form caffeine was detected only in *Camellia ptilophylla* (Table 9). The theobromine synthase appears to participate principally in the synthesis of **5** in theobromine-accumulating plants. The genes encoding theobromine synthase have been isolated from some *Camellia* plants that do not accumulate **5**. In vitro, the recombinant enzymes principally catalyzed the conversion of **2** to **5** [31]. The role of theobromine synthase in these species is unknown, although it would appear not to be involved in the biosynthesis of xanthine alkaloids.

Table 9 Comparison of substrate specificity of theobromine synthase from different plant sources[a]

Enzyme source(s)		Substrate			Ref.
		7mX (**2**)	Paraxanthine (**7**)	Theobromine (**5**)	
Coffee	CTS1	100	1.4	nd	[121]
	CaMXT1	100	5.0	nd	[124]
	CaMXT2	100	5.3	nd	[124]
Cacao	BTS1	100	nd	nd	[131]
Camellia irrawadiensis	ICS1	100	10.9	0.7	[131]
Camellia ptilophylla	PCS1	100	11.0	nd	[131]

[a]Theobromine synthase activity is expressed as a percentage of the activity with 7-methylxanthine (**2**); nd = not determined

3.2.3 Evolutionary Relationship of the Caffeine Synthase Family Proteins

To investigate the evolutionary relationship, a comparison of the amino acid sequences of the caffeine synthase family of proteins was undertaken. Figure 7 shows the amino acid sequences of caffeine synthases from coffee, cacao, maté, tea, and guaraná. There are four highly conserved regions, namely, motif A, motif B′, motif C, and the YFFF-region [132]. Three conserved motifs, A, B, and C, of the binding site of the methyl donor of SAM, have been reported in most plant SAM-dependent *O*-methyltransferases [133].

The motif B′ and the YFFF region are specific to the motif B′ methyltransferase family. The consensus sequence of the motif B′ region, which is located between motif A and motif C, contains a large number of hydrophilic amino acids such as asparagine and aspartic acid, while the consensus sequence of the YFFF region, located downstream to motif C, contains aromatic amino acid residues [121]. These four regions may have important roles both in the N-methylation reaction and in discriminating the position of N-methylation on the purine ring.

In addition to the caffeine synthase family, members of the motif B′ methyltransferase include some *C*-methyltransferases, such as salicylic acid carboxyl methyltransferase [134], benzoic acid carboxyl methyltransferase [135],

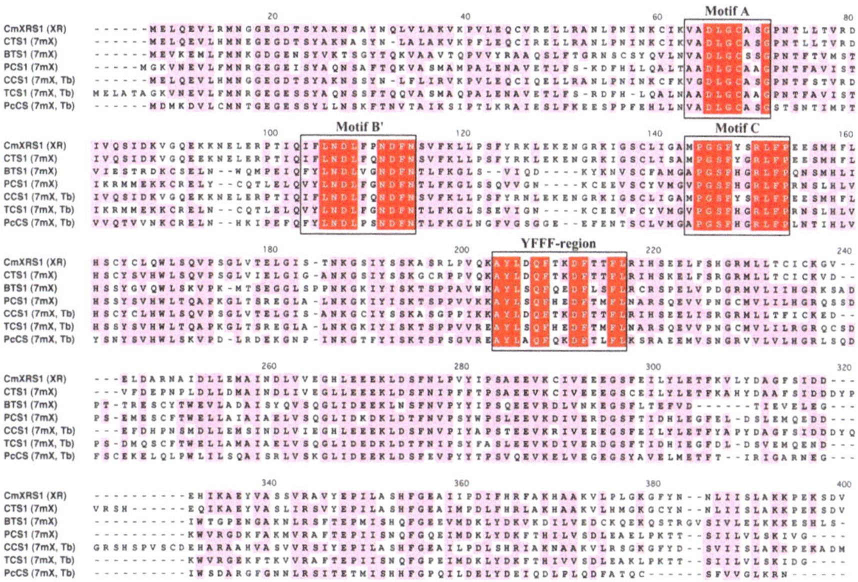

Fig. 7 Comparison of the amino acid sequences of caffeine synthases from several caffeine-containing plants. Alignment of the amino acid sequences for CmXRS1, CTS1, and CCS1 (accession numbers AB034699, AB034700, and AB086414, respectively) from coffee, BTS1 (accession number AB096699) from cacao, PCS1 (accession number AB207817) from cocoa tea (*Camellia ptilophylla*), TCS1 (accession number AB031280) from tea, and PcCS (accession number DAA64605) from guaraná. The identified substrates are shown in parentheses. The core SAM binding domain is indicated in red. The SAM binding motifs (A, B′, and C) and YFFF conserved region are shown by black boxes

jasmonic acid carboxyl methyltransferase [136], farnesoic acid carboxyl methyltransferase [137], indole-3-acetic acid methyltransferase [138], gibberellic acid methyltransferase [139], and loganic acid carboxyl methyltransferase [140] (Table 10). The motif B′ methyltransferase family is sometimes referred to as the

Table 10 Examples of the motif B′ methyltransferase family enzymes

Species	Enzyme name	Gene name	Accession number
Coffee (*Coffea arabica*)	7-Methylxanthine synthase	*CmXRS1*	AB034699
		CaXMT1	AB048793
		CaXMT2	JX978522
	Theobromine synthase	*CTS1*	AB034700
		CaMXMT1	AB048794
		CaMXMT2	AB084126
	Caffeine synthase	*CCS1*	AB086414
		CaDXMT1	AB084125
	Tentative caffeine synthase	*CtCS7*	AB086415
	Methyltransferase-like protein	*CaMTL1*	AB039725
		CaMTL2	AB048792
	Trigonelline synthase	*CTgS*	AB054842
Coffee (*Coffea canephora*)	7-Methylxanthine synthase	*CcXMT1*	JX978518
	Theobromine synthase	*CcMXMT1*	JX978517
	Caffeine synthase	*CcDXMT*	JX978516
	Methyltransferase-like protein	*CcMTL*	JX978528
Cacao (*Theobroma cacao*)	Theobromine synthase	*BTS1*	AB096699
Tea (*Camellia sinensis*)	Caffeine synthase	*TCS1*	AB031280
		TCS2	AB031281
Cocoa tea (*Camellia ptilophylla*)	Theobromine synthase	*PCS1*	AB207817
		PCS2	AB207818
Camellia irrawadiensis	Theobromine synthase	*ICS1*	AB056108
		ICS2	AB207816
Camellia (*Camellia japonica*)	Theobromine synthase	*CjCS1*	AB297451
Guaraná (*Paullinia cupana*)	Caffeine synthase	*PcCS*	DDA64605
Fairy fans (*Clarkia breweri*)	Salicylic acid carboxyl methyltransferase	*CbSAMT*	AF133053
Snapdragon (*Antirrhinum majus*)	Benzoic acid carboxyl methyltransferase	*AmBAMT*	AF198492
Rice (*Oryza sativa*)	Indole-3-acetic acid methyltransferase	*OsIAMT*	EU375746
Thale cress (*Arabidopsis thaliana*)	Benzoic acid/salicylic acid carboxyl methyltransferase	*AtBSMT1*	NM111981
	Jasmonic acid carboxyl methyltransferase	*AtJAMT*	AY008434
	IAA carboxylmethyltransferase	*AtIAMT1*	NM124907
	Gibberellin methyl transferase	*AtGAMT1*	NM118775

Adapted from Nakayama et al. [262]

SABATH family, based on the initial letters of the names of the substrates of methyltransferases initially named by D'Auria et al. [141].

Crystallographic data on salicylic acid carboxyl methyltransferase from *Clarkia breweri* suggest that members of this family exist as dimers in solution [142]. Further structural analysis of 7-methylxanthosine synthase and caffeine synthase from *Coffea canephora* has also revealed a dimeric structure [129]. Despite the marked similarity in amino acid sequences of *N*-methyltransferases, each enzyme catalyzes the methylation of specific substrate(s). Some reports suggest that a single amino acid residue of the *N*-methyltransferases decides the substrate specificity [123, 131].

Amino acid sequences of the caffeine synthase family derived from coffee are more than 80% homologous but share only 40% homology with caffeine synthase from tea. Figure 8 shows the phylogenetic tree analysis of the motif B′ methyltransferase family. The figure indicates that the amino acid sequences of several methyltransferases share a high degree of sequence identity within the same genus. This implies that the xanthine alkaloid biosynthesis pathways in coffee, tea,

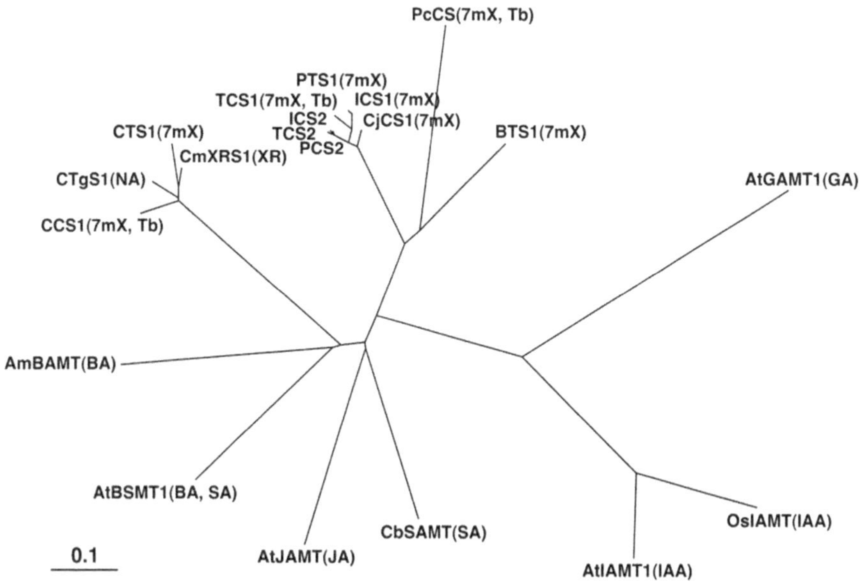

Fig. 8 Molecular phylogenetic tree of motif B′ methyltransferase family. The identified substrates are shown in parentheses. Abbreviation of substrates are as follows: BA, benzoic acid; GA, gibberellic acid; IAA, indole-3-acetic acid; JA, jasmonic acid; 7mX, 7-methylxanthine; NA, nicotinic acid; SA, salicylic acid; Tb, theobromine; XR, xanthosine. The unrooted tree is created by a neighbor-joining method using ClustalW software. The sources of sequences given by the accession numbers are as follows: TCS1, AB031280; TCS2, AB031281; ICS1, AB056108; ICS2, AB207816; PCS1, AB207817; PCS2, AB207818; CjCS1, AB297451; CmXRS1, AB034699; CTS1, AB034700; CCS1, AB086414; CTgS1, AB054842; BTS1, AB096699; CbSAMT, AF133053; AmBAMT, AF198492; OsIAMT, EU375746; AtBSMT1, NM111981; AtJAMT, AY008434; AtIAMT1, NM124907; AtGAMT1, NM118775; PcCS, DDA64605

cacao, and guaraná might have evolved in parallel with one another. Denoeud et al. [143] also have suggested that convergent evolution in caffeine biosynthesis occurs in different plant species. Caffeine (**8**)-forming plants belong to several unrelated families, but they accumulate caffeine synthesized by a similar, if not identical, biosynthesis pathway. The genome sequence of caffeine biosynthesis indicates that the methyltransferase genes in some lineages have evolved independently from other branches of the SABATH methyltransferase gene family [144]. Ishida et al. reported the occurrence of theobromine synthase genes in species of *Camellia* that do not accumulate xanthine alkaloids [31]. Arguably this represents additional evidence that monophyletic genes occur in *Camellia* plants.

Perrois et al. [145] proposed that the different *N*-methyltransferases involved in caffeine biosynthesis in coffee belong to three different clusters that align with the function of each enzyme. Clusters I, II, and III correspond to 7-methylxanthosine synthase, theobromine synthase, and caffeine synthase (Fig. 9). Recently, Mizuno

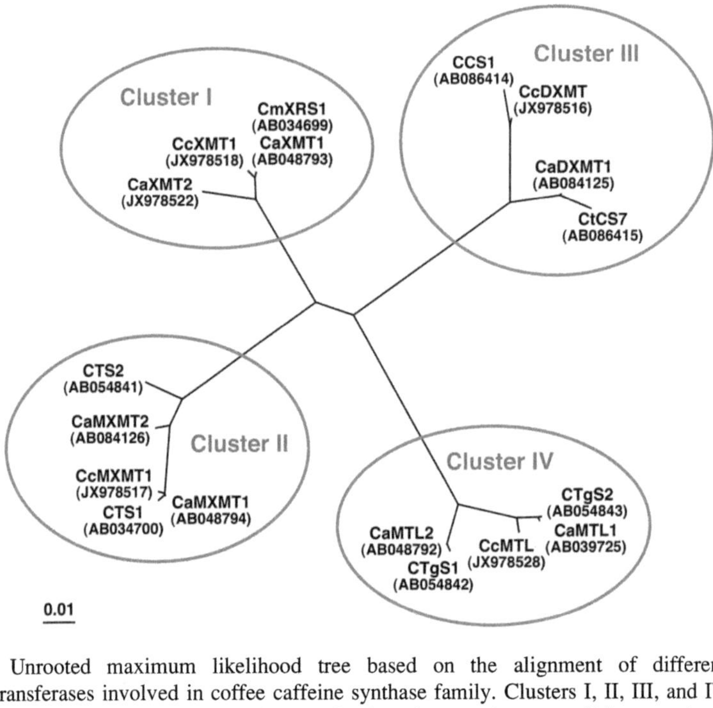

Fig. 9 Unrooted maximum likelihood tree based on the alignment of different 18*N*-methyltransferases involved in coffee caffeine synthase family. Clusters I, II, III, and IV correspond to 7-methylxanthosine synthases, theobromine synthases, caffeine synthases, and trigonelline synthases, respectively. Those accession numbers are as follows: CmXRS1 (AB034699); CaXMT1 (AB048793); CcXMT1 (JX978518); CaXMT2 (JX978522); CcMXMT1 (JX978517); CaMXMT1 (AB048794); CaMXMT2 (AB084126); CTS1 (AB034700); CTS2 (AB054841); CcDXMT (JX978516); CaDXMT1 (AB084125); CtCS7 (AB086415); CCS1 (AB086414); CTgS1 (AB054842); CaMTL1 (AB039725); CcMTL (JX978528); CTgS2 (AB054843); CaMTL2 (AB048792)

et al. [146] revealed that the conversion of nicotinic acid (**18**) to trigonelline (**20**) is also catalyzed by *N*-methyltransferases belonging to the motif B′ methyltransferase family in *Coffea arabica*. Genes previously cloned from *C. arabica*, and tentatively named as *CtCS3* and *CtCS4*, were shown to encode trigonelline synthase 1 (CTgS1) and 2 (CTgS2). The sequence of these isozymes exhibited >95% homology while homology between the trigonelline synthase (CTgS) and caffeine synthase (CCS1) was 82%. The CTgS can be classified as Cluster IV as shown in Fig. 9. Trigonelline (**20**) is accumulated in coffee seeds in similar concentrations to caffeine [147, 148]. It is interesting that both biosynthetic enzymes possess high homology.

3.3 Pathways that Supply Xanthosine for Caffeine Biosynthesis

Caffeine biosynthesis is initiated by the methylation of xanthosine (**16**). The direct precursor, **16**, is supplied by several different pathways: (i) de novo purine biosynthesis (de novo route); (ii) degradation of adenine nucleotides (AMP route); (iii) the SAM cycle (SAM route); (iv) degradation of NAD or NADP (NAD route), and (v) guanine nucleotides (GMP route) (Fig. 5).

3.3.1 De Novo Route

Plants, as well as most other organisms, produce inosine-5′-phosphate (IMP) (**31**) by de novo purine nucleotide biosynthesis. In the de novo pathway, IMP is synthesized from glycine (**35**), glutamine (**38**), and aspartic acid (**39**), 5-phosphoribosyl-1-pyrophosphate (PRPP) (**62**), 10-formyl tetrahydrofolate (**40**), and carbon dioxide. The genome information from *Arabidopsis* [149] and rice [150] indicates that plants that synthesize **31** use similar reactions to those found in microorganisms and animals [151]. Purine biosynthesis starts with the formation of phosphoribosylamine (PRA) (**63**) from **62** and **38** (step 1, Fig. 10). This reaction is catalyzed by PRPP amidotransferase. GAR synthetase catalyzes the ATP-dependent formation of glycine amide ribonucleotide (GAR) (**64**) by attaching **35** via an amide bond to **63** (step 2, Fig. 10). Compound **64** is metabolized subsequently in a reaction catalyzed by GAR transformylase involving **40**, to generate formylglycinamide ribonucleotide (FGAR) (**65**) (step 3, Fig. 10). The next step, catalyzed by formylglycinamidine ribonucleotide synthetase, consumes ATP and glutamine (**38**), and leads to the formation of formylglycinamidine ribonucleotide (FGAM) (**66**) (step 4, Fig. 10). Compound **66** then undergoes ring closure to form 5-aminoimidazole ribonucleotide (AIR) (**67**) by consuming an additional molecule of ATP (step 5, Fig. 10). This reaction is catalyzed by AIR-synthase. To build the second ring of the purine skeleton, CO_2, **39** and a second molecule of **40** are inserted. Compound **67** is carboxylated in a reaction

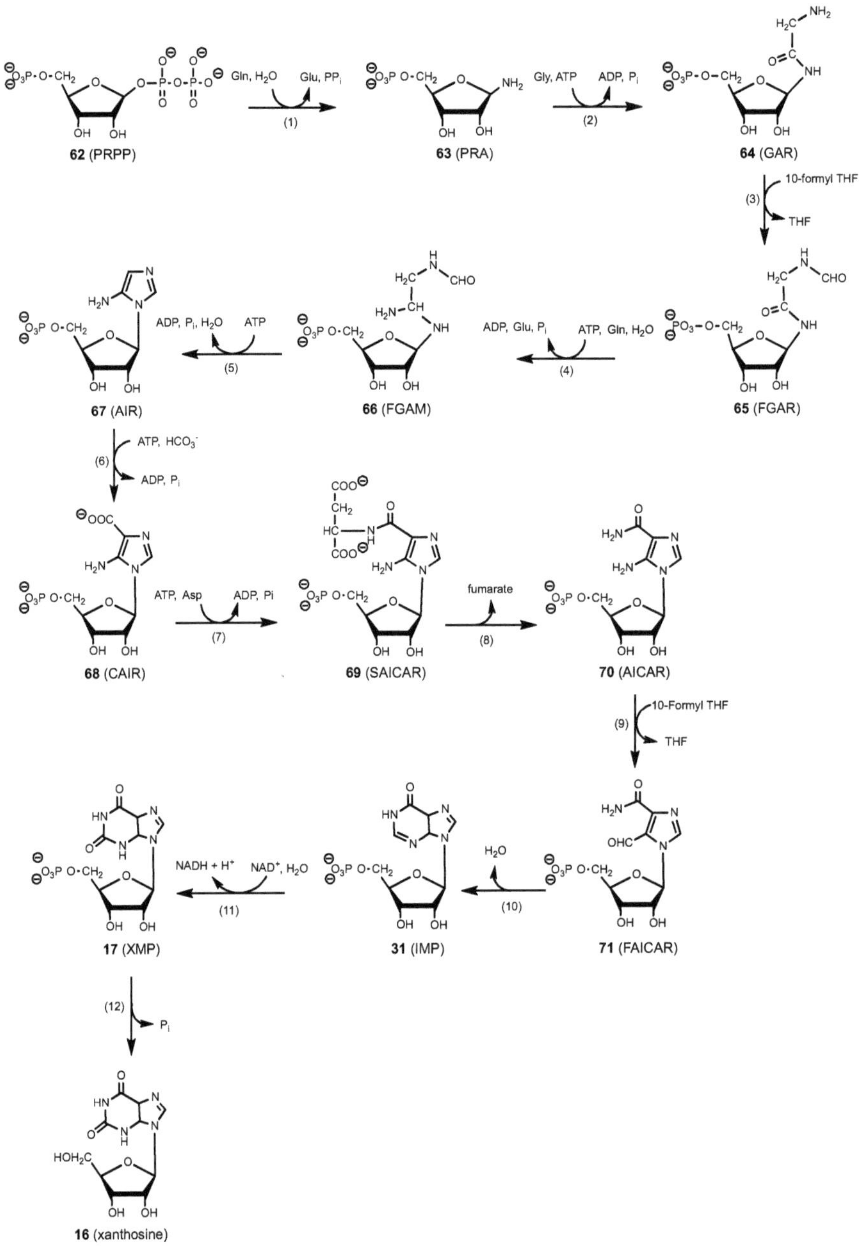

Fig. 10 The de novo biosynthesis pathway of caffeine initiated from 5-phosphoribosyl-1-pyrophosphate. Abbreviations of metabolites are as follows: PRA, 5-phosphoribosyl amine; GAR, glycineamide ribonucleotide; FGAR, formylglycineamide ribonucleotide; FGRAM, formylglycine amidine ribonucleotide; AIR, 5-aminoimidazole ribonucleotide; CAIR, 5-aminoimidazole 4-carboxylate ribonucleotide; SCAIR, 5-aminoimidazole-4-*N*-succinocarboxyamide ribonucleotide; AICAR, 5-aminoimidazole-4-carboxyamide

catalyzed by AIR carboxylase to 4-carboxy aminoimidazole ribonucleotide (CAIR) (**68**) (step 6, Fig. 10). On adding **39** and using a further molecule of ATP, *N*-succinyl-5-aminoimidazole-4-carboxamide ribonucleotide (SAICAR) (**69**) is formed (step 7, Fig. 10). This step is catalyzed by SAICAR synthase. Fumarate is released to build 5-aminoimidazole-4-carboxamide ribonucleotide (AICAR) (**70**) (step 8, Fig. 10), catalyzed by adenylosuccinatelyase. The last two steps that form the first complete purine molecule, **31**, are catalyzed by the bifunctional enzyme 5-aminoimidazole-4-carboxamide ribonucleotide formyltransferase/inosine monophosphate cyclohydrolase. In the first part of this reaction, the final carbon of the purine ring is provided by **40** to form 5-formaminoimidazole-4-carboxamide ribonucleotide (FAICAR) (**71**) (step 9, Fig. 10). Compound **71** undergoes dehydration and a ring closure to generate **31** (step 10, Fig. 10). Xanthosine (**16**) is produced from IMP (**31**) by a two-step reaction catalyzed by IMP dehydrogenase and 5′-nucleotidase (steps 11 and 12 in Fig. 10).

The contribution of de novo purine biosynthesis to caffeine (**8**) biosynthesis is indicated by the incorporation of some small-molecule precursors of purine biosynthesis into **8** in coffee [74] and tea leaves [75–77]. Ito and Ashihara [152] examined whether the de novo pathway of purine nucleotide biosynthesis contributed directly to the biosynthesis of caffeine in tea leaves. The ^{15}N atom from [^{15}N] glycine (**35**) was incorporated into theobromine (**5**) and **8**, and its incorporation was reduced markedly by azaserine (**36**) and aminopterin (**37**), known inhibitors of purine biosynthesis de novo. The radioactivity from [2-^{14}C]5-aminoimidazole-4-carboxamide ribonucleoside (AICA) (**41**), a precursor of AICAR (**70**), was also incorporated into theobromine and caffeine. Pulse-chase experiments with [^{15}N] glycine (**35**) and [2-^{14}C]AICA (**41**) suggested that **5** is the immediate precursor of **8**. Incorporation of radioactivity from [2-^{14}C]AICA (**41**) into purine alkaloids was not influenced by treatment with coformycin (**43**), an inhibitor of plant AMP deaminase. The rate of the incorporation of radioactivity into **5** from [8-^{14}C]inosine (**42**), an immediate precursor of IMP (**31**), was faster than from [8-^{14}C]adenine (**13**), a precursor of AMP (**30**). From these results, it was concluded that most of the purine alkaloids originate from newly produced **31** formed by the de novo pathway for purine nucleotide biosynthesis. Although it is difficult to totally exclude the participation of other routes, the contribution of the de novo pathway to caffeine biosynthesis appears to be important, especially in young tea leaves in which a very rapid accumulation of purine alkaloids is observed.

Fig. 10 (continued) ribonucleotide; FAICAR, 5-formamidoimidazole-4-carboxyamide ribonucleotide; XMP, xanthosine-5′-monophosphate. Enzymes (EC numbers) shown are: (1) PRPP amidotransferase (2.4.2.14); (2) GAR synthetase (6.3.4.13); (3) GAR formyl transferase (2.1.2.2); (4) FGAM synthetase (6.3.5.3); (5) AIR synthetase (6.3.3.1); (6) AIR carboxylase (4.1.1.21); (7) SAICAR synthetase (6.3.2.6); (8) adenylosuccinate lyase (4.3.2.2); (9) AICAR formyltransferase (2.1.2.3); (10) IMP cyclohydrolase (3.5.4.10); (11) IMP dehydrogenase (1.1.1.205); (12) 5′-nucleotidase (3.1.3.5)

3.3.2 AMP Route

A portion of the xanthosine (**16**) used for caffeine (**8**) biosynthesis is derived from the adenine nucleotide pools (route ii in Fig. 5). The cellular concentration of adenine and guanine nucleotides was, respectively, 225 and 26 nmol/g f.w. in young tea leaves [13] and 303 and 60 nmol/g f.w. in young maté leaves [103]. Thus, the biosynthesis of **8** from adenine nucleotides would appear to be quantitatively more important than that from guanine nucleotides. There are several potential pathways for xanthosine (**16**) synthesis from AMP (**30**), but the AMP (**30**) → IMP (**31**) → XMP (**17**) → xanthosine (**16**) route is likely to predominate (route ii in Fig. 5). Coformycin (**43**), an inhibitor of AMP deaminase in plants [153], inhibits the caffeine biosynthesis in tea flower buds [154] and maté leaves [103]. This suggests an involvement of AMP deaminase in the biosynthesis of the purine alkaloids. All three enzymes participating in the conversion, namely, AMP deaminase, IMP dehydrogenase, and 5′-nucleotidase have been detected in tea leaves [106].

3.3.3 SAM Cycle Route

The SAM cycle route (route iii in Fig. 5) is a variation in the route from adenine nucleotides discussed above. The four-step caffeine biosynthesis pathway includes three methylation steps that utilize SAM (**28**) as the methyl donor. In this process, **28** is converted to SAH (**73**), which in turn is hydrolyzed to L-homocysteine (**74**) and adenosine (**44**) (Fig. 11). Koshiishi et al. [106] demonstrated that significant amounts of radioactivity from [methyl-^{14}C]methionine (**72**) and [methyl-^{14}C]SAM (**28**) were incorporated into theobromine and caffeine in young tea leaf segments. In addition, very high SAH hydrolase activity was found in cell-free extracts from young tea leaves. Substantial amounts of radioactivity from [adenosyl-^{14}C]SAH (**73**) were also recovered as theobromine (**5**) and caffeine (**8**) in tea leaf segments, demonstrating that **44** derived from **73** is utilized for the synthesis of the purine ring of **8**. From the profiles of activity of related enzymes in tea leaf extracts, it is proposed that the major route from SAM to **8** is a SAM (**28**) → SAH (**74**) → adenosine (**44**) → adenine (**13**) → AMP (**30**) → IMP (**31**) → XMP (**17**) → xanthosine (**16**) → 7-methylxanthosine (**23**) → 7-methylxanthine (**2**) → theobromine (**5**) → caffeine (**8**) pathway. In addition, direct adenosine kinase-catalysed formation of **30** from **44** may participate as an alternative minor route. From these results, Koshiishi et al. [106] proposed that in young tea leaves, not only the methyl groups, but also the purine ring of caffeine is derived from SAM (**28**). The capacity of the pathway is such that it is possible that the purine ring of caffeine is produced exclusively by this route.

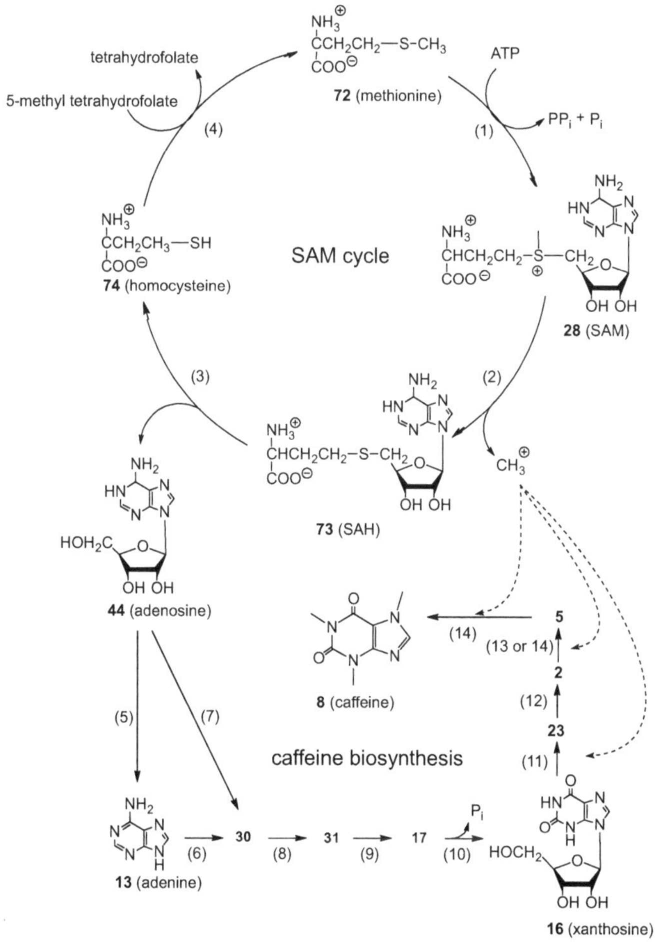

Fig. 11 A possible caffeine biosynthesis pathway from adenosine released by the *S*-adenosyl-L-methionine cycle. Abbreviations of metabolites are as follows: SAM, *S*-adenosyl-L-methionine; SAH, *S*-adenosyl-L-homocysteine. Enzymes (EC numbers) shown are: (1) SAM synthetase (2.5.1.6), (2) SAM-dependent *N*-methyltransferases (2.1.1), (3) SAH hydrolase (3.3.1.1), (4) methionine synthase (2.1.1.13), (5) adenosine nucleosidase (3.2.2.7), (6) adenine phosphoribosyltransferase (2.4.2.7), (7) adenosine kinase (2.7.1.20), (8) AMP deaminase (3.5.4.6); (9) IMP dehydrogenase (1.1.1.205), (10) 5-nucleotidase (3.1.3.5), (11) 7-methylxanthosine synthase (2.1.1.158), (12) *N*-methylnucleosidase (3.2.2.25), (13) theobromine synthase (2.1.1.159), and (14) caffeine synthase (2.1.1.160)

3.3.4 NAD Route

The NAD route (route iv in Fig. 5) is also a variation in the AMP route. If NAD (**33**) is degraded to nicotinamide mononucleotide (NMN) (**34**) and AMP (**30**) by NAD pyrophosphatase, the product, **30**, enters the AMP pathway of caffeine biosynthesis.

The NAD route appears to be important in coffee plants, because another alkaloid, trigonelline (**20**), is formed from nicotinic acid (**18**) derived from **33** [147, 148]. In a recent paper, Baumann [155] considered a hypothesis originally proposed by Kremers in the 1950s [156], and speculated that **33** is a precursor for both **8** and **20**. In this pathway, **33** would be transformed to nicotinic acid-hypoxanthine dinucleotide (**45**) by the simultaneous action of nicotinamidase and a deaminase. In a second step, the "hypoxanthine part" is oxidized by an IMP dehydrogenase to yield nicotinic acid xanthine dinucleotide (**46**). Finally, by the dual action of the same *N*-methyltransferase, **20** and 7-methylxanthine (**2**) are liberated. Although this is an interesting speculation, there is an absence of supporting experimental evidence for this hypothesis.

3.3.5 GMP Route

Xanthosine for caffeine biosynthesis is also produced from guanine nucleotides by a GMP (**29**) → guanosine (**32**) → xanthosine (**16**) pathway (route v in Fig. 5). Guanosine deaminase (EC 3.5.4.15) activity has been found in enzyme extracts from young tea leaves [157]. The synthesis of caffeine (**8**) from [8-^{14}C]guanosine (**32**) in tea leaves [158] has also confirmed the operation of this pathway. Although GMP (**29**) is converted to IMP (**31**) by the action of GMP reductase in bacteria and mammals [159–161], there is no evidence for the presence of GMP reductase in higher plants [158, 162, 163]. Therefore, a potential alternative route, the GMP (**29**) → IMP (**31**) → XMP (**17**) → xanthosine (**16**) pathway, would appear not to be functional in plants.

3.4 Localization of Xanthine Alkaloid Biosynthesis

Secondary metabolites, such as alkaloids, accumulate in the vacuoles of cells of vegetative tissues and seeds and seed coats of fruits. However, biosynthesis is often carried out at other sites and translocated from source cells to sink tissues. For example, nicotine (**47**), a pyridine alkaloid, is synthesized exclusively in the roots, loaded into the xylem, and transported upwards in with the transpiration stream to the leaves [164]. For such a translocation, at least three membrane transport events would seemingly have to be involved, namely, export from the plasma membrane in root cells, import at the plasma membrane in leaf cells, and additional transport into the vacuoles in leaf cells [165]. In contrast to **47**, long-distance transport of caffeine (**8**) has not yet been observed, although intracellular transport to vacuoles may occur.

3.4.1 Organs and Tissues

Xanthine alkaloid biosynthetic activity in different organs and tissues was investigated from the ^{14}C-feeding experiments, determination of enzyme activity and/or expression of caffeine synthase genes. Activity has been detected in leaves, flowers and fruits. In general, high biosynthetic activity was found in young tissues and disappeared with maturation. The activity in roots is low. Details of each plant are described below.

Camellia

Feeding experiments investigating methylxanthine biosynthesis with ^{14}C-labeled substrates have often used leaf tissues and/or fruits. In tea, caffeine biosynthesis from [8-^{14}C]adenine (**13**) was found to occur exclusively in the green chlorophyll-containing buds, leaves, and shoots. The biosynthetic activity of theobromine and caffeine was extremely high in younger leaves, and decreased with the age of the tissue of 4-month-old seedlings. In contrast, very limited incorporation of [8-^{14}C] adenine into these alkaloids was detected in the roots and cotyledons [33]. Similar feeding experiments also indicated that the biosynthesis pathway to caffeine (**8**) was operative in the stamens and petals of flowers and theobromine (**5**) biosynthesis was also detected in flowers of theobromine-accumulating *Camellia irrawadiensis*. The biosynthetic activity was higher in the flower buds than in the flowers [166]. A detailed investigation of caffeine biosynthesis in the stamens isolated from tea flower buds has been also reported [154]. Caffeine biosynthesis is also found in fruits of tea plants. The activity was detected in the pericarp, cotyledons, and seed coats. The biosynthetic activity was high in young tissues of all parts of the tea fruits, especially in the pericarp [167]. The studies on the expression of genes involved in caffeine biosynthesis in the leaves, stems, and roots of 4-month-old tea seedlings also indicated that expression occurred in all organs, but the amounts of transcripts of caffeine synthase (*TCS1*) were much higher in young leaves than in other tissues [168]. The activity of *N*-methyltransferases involved in caffeine biosynthesis was highest in young leaves, but disappeared with subsequent development and growth [128]. Using an RNA in situ hybridization technique, Li et al. [169] reported that the caffeine synthase gene is expressed mainly in the palisade parenchyma and the epicuticle of tea leaves and less so in the spongy parenchyma and hypoderm.

Coffea

In coffee, biosynthetic activity has been detected in leaves and fruits. Transcript accumulation for the identified *N*-methyltransferase genes for caffeine biosynthesis was analyzed by the RT-PCR using gene-specific primer sets. The results indicated plural genes were expressed in different organs and tissues of coffee plants. Mizuno et al. [111, 121] reported that the transcripts of *CmXRS1*, *CTS2*, and *CCS1* were detected in the developing endosperm, young leaves, and flower buds, and the amount of transcripts was highest in the developing endosperm. In the flower buds, marked expression of *CTS2* was detected but the expression of *CmXRS1* and *CCS1* was weak. In young leaves, the amounts of *CTS2* transcript were much smaller than those of *CmXRS1* and *CCS1*. Similarly, Uefuji et al. [124] reported that transcripts for *CaXMT1* were found in all tissues except mature fruits. Transcripts for *CaMXMT1* and *CaMXMT2* were identified at high levels in young leaves, floral buds, and immature fruits. In contrast, those of *CaDXMT1* were detected predominantly in immature fruits.

Feeding experiments revealed that the incorporation of [8-^{14}C]adenine (**13**) into theobromine and caffeine was found in small, young leaves of *Coffea arabica* but it disappeared in fully developed leaves [43]. In 6-month-old *C. arabica* seedlings, theobromine and caffeine were synthesized from [8-^{14}C]adenine (**13**) only in young leaves and young shoots including the buds, but no biosynthetic activity was found in the roots or aged cotyledons [41]. Caffeine biosynthesis activity was found in the pericarp and seeds of *Coffea arabica* and *Coffea canephora* fruits. A high biosynthetic activity, estimated from in situ [8-^{14}C]adenine (**13**) metabolism experiments, was found in small, immature fruits, which consist of perisperm and pericarp, and in developing seeds (endosperm). The biosynthetic activities of caffeine (**8**) were reduced in both the pericarp and seeds with the growth and ripening of fruits [45]. Gene expression studies revealed that the transcripts of three *N*-methyltransferase genes for caffeine biosynthesis *CmXRS1*, *CTS2*, and *CCS1* were detected in every stage of growth, although the amounts of these transcripts were significantly lower in later stages of fruit ripening. The pattern of expression of genes for caffeine synthesis during growth is broadly related to the in situ synthesis of **8** from adenine nucleotides. The amounts of the transcripts of three *N*-methyltransferase genes were higher in the seeds than in the pericarp [45]. These observations strongly suggest that **8** accumulated in coffee seeds is mainly synthesized in the seeds. This contradicts an early paper in which Baumann and Wanner [170] argued that in *Coffea arabica* fruits **8** in the seeds was derived from the pericarp.

Theobroma

The biosynthetic activity of theobromine (**5**) has been investigated in the leaves and fruits of *Theobroma cacao* [56, 61]. Koyama et al. [56] examined purine metabolism in young small leaves, developing, intermediate-sized leaves, and fully developed leaves from flush shoots of cacao trees, and aged leaves from 1-year-old shoots. Biosynthesis of **5** from various purine precursors was found only in young leaves. Although no caffeine biosynthesis from purine bases and nucleosides was observed in the short-term feeding experiments, [8-^{14}C]theobromine (**5**) was converted to caffeine (**8**) in young cacao leaves. The unique accumulation of

5 observed in cacao leaves can be explained by the fact that conversion of **5** to **8** is very slow; **5** is rapidly catabolized in the leaves.

Xanthine alkaloid biosynthesis was studied using three different growth stages of *Theobroma cacao* fruits [61]. Feeding experiments with [8-^{14}C]adenine (**13**) showed that the theobromine (**5**) biosynthesis occurred in small, immature fruits, the pericarp of developing fruits, and cotyledons of large size fruits. No theobromine biosynthesis was detected in the pericarp, placenta, and seed coats. Therefore, very limited amounts of xanthine alkaloids may be transported from the pericarp to seed tissue in young fruits, but most alkaloids appeared to be synthesized in the cotyledons of cacao seeds.

Maté and Guaraná

Biosynthesis of xanthine alkaloids in maté was investigated using ^{14}C-labeled adenine (**13**), guanosine (**32**), and hypoxanthine (**15**), which were incorporated into theobromine (**5**) and caffeine (**8**) in the young, pale-green leaves but not in the mature, dark-green leaves [103].

In guaraná, the expression of caffeine synthase gene (*PcCS*) has been compared with different organs. The gene *PcCS* is expressed in all tissues. Transcript levels in young leaves and the apical stem were much higher than in mature leaves and roots. In fruits, expression was observed in both the seeds and pericarp. The highest expression occurred in immature seeds. Transcript levels decreased markedly as the fruits matured. These results obtained from RT-PCR were confirmed by in situ hybridization analysis; *PcCS* was expressed in the cotyledon of the seeds of immature and intermediate stages of fruit maturation but not in mature fruits.

3.4.2 Subcellular Accumulation of Xanthine Alkaloids

Currently, there is no conclusive evidence on the subcellular localization of methylxanthine alkaloids, although it has been speculated that caffeine (**8**) accumulates in vacuoles of leaf tissue. It has also been proposed that in coffee, **8** occurs in the vacuoles as a complex with chlorogenic acids [171].

3.4.3 Subcellular Localization of Caffeine Biosynthesis Enzymes

Caffeine Synthase

The subcellular localization of caffeine synthase has been the subject of controversy. The results of traditional biochemical fractionation of intracellular organelles demonstrated that caffeine synthase was associated with purified chloroplasts [106, 172]. In contrast, more recent molecular investigations on the localization of caffeine synthase suggest that it is probably located in the cytosol [123] or the external surface of the vacuole [173].

Kato et al. [172] examined the subcellular distribution of caffeine synthase using a Percoll density gradient. The vast majority of the caffeine synthase activity appeared as a sharp peak corresponding with the distribution of chlorophyll. The distribution of triose phosphate isomerase activity, which is a marker for both chloroplasts and cytosol, indicated that little or no caffeine synthase activity was present in the cytosol. They also obtained a crude chloroplast preparation that was washed then purified using a Percoll density gradient. The purified chloroplasts were frozen with liquid N_2 and then stored at −80°C overnight. The thawed samples were centrifuged at 10,000 g and the supernatant was obtained. Since caffeine synthase activity was found in the supernatant, it was concluded that the enzyme is located in the stroma of the chloroplasts [172]. This is also supported by the evidence that the optimum pH for tea caffeine synthase activity was 8.5, which is the typical pH optimum of various enzymes located in the stroma of chloroplasts [174].

A molecular approach to investigate the localization of caffeine synthase in coffee has been carried out by two groups. Ogawa et al. [123] used a cDNA fragment covering the entire coding region of the caffeine synthase gene (*CaMXMT1*) fused to pGFP2. When the resulting plasmid was introduced into the epidermal layer of onions by particle bombardment, green fluorescence was detected in the cytoplasm (cytosol). Kodama et al. [175] confirmed this result and demonstrated that all three *N*-methyltransferases involved in caffeine biosynthesis were localized in the cytosol.

Kumar et al. [173] investigated the localization of caffeine synthase in coffee endosperm cells using the promoter for one of the *N*-methyltransferase gene families involved in caffeine biosynthesis. These constructs and *pCAMBIA 1301* bearing the intron *uidA* gene driven by the cauliflower mosaic virus 35S promoter, were electroporated into coffee endosperm, and the activity of β-glucuronidase (GUS) localized. In tissues transformed with the construct-containing promoter and first exon, enzymatic activity was localized on the outer surface of the vacuole. Antibodies to coffee caffeine synthase were also specifically localized in the same region. In tissues bearing either the caffeine synthase-GUS construct without the first exon or *pCAMBIA 1301* with intron GUS, GUS activity was spread throughout the cytoplasm. These findings suggest that *N*-methyltransferase is located on the external surface of the vacuole.

Enzymes Involved in the de Novo Route

There are several reports on the subcellular localization of enzymes involved in xanthosine (**16**) formation from the de novo route. Although no data are available on the cellular localization of de novo purine nucleotide synthesis in xanthine alkaloid-forming plants, it has been shown in *Arabidopsis thaliana* that all genes encoding enzymes required for the ten-step synthesis of IMP (**31**) from PRPP (**62**) (steps 1–10 in Fig. 10), contain sequences that are predicted to encode *N*-terminal plastid-transit peptides [151, 176]. This observation suggests chloroplast localization of de novo IMP synthesis. Experimental evidence for a chloroplastic localization of these enzymes has been obtained but is fragmentary, showing only localization of the PRPP amidotransferase in *A. thaliana* [177]. In *Vigna unguiculata* (cowpea), AIR synthase was immunolocalized in chloroplasts and mitochondria [178]. The genes encoding IMPDH, which are involved in the conversion of IMP (**31**) to XMP (**17**), do not contain *N*-terminal transit peptides; this observation suggests that the synthesis of **17** from **31** may occur in the cytosol, although experimental evidence supporting this possibility is, as yet, lacking in plants.

Although the situation may be somewhat different from non-symbiotic plants, Atkins et al. [179], investigated intracellular localization of de novo purine synthesis in ureide-forming cow-pea nodules using sucrose (**48**) and Percoll density gradient centrifugation. Enzyme activity analysis showed that both plastids and mitochondria had a full complement of enzymes for de novo purine synthesis, from ribose-5-phosphate (**49**) to IMP (**31**). On the other hand, the less soluble enzymes of purine oxidation including IMP dehydrogenase, were not associated with either plastids or mitochondria.

Enzymes Involved in the AMP Route

AMP (**30**) is synthesized from IMP (**31**) via adenylosuccinate (**50**). The cDNAs encoding the enzymes for this conversion, adenylosuccinate lyase and adenylosuccinate synthase were cloned from several plant species including *A. thaliana* and shown to contain putative plastid transit sequences [180]. Therefore, all steps of AMP biosynthesis from PRPP (**62**), namely, the de novo purine biosynthetic pathway and the AMP branch point pathway from **31** are located in the chloroplasts. Although **30** is mainly produced in the chloroplasts, it is exported to the cytosol and other organelles.

As discussed above, the AMP route of xanthosine synthesis was catalyzed by AMP deaminase, IMP dehydrogenase, and 5′-nucleotidase. Since amino acid sequences of these enzymes do not contain a *N*-terminal transit peptide, these enzymes are probably located in the cytosol.

Enzymes Involved in the SAM Route

Using traditional biochemical isolation techniques, Koshiishi et al. [106] showed that caffeine synthase activity and part of the activities of SAH hydrolase, adenosine nucleosidase, adenine phosphoribosyltransferase and adenosine kinase are located in tea chloroplasts. In contrast, no SAM synthetase activity was detected in the purified chloroplast fraction. Although localization of caffeine synthase was not part of the study, chloroplastic localization of adenosine kinase, SAH hydrolase, and adenine phosphoribosyltransferase has been reported; Schoor et al. [181] used immunogold labeling to detect ADK and SAH hydrolase isoforms in the cytosol, chloroplast and nucleus of *Arabidopsis*. Chloroplastic adenine phosphoribosyltransferase 1 (P31166, AT1G27450) is also registered.

3.4.4 Subcellular Localization and Transport of Intermediates

The precursors of xanthine alkaloid biosynthesis, namely, purine nucleotides may be synthesized in chloroplasts and the final products, typically caffeine (**8**) or theobromine (**5**), accumulated in vacuoles. All enzymes involved in caffeine biosynthesis may be organized in the cells. In some natural product biosynthesis pathways, metabolic channeling with a multi-enzyme complex has been postulated [182].

Using bimolecular fluorescence complementation analyses, Kodama et al. [175] revealed that each of the three *N*-methyltransferases involved in caffeine biosynthesis in coffee form not only homo-dimers, but also hetero-dimers. They speculated that if dimerization among three closely related proteins resulted in minimal functional interference there may be a rapid conversion of the initial substrate to the final product, effectively as a single step. Further studies are necessary to confirm the possible metabolic channeling of caffeine biosynthesis.

3.5 Regulation of Caffeine Biosynthesis

Similar to the biosynthesis of other secondary metabolites, caffeine and theobromine biosynthesis is regulated primarily by gene expression and the subsequent synthesis of enzymes. Changes in the expression rates of genes (transcript level), quantity (activity) of key enzymes of caffeine biosynthesis and accumulation have often been associated with various physiological changes. The expression of *N*-methyltransferases is, for instance, closely related to the accumulation of caffeine (**8**) in tea leaves [168, 183–185], coffee fruits [45, 145], stressed leaves of cacao [186, 187], and guaraná plants [64]. In vitro activity of *N*-methyltransferases is correlated to the expression of enzymes in tea leaves [128].

There are several examples of regulation of caffeine biosynthetic genes in response to salicylic acid (**51**) and methyl jasmonate (**52**) [188], light [189], drought, and salinity [190]. The data suggest that transcriptional regulation of caffeine biosynthesis genes has a major role in the accumulation of caffeine (**8**).

Minor modification of the rate of caffeine biosynthesis may also be regulated by "fine control" comprising the supply of substrates and/or methyl donor and feedback control of enzyme activity. The results obtained with tea tissue culture have shown that adding precursors of purine nucleotides, adenosine (**44**), guanosine (**32**) or hypoxanthine (**15**) to the culture medium, influences the rate of growth but not the accumulation of caffeine (**8**). This suggests that caffeine biosynthesis is not influenced by the availability of purine precursors [191]. In contrast, addition of paraxanthine (**7**), a preferred substrate of caffeine synthase, doubled the level of **8** compared to controls. This indicates that availability of SAM (**28**) is not a principal factor in the control of caffeine biosynthesis and points to 7-methylxanthosine synthase activity being the most plausible rate-limiting factor of caffeine biosynthesis [191].

4 Metabolism of Xanthine Alkaloids

In most xanthine alkaloid-forming plants, caffeine (**8**) is the end product of xanthine alkaloid biosynthesis. However, in a limited number of species, **8** is either further metabolized to methyluric acids or rapidly catabolized via the conventional purine catabolic pathway after following cleavage of the three methyl groups on the purine ring.

4.1 Methyluric Acid Biosynthesis

Sizeable amounts of theacrine (**10**), which was discovered as a minor component of *Camellia sinensis* leaves [27], accumulate in kucha (*C. assamica* var. *kucha*) [28]. The biosynthesis of **10** was investigated using ^{14}C-labeled adenosine (**44**), caffeine (**8**), and SAM (**28**). The feeding experiments indicated that **8** is synthesized in kucha leaves by the same pathway that is utilized in *C. sinensis*, and it is then converted to **10**, probably by successive oxidation and methylation with 1,3,7-trimethyluric acid (**75**) as the intermediate [29] (Fig. 12).

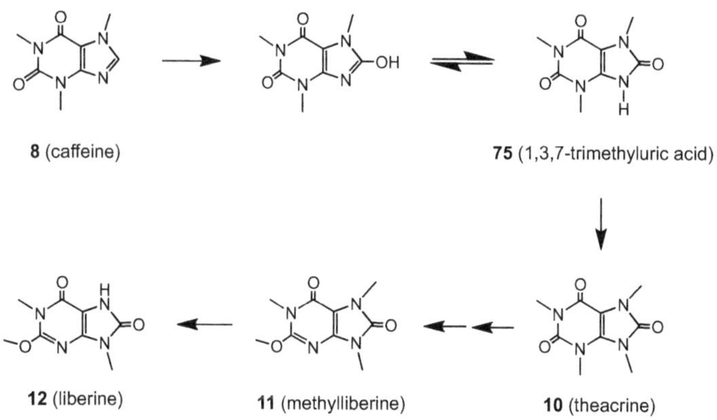

Fig. 12 Biosynthesis pathway of methyluric acids in plants. Enzymes participating in the reactions have not yet been identified

Methyluric acid synthesis in the leaves of *Coffea dewevrei*, *Coffea liberica*, and *Coffea abeokutae* was investigated by Petermann and Baumann [40] using ^{14}C-labeled theobromine (**5**), caffeine (**8**), and theacrine (**10**). In all three species, **5** was converted to **8**, which, in turn, was metabolized to libertine (**12**) via methylliberine (**11**) (Fig. 12).

No enzyme activity that converts caffeine (**8**) to methyluric acids has yet been isolated from plants. Interestingly the bacterium *Pseudomonas* contains a novel caffeine dehydrogenase [192]. This enzyme oxidizes caffeine to trimethyluric acid (**75**) stoichiometrically and hydrolytically, without producing hydrogen peroxide. The enzyme is not $NAD(P)^+$-dependent but coenzyme Q_0 (**53**) is the preferred electron acceptor. Arguably, a similar enzyme may participate in the formation of methyluric acids in plants.

4.2 Biodegradation and Inter-conversion of Xanthine Alkaloids

4.2.1 The Major Pathway of Caffeine Degradation

Compared with the biosynthesis of caffeine (**8**), few studies have investigated the catabolism of xanthine alkaloids. The available data were obtained by feeding radiolabeled substrates to leaves and no activity of the participating enzymes been detected in cell-free preparations.

Biodegradation of caffeine (**8**) to xanthine (**1**) was demonstrated initially in *C. arabica* leaves by Kalberer in the 1960s [193, 194]. He found that radioactivity from ^{14}C-labeled caffeine (**8**) applied to old coffee leaves was incorporated into uric acid and CO_2. In the 1980s, Suzuki and Waller analyzed the radiolabeled products with paper chromatography and showed that both immature and mature coffee fruits degraded [2-^{14}C]caffeine (**8**) to theobromine (**5**), theophylline (**6**), 3-methylxanthine (**3**), 7-methylxanthine (**2**), allantoin (**76**), allantoic acid (**77**), and urea (**78**) [195]. They suggested that biodegradation of **8** occurs in mature, ripened coffee fruits through **5** and **6** as the first biodegradation products. Theobromine (**5**) is involved in both biosynthesis and biodegradation of **8**, whereas **6** is associated primarily with biodegradation [196].

Subsequently, Ashihara et al. [197] performed more sophisticated analysis with HPLC and on-line radioactivity detection with the identity of some metabolites being confirmed by gas chromatography-mass spectrometry. It was concluded that degradation via theophylline (**6**) is the major route for caffeine (**8**) catabolism in leaves of *Coffea arabica*. This study also demonstrated that theobromine (**5**) is not a direct catabolite of caffeine and that catabolism proceeds via a caffeine (**8**) → theophylline (**6**) → 3-methylxanthine (**3**) → xanthine (**1**) pathway (Fig. 13). Catabolism of **8** via **6** has also been demonstrated in *Camellia sinensis* [130], but the pathway was not been detected in maté leaves [198]. Several unknown demethylases may participate in the sequential demethylation reactions. Xanthine

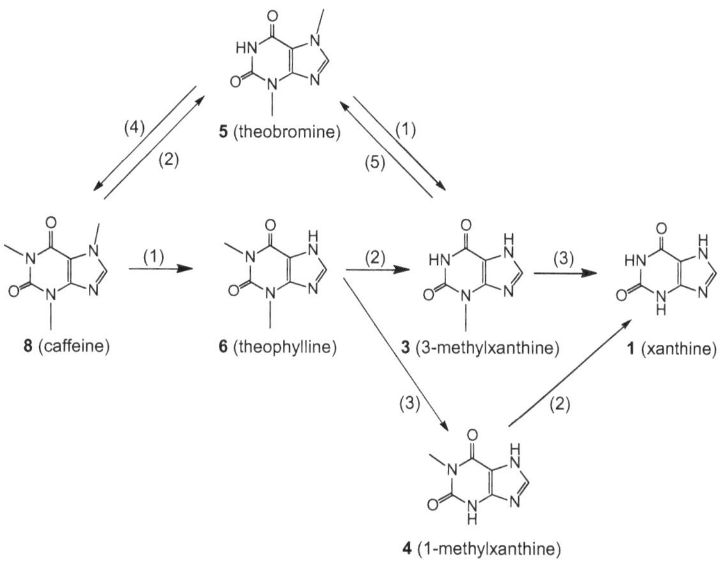

Fig. 13 Inter-conversion of xanthine alkaloids in plants. Possible enzymes: (1) 7*N*-demethylase, (2) 1*N*-demethylase, (3) 3*N*-demethylase, (4) caffeine synthase (1*N*-methyltransferase), and (5) 7*N*-methyltransfease

(**1**), formed by demethylation of **8**, is converted to uric acid (**9**) and then the purine ring is further degraded via allantoin (**77**). Two distinct pathways have been proposed in plants (Fig. 14).

4.2.2 Conventional Purine Catabolic Pathways in Plants

Although detailed enzymatic studies on xanthine (**1**) catabolism have not been performed with caffeine-accumulating plants, the catabolic pathway of **1** may be operative in all plant species via what is termed the conventional purine catabolic pathway [199]. In this pathway, **1** is converted to uric acid (**9**) by xanthine dehydrogenase (step 1, Fig. 14). Allopurinol (**54**) inhibits this reaction. Compound **9** is converted to allantoin (**77**) by uricase (step 2, Fig. 14). Allantoin amidohydrolyase (allantoinase) catalyzes the hydrolysis of the internal amide bond of **76** resulting in its conversion to allantoic acid (**77**) (step 3, Fig. 14). Different metabolic fates of **77** are proposed in plants. In the allantoicase pathway [200], **77** is degraded to CO_2, NH_3 and glyoxylate (**80**) via urea (**78**) and ureidoglycolate (**79**) (steps 4–5 and 9, Fig. 14). An alternative route, the allantoic acid amidohydrolase pathway, has been proposed by Winkler et al. [201] in which **77** is initially converted to ureidoglycine (**81**), CO_2, and NH_3 (steps 6–8, Fig. 14). The NH_3 is released directly and urea formation is not involved in the subsequent catabolism to glyoxylate (**80**). Since ^{14}C-urea (**78**) was found when [8-^{14}C]xanthine

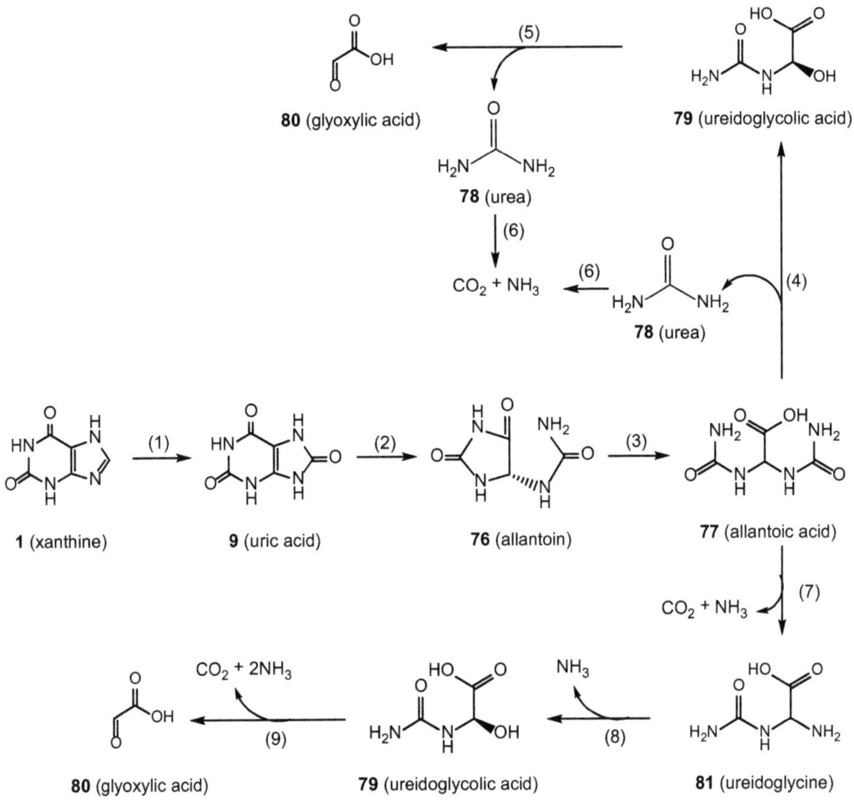

Fig. 14 Catabolic pathways of xanthine in plants. Enzymes: (1) xanthine dehydrogenase (1.1.1.204), (2) uricase (1.7.3.3), (3) allantoinase (3.5.2.4), (4) allantoicase (3.5.3.4), (5) ureidoglycolate lyase (4.3.2.3), (6) urease (3.5.1.5), (7) allantoin deaminase (3.5.3.9), (8) ureidoglycine amidohydrolase (3.5.3.26), and (9) ureidoglycolate amidohydrolase (3.5.1.116). The allantoicase pathway (steps 4–6) and the allantoic acid amidohydrolase pathway (steps 7–9) have been proposed (see text)

(**1**) was administered to leaves of *Camellia sinensis* [202] and *Coffea arabica* [203], the former pathway may operate in tea and coffee plants.

4.2.3 Diversity of Xanthine Alkaloid Metabolism in Plants

Coffea

In *Coffea arabica*, endogenous caffeine levels decrease as tissues age, although substantial quantities remain in old leaves and in fully ripened fruits. Ashihara et al. [197] reported a detailed investigation of caffeine catabolism in which *C. arabica* leaves were incubated with a number of ^{14}C-labeled purine alkaloids. The release of $^{14}CO_2$ from [1-methyl-^{14}C]-, [3-methyl-^{14}C]- and [2-^{14}C]caffeine (**8**) by mature leaves was only 1.0, 0.1 and 0.03%, respectively, of total radioactivity taken up by the leaves, when they were incubated for an 18-h pulse followed by a 24-h chase. Compound **8** is clearly catabolized very slowly by the leaves of *C. arabica*.

Catabolism of **8** has been compared in leaves of three low caffeine-containing coffee species, *Coffea salvatrix*, *Coffea eugenioides* and *Coffea bengalensis* [204]. Like *Coffea arabica* leaves, little or no catabolism was observed in leaf segments of *C. salvatrix* and *C. bengalensis*. However, very rapid catabolism of **8** was observed in *C. eugenioides*. More than 75% of [8-^{14}C]caffeine (**8**) was catabolized in both young and mature leaves, with radioactivity recovered as theophylline (**6**), 3-methylxanthine (**3**), 1-methylxanthine (**4**), xanthine (**1**), ureides (allantoin (**76**) plus allantoic acid (**77**)), urea (**78**), and CO_2. This suggests that [8-^{14}C]caffeine (**8**) undergoes demethylation via theophylline (**6**). In addition to **3**, **4** may also be involved in caffeine catabolism in *C. eugenioides* (Fig. 13). Among three low-caffeine-containing *Coffea* species, only *C. eugenioides* was found to possess a strong capacity for **8** catabolism, with the demethylation of **6** to **1** proceeding mainly via **3** and to a lesser extent via **4**. In contrast, catabolism of **8** is low in *C. salvatrix* and *C. bengalensis*. The low **8** content of these *Coffea* species, thus, appears to be due primarily to their reduced biosynthetic activity.

Silvarolla et al. [38] reported that a natural decaffeinated *Coffea arabica* from Ethiopia (AC1) degraded radiolabeled caffeine (**8**), at a rate similar to that which occurring in *Coffea arabica* cv. Mundo Novo. Therefore, the naturally low **8** content in the AC1 plants is likely to be due to a reduced rate of biosynthesis rather than an enhanced degradation of **8**.

In contrast to the slow metabolism of caffeine (**8**), [8-^{14}C]theophylline (**6**) was catabolized very rapidly in *Coffea arabica* leaves [197]. Most of the [8-^{14}C]theophylline (**6**) taken up by young, mature and aged *C. arabica* leaves was recovered as $^{14}CO_2$. Radioactivity was also associated with allantoic acid (**77**), allantoin (**76**), xanthine (**1**), 3-methylxanthine (**3**), and 7-methylxanthine (**2**). Incorporation of radioactivity into **3** was high in young and mature leaves while the accumulation of **1** and **2** was greatest in aged leaves. The inclusion of 5 m*M* allopurinol (**54**) in the incubation medium had major effects on [8-^{14}C]theophylline (**6**) metabolism. The production of $^{14}CO_2$ declined dramatically, as a consequence of the degradation of **1** being blocked, and there were concomitant increases in the incorporation of label into **1**, **2**, and **3**. It is noteworthy that >70% of the radioactivity recovered from aged leaves treated with **54** was incorporated into **2**. While the presence of **3** and **1** as catabolites of **6** is to be anticipated, the detection of **2**, especially in such large amounts, is unexpected, because previously unknown step(s) are necessary to convert **6** to **2**.

In contrast to theophylline (**6**), theobromine (**5**) was preferentially utilized for caffeine biosynthesis and only small quantities were catabolized. In mature *Coffea arabica* leaves, catabolism of **5** was slightly higher than that of **8**; the release rates of $^{14}CO_2$ from [2-^{14}C]theobromine (**5**) and [2-^{14}C]caffeine (**8**) were, respectively, 5% and 0.03% of total radioactivity after incubation of an 18-h pulse followed by a 24-h chase [197]. This implies that a minor portion of the [2-^{14}C]theobromine (**5**) was subjected to demethylation, yielding xanthine (**1**), which was further

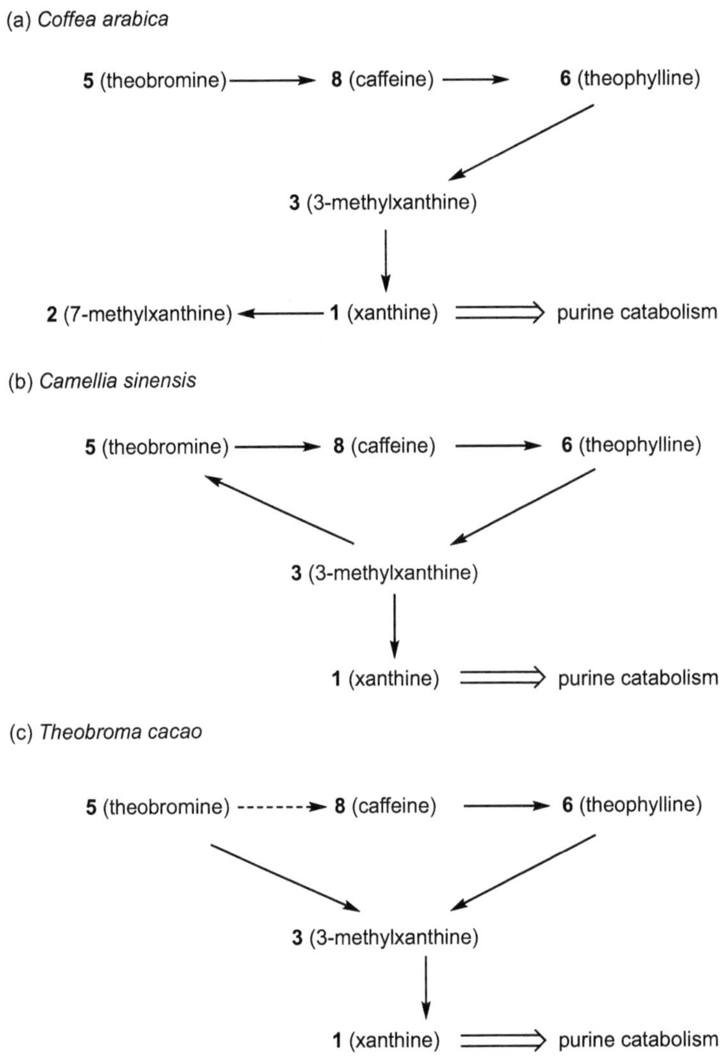

Fig. 15 Diversity of xanthine alkaloid interconversion in plants. (a) *Coffea arabica* [197]; (b) *Camellia sinensis* [130]; (c) *Theobroma cacao* [56, 61]

catabolized to $^{14}CO_2$. Catabolism of **5** to **1** would be via either 3-methylxanthine (**3**) or 7-methylxanthine (**2**), although radioactivity was not incorporated into either compound in detectable quantities during the experimental period. An outline of caffeine (**8**) metabolism in *Coffea arabica* is illustrated in Fig. 15a.

Tea

Ashihara et al. [130] found that only a small amount of $^{14}CO_2$ was released from [8-^{14}C]caffeine (**8**) when young, mature, and aged leaves of *Camellia sinensis* were incubated over a 48-h period and no methanol-soluble radiolabeled metabolites were detected. This finding indicates that the rate of catabolism of **8** is extremely low. Caffeine (**8**) metabolism in high and low caffeine-containing cultivars of *C. sinensis* have been compared [205]. The results suggest that different levels of caffeine (**8**) accumulation is mainly a consequence of caffeine (**8**) biosynthetic activity, although catabolic activity, namely, the rate of release of $^{14}CO_2$ from [2-^{14}C]caffeine (**8**), was about two times higher in the low-caffeine cultivars. Therefore, catabolic activity of **8** may also impact on the content of **8** in tea [205].

Unlike caffeine (**8**), theophylline (**6**) was found to be rapidly metabolized to a variety of compounds. There were marked differences in the metabolic fate of [8-^{14}C]theophylline (**6**) in young leaves compared to mature and aged leaves (Fig. 16). Similar to coffee leaves [197], the catabolism of [8-^{14}C]theophylline (**6**) by leaf disks from mature and aged leaves was rapid with more than 75% of the total radioactivity being recovered as CO_2, 3-methylxanthine (**3**), xanthine (**1**), and allantoin (**76**), with trace amounts of label also being incorporated into theobromine (**5**) and **8** (Fig. 16b, c). The addition of 5 m*M* allopurinol (**54**) caused a marked decline in purine catabolism and with a concomitant increase in the accumulation of [^{14}C]xanthine (**1**). However, conversion of **6** to **8** was not influenced by **54** (Fig. 16). These results demonstrate that theophylline is mainly a precursor of methylxanthine catabolism and that it is not involved in caffeine (**8**) biosynthesis in mature and aged tea leaves.

In young tea leaves, catabolism of [8-^{14}C]theophylline (**6**) was slower than in mature and aged leaves. A marked reduction in $^{14}CO_2$ output suggests that the purine catabolism is less active in young leaves than in the older leaves. A considerable amount (~20%) of radioactivity was incorporated into caffeine (**8**). Since radioactivity from [8-^{14}C]theophylline (**6**) is also incorporated into 3-methylxanthine (**3**) and theobromine (**5**), a theophylline (**6**) $\rightarrow$ 3-methylxanthine (**3**) $\rightarrow$ theobromine (**5**) $\rightarrow$ caffeine (**8**) pathway (Fig. 13), may be functioning in young tea leaves.

In *Camellia sinensis* leaves, theobromine (**5**) is utilized predominantly for caffeine (**8**) biosynthesis [130]. The highest rate of [^{14}C]caffeine (**8**) production from [2-^{14}C]theobromine (**2**) was observed in the mature leaves. Although the results do not agree with the finding that caffeine biosynthesis is most active in young tea leaves [33], it may be a consequence of the conversion of **5** to **8**, the last step of caffeine biosynthesis, being more rapid in mature leaves than in young leaves. The rate-limiting step in the overall caffeine biosynthesis pathway may be earlier, possibly the 7-methylxanthosine synthase reaction, so that the overall capacity for the biosynthesis of caffeine (**8**) is highest in young leaves. No detectable incorporation of radioactivity from [2-^{14}C]theobromine (**5**) into CO_2 or any other metabolites was detected in young leaves [130]. An outline of caffeine (**8**) metabolism in *Camellia sinensis* is illustrated in Fig. 15b.

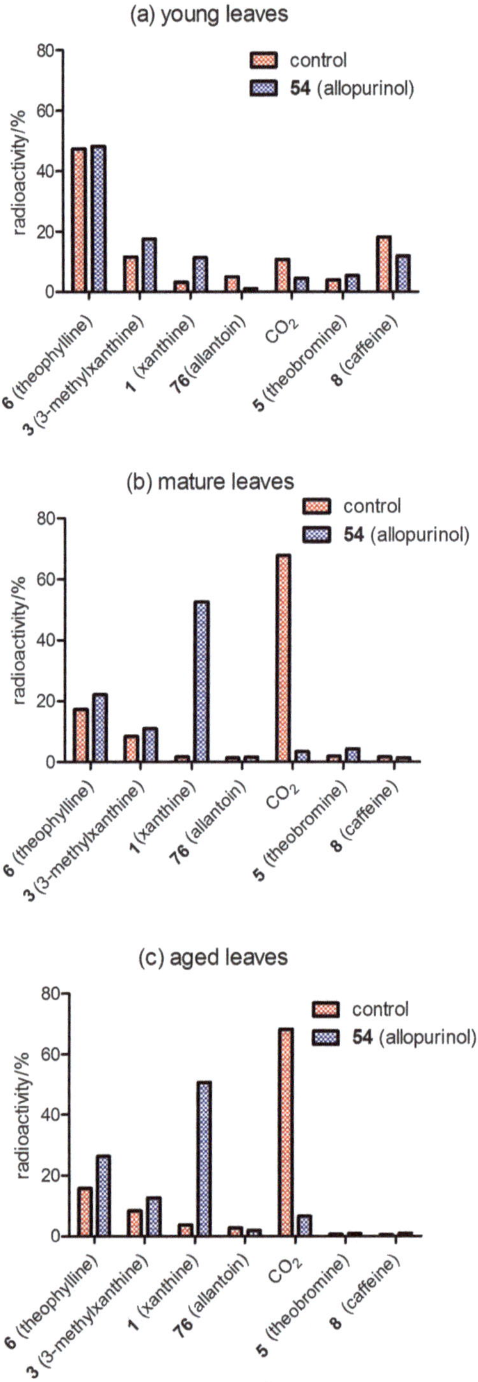

Fig. 16 Summary of the metabolism of [8-^{14}C]theophylline in young, mature and aged leaves of *Camellia sinensis* in the presence and absence of 5 m*M* allopurinol for 24 h. Based on data from Ashihara et al. [130]

Metabolism of xanthine alkaloids has been examined in three other *Camellia* plants; theobromine-accumulating *Camellia ptilophylla* (cacao tea) and *Camellia irrawadiensis* and theacrine-accumulating *Camellia assamica* var. *kucha* (kucha). In *C. ptilophylla* leaf sections only a limited amount of [8-^{14}C]caffeine (**8**) was converted to theobromine (**5**) (6–8% of total radioactivity) and CO_2 (0.2–2%). Most of the [2-^{14}C]theobromine (**5**) (>93%) was recovered unchanged after a 21-h incubation [206].

The metabolism of [8-^{14}C]theophylline (**6**) was investigated in *Camellia irrawadiensis*, and, as shown in *Camellia sinensis* (Fig. 16), **6** was metabolized to 3-methylxanthine (**3**), xanthine (**1**), allantoin (**76**), and CO_2 with a small amount of radioactivity also being incorporated into theobromine (**5**) and caffeine (**8**) (Fig. 17). However, in contrast to *C. sinensis*, complete degradation of the theophylline (**6**) purine ring declined significantly with leaf age and this was accompanied by a relatively high recovery of unmetabolized [8-^{14}C]theophylline from aged leaves (Fig. 17c).

In kucha, caffeine (**8**) was converted to theacrine (**10**) by expanding buds, and young and mature leaves; the conversion to methyluric acid was highest in young tissues [29]. However, most [8-^{14}C]caffeine (8) remained unmetabolized. Small amounts of radioactivity, <3% of total activity, were recovered as theobromine (**5**), ureides (**76** and **77**), CO_2, and an unknown metabolite that arguably could be an intermediate between **8** and **10**.

Almost no [8-^{14}C]caffeine catabolism was detected in young and mature maté leaves during a 24-h incubation period [198]. In young leaves, considerable amounts of radioactivity from [8-^{14}C]theophylline were incorporated in caffeine (**8**) (31%), theobromine (**5**) (5%), and 3-methylxanthine (**3**) (1%). Smaller amounts (<1%) were found in xanthine (**1**), ureides [the sum of allantoin (**76**) and allantoic acid (**77**)], and CO_2. In mature leaves, only <20% of theophylline (**6**) was metabolized and the radioactivity was found in theobromine (**5**) (7%), 3-methylxanthine (**3**) (3%), ureides (**76** and **77**) (7%), and CO_2 (2%).

In young maté leaves, more than 30% of the radioactivity from [8-^{14}C]theobromine (**5**) was incorporated into metabolites, with the most heavily labeled being caffeine, followed by 3-methylxanthine (**3**). In mature leaves, a small proportion of [8-^{14}C]theobromine was converted to **3** and CO_2. In addition to theobromine (**5**), exogenous theophylline (**6**) is utilized for caffeine biosynthesis in maté leaves. However, it has been shown that **6** is not an intermediate in the main caffeine biosynthesis pathway from purine nucleotides [103].

In contrast to coffee and tea leaves, [8-^{14}C]theobromine (**5**) was degraded to CO_2 via 3-methylxanthine (**3**), xanthine (**1**) and allantoic acid (**77**) by *Theobroma cacao* leaves [56]. In young leaves, 12% of the total radioactivity taken up by the leaf segments was found in caffeine (**8**) and less than 4% in **3**. However, in developed and mature leaves, **5** was not utilized for caffeine (**8**) biosynthesis but was degraded to CO_2 via **3**. Thus, **5** is degraded by the conventional purine catabolic pathway.

[8-^{14}C]Caffeine (**8**) was converted to theobromine (**5**) and theophylline (**6**). In young cacao leaves, metabolism of **8** to **5** exceeded conversion to theophylline (**6**).

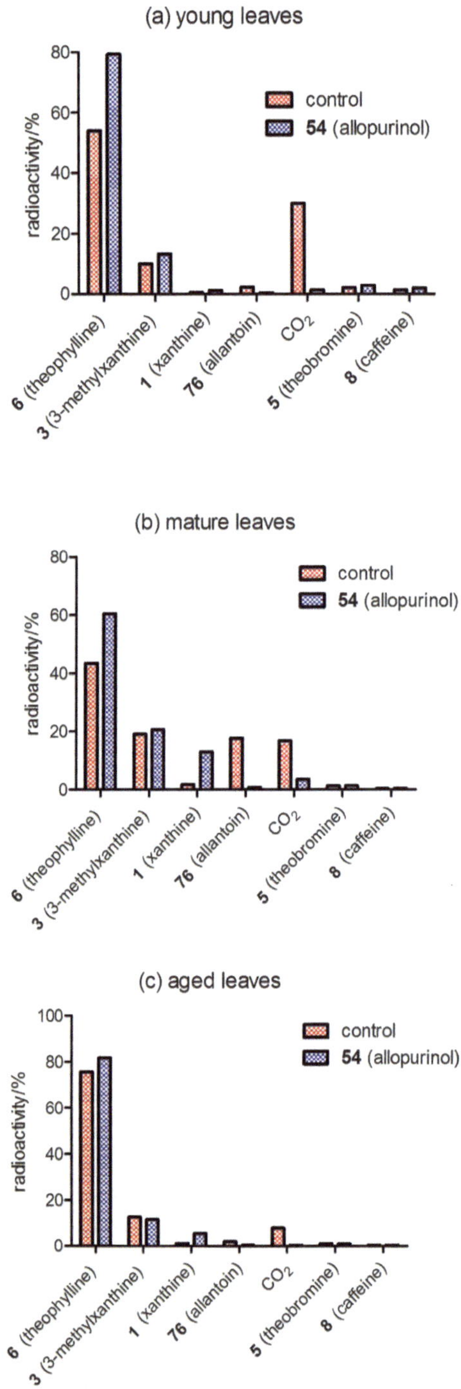

Fig. 17 Summary of the metabolism of [8-^{14}C]theophylline in young, mature and aged leaves of *Camellia irrawadiensis* in the presence and absence of 5 m*M* allopurinol for 24 h. Based on data from Ito et al. [207]

However, in fully developed leaves, where the catabolic activity of xanthine alkaloids was the highest, radioactivity from [8-^{14}C]caffeine (**8**) was incorporated principally in **6** and CO_2. In the oldest leaves, the metabolism of **8** was reduced [56].

Although conversion of caffeine (**8**) to theobromine (**5**) was detected in younger cacao leaves, **8** is mainly catabolized to CO_2 via theophylline (**6**) in the later stages of growth as occurs in coffee and tea leaves (Fig. 15c).

The metabolic fate of [8-^{14}C]theobromine was also examined in cacao fruits at different stages of development [61]. Theobromine (**5**) metabolism was only detected in the pericarp of mature fruits. The radioactivity was recovered in 3-methylxanthine (**3**) (2%), ureides (**76** and **77**) (3%), and CO_2 (29%) (Table 4). The degradation of [8-^{14}C]caffeine (**8**) in the pericarp was also detected; the radioactivity was found in theophylline (**6**) (2%), theobromine (**5**) (1%) and CO_2 (6%). An outline of caffeine (**8**) metabolism in cacao leaves and pericarp is illustrated in Fig. 15c.

Plants as well as other organisms can metabolize exogenously supplied compounds as xenobiotics, which may be converted to non-toxic compounds. Ito et al. [207] reported [8-^{14}C]theophylline (**6**) metabolism in three non-purine alkaloid containing species.

It is noteworthy that compared to species that synthesize purine alkaloids, the uptake and metabolism of [8-^{14}C]theophylline (**6**) by leaf segments of *Avena sativa*, roots of *Vigna mungo*, and suspension cell cultures of *Catharanthus roseus* was extremely low. Relatively small amounts of radioactivity were associated with 3-methylxanthine (**3**) (1–5% of total activity), xanthine (**1**) (0–2%), and CO_2 (0–4%). The effects of allopurinol (**54**) treatment were minimal, confirming that relatively minor amounts of the theophylline were being catabolized to xanthine and entering the purine catabolism pathway.

4.2.4 Xanthine Alkaloid Metabolism in Bacteria and Animals

Bacteria

Although xanthine alkaloid biosynthesis from purine nucleotides does not occur in bacteria and animals, they nonetheless have the ability to metabolize exogenously supplied xanthine alkaloids. Comprehensive details of this topic are outside the scope of this review, but a summary is provided below.

Although high concentrations of caffeine seem to be toxic for microorganisms, several bacteria including *Pseudomonas putida* and *Serratia marcescens* and fungi, such as *Aspergillius niger* and *Penicillium roqueforti*, are able to break down caffeine (**8**) and its catabolites and use them as a nitrogen source for growth [208]. Catabolism of **8** has been extensively studied in bacteria including *P. putida* and *S. marcescens*.

Two distinct pathways of caffeine metabolism occur in bacteria: (i) oxidative and (ii) N-demethylation pathways. Oxidation of **8** to form 1,3,7-trimethyluric acid (**76**) (step 1 in Fig. 18) occurred in a *Rhodococcus* sp.–*Klebsiella* sp. mixed-culture

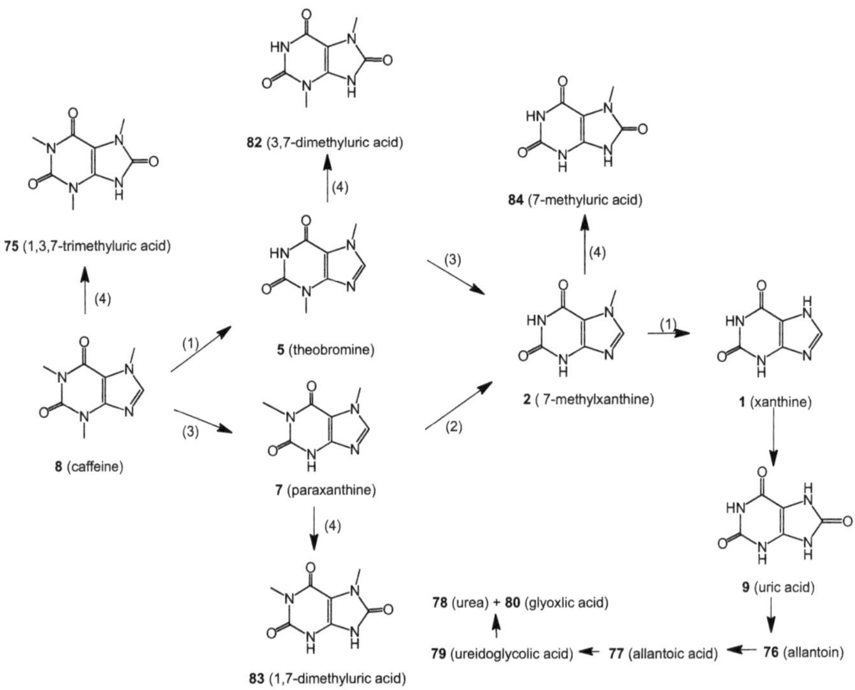

Fig. 18 Caffeine metabolism proposed in bacteria (*Pseudomonas putida*). Possible enzymes: (1) 7*N*-demethylase; (2) 1*N*-demethylase; (3) 3*N*-demethylase, and (4) caffeine dehydrogenase, caffeine oxidase or related enzymes. Adapted from Yu et al. [210] and Mazzafera [208]

consortium [209]. The enzyme involved in the oxidation seems to be a non-$NAD(P)^+$-dependent caffeine dehydrogenase that was observed in *Pseudomonas* sp. [192]. Several caffeine-degrading bacteria metabolize caffeine (**8**) via the N-demethylation pathway and produce theobromine (**5**) and minor amounts of paraxanthine (**7**) as the initial products (steps 2 and 3 in Fig. 18). Subsequently, N-demethylation of **5** or **7** to xanthine (**1**) occurs via 7-methyxanthine (**2**). Xanthine (**1**) is further oxidized to uric acid (**9**) by xanthine dehydrogenase/oxidase as shown in higher plants [210]. In contrast to plants, theophylline (**6**) has not been reported to be a bacterial metabolite of caffeine (**8**). Yu et al. [210] reported that *Pseudomonas putida* strain CBB5 was also capable of utilizing **6** as the sole carbon and nitrogen source. This bacterium N-demethylated theophylline to 1-methylxanthine (**4**) and 3-methylxanthine (**3**), which were both further demethylated to **1**. Two distinct pathways for metabolism of **8** and **6** operate in *Pseudomonas putida*.

A novel methylxanthine *N*-demethylase with broad substrate specificity was purified from *Pseudomonas putida* [211]. It was characterized as a soluble oxygenase composed of two subunits (A and B). The N-demethylation activity was dependent on a specific reductase present in the bacteria, which oxidized NAD (P)H and presumably transferred electrons to the *N*-demethylase for N-demethylation. Genes encoding the two *N*-demethylase subunits and the reductase were identified within a gene cluster. Functional expression of these genes with the recombinant enzymes established *N*-demethylase A and B as individual Rieske oxygenases with highly specific N-1- and N-3-demethylation activities, respectively [212].

Animals

Metabolism of caffeine (**8**) and related xanthine alkaloids has been studied extensively in animals especially in mammals, including humans [213, 214]. Although absorption, bioavailability and the route of excretion were generally similar between humans, dogs, rabbits, rats, and mice, there are interspecies differences in the route of metabolism and enzymes involved in this process. Humans metabolize caffeine mainly via N-demethylation catalyzed by the hepatic cytochrome P450s, CYP1A2, and CYP2E1 [214]. Caffeine (**8**) catabolic pathways in humans are illustrated in Fig. 19.

5 Ecological Roles of Xanthine Alkaloids

There are hypotheses concerning the ecological roles of caffeine (**8**) in plants. The "allelopathic or autotoxic function theory" proposes that **8** in seeds and falling leaves is released into the soil where it inhibits germination of seeds around the parent plants. In caffeine-synthesizing plant tissues, this methylxanthine probably accumulates in compartments, such as vacuoles, which are different from active cellular metabolism. In contrast, **8** absorbed by roots may inhibit various aspects of metabolism in the plant cells.

The "chemical defense theory" proposes that caffeine in young leaves, fruits, and flower buds of species such as tea and coffee acts as a defense to protect young soft tissues from pathogens and herbivores. There are currently several examples indicating that caffeine (**8**) can have a toxic effect. The effective caffeine (**8**) doses for inhibiting the development of various organisms are summarized in Table 11.

5.1 Allelopathic Function Theory

The term "allelopathy" was first used by Hans Molisch in 1937 [215]. It is utilized for chemical interactions whereby a plant inhibits the growth of neighboring plants. Recently, the definition was enlarged to include interaction of plants with microorganisms. This topic is included in this chapter in Sect. 5.2 on chemical defense. Allelochemicals include phenolic acids, glucosinalates, terpenes, and flavonoids. Compared to these secondary metabolites, there is little information on the allelopathic activity of alkaloids [216, 217]. A high allelopathic effect of alkaloids has

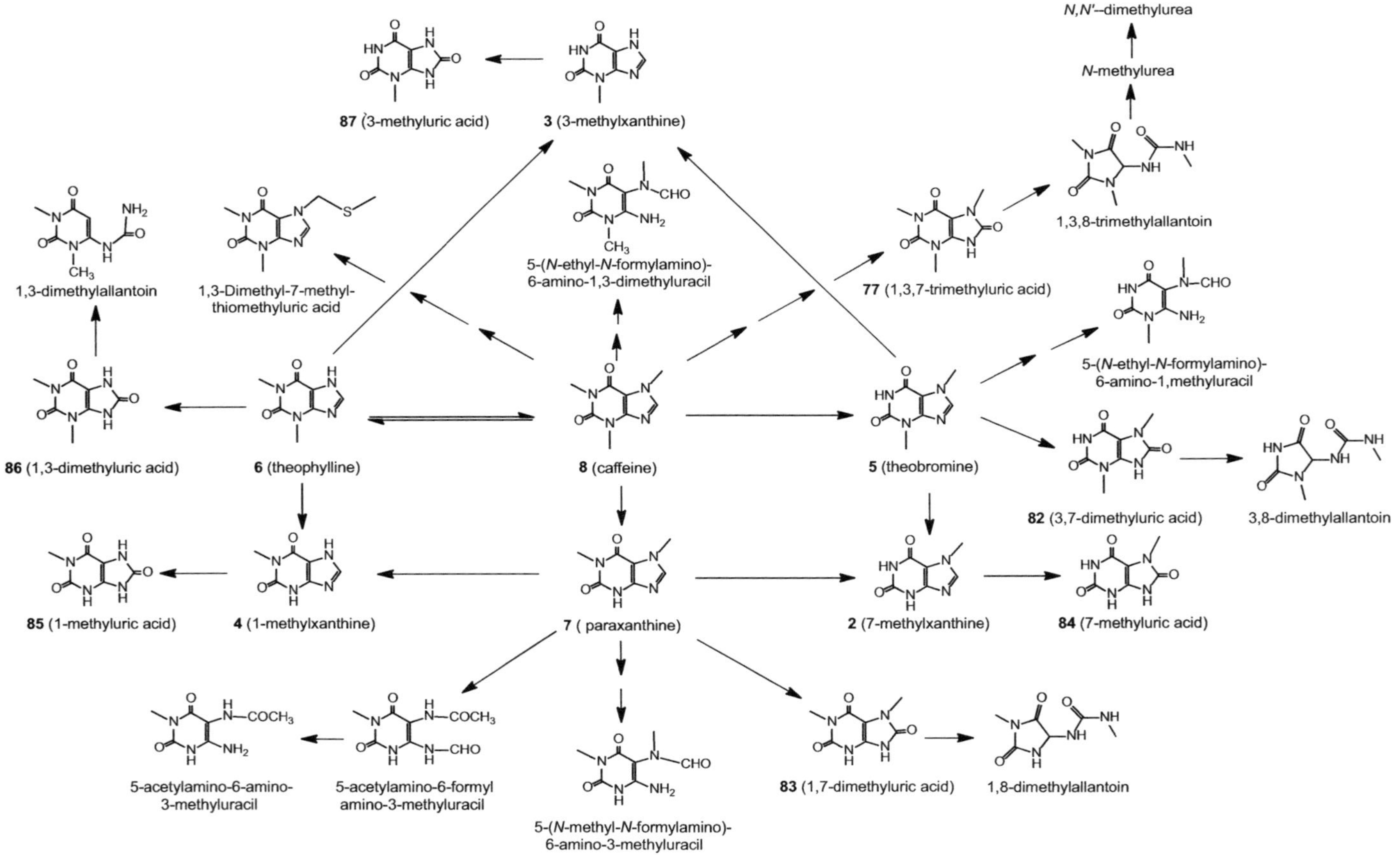

Fig. 19 Caffeine metabolism of humans. Adopted from Arnaud [214]

Table 11 Effective caffeine (**8**) dose for biological activity

Organism	Species	Dose/%
Bacterium	*Pseudomonas syringae*	0.04
Fungus	*Crinipellis perniciosa*	0.05
	Monacrosporium ambrosium	0.05–0.5
	Monacrosporium ambrosium	0.5–1.0
Insect (larvae)	*Pieris rapae* (small white butterfly)	0.05
	Manduca sexta (tobacco horn moth)	0.03–10
	Vanessa cardui (painted lady butterfly)	0.1–0.3
	Tenabrio ssp (mealworm)	0.1–0.3
	Oncopeltus fasciatus (milkweed bug nymph)	0.3
	Culex spp. (mosquito)	0.0007
	Tribolium confusum (confused flour beetle)	0.2
	Tribolium castaneum (red flour beetle)	0.2
Mollusk	*Zonitoides arboreus* (orchid snail)	0.1
	Veronicella cubensis (slug)	0.01–0.1
Plant	*Amaranthus spinosum*	0.06–0.12
	Lactuca sativa (lettuce)	0.03
	Lactuca sativa (lettuce) protoplasts	0.01–0.02
	Avena fatua	0.2–1
	Vicia sativa	0.5–1

Adapted from Kim et al. [252]

been observed in leachates of coffee leaves and roots and extracts from the soil of coffee plantations [218]. Waller et al. [219] reported on the presence of caffeine (**8**) in soils of long-established coffee plantations. As young roots of coffee plants are highly susceptible to **8**, this may cause autotoxicity.

5.1.1 Effect of Xanthine Alkaloids on Germination and Growth of Plants

Around 20% of the caffeine (**8**) in the seeds of germinating *Coffea arabica* is excreted when radicles emerge [220, 221]. Friedman and Waller [220] reported that the growth of radicles was almost completely inhibited by >10 m*M* caffeine (**8**). The tips of radicles were darkened within five days and subsequently died. In addition, 5–20 m*M* caffeine (**8**) reduced hypocotyl extension by ~50%. A similar but milder inhibition was observed with theophylline (**6**). Xanthine alkaloids excreted by coffee seeds, thus, appear to have autotoxic properties.

The effect of xanthine alkaloids on the growth of plant species that do not produce the alkaloids has been investigated. Chou and Waller [222] reported inhibitory effects of alkaloids on the growth of radicles of lettuce seedlings at concentrations of 0.5–2.0 μ*M*. Inhibition by caffeine (**8**), theophylline (**6**), paraxanthine (**7**) and theobromine (**5**) ranged from 54–90%, 58–81%, 62–85%, and 32–64%, respectively. Rizvi et al. [223] reported the selective phytotoxicity of **8** between leguminous plants and some weeds. Germination of *Amaranthus spinosus* and *Echinochloa colonum* seed was inhibited by less than ~6 m*M* caffeine (**8**), while in contrast, it had no significant effect on the germination of *Phaseolus mungo*. Smyth [224] found that 2.5 m*M* caffeine (**8**) caused respectively, 90% and 50% inhibition in the elongation of roots and shoots of rice seedlings.

5.1.2 Effect of Xanthine Alkaloids on Proliferation of Plant Cells

Little is known as to whether or not xanthine alkaloids function as allelochemicals and directly affect cell proliferation of non-purine alkaloid producing plants. Sasamoto et al. [225] reported the effect of four xanthine alkaloids on the proliferation of lettuce cells using a method specifically designed to investigate allelopathy [226]. The effects of caffeine (**8**), theophylline (**6**), paraxanthine (**7**) and theobromine (**5**) on the division and colony formation of lettuce cells were assessed at concentrations up to 1 m*M*. Cell division of protoplasts was significantly inhibited by 0.5 m*M* caffeine (**8**), theophylline (**6**) and paraxanthine (**7**), five days after treatment. After 13 days treatment, 0.25 m*M* caffeine (**8**) had a marked inhibitory effect on colony formation of cells derived from the protoplasts. Other purine alkaloids also acted as inhibitors. The order of the inhibition was caffeine (**8**) > theophylline (**6**) > paraxanthine (**7**) > theobromine (**5**). These observations suggest that a relatively low concentration of **8** is toxic for the proliferation of plant cells. In contrast, **5** is a weak inhibitor. Although the concentration of **8** used with intact plants is much higher than that used for these protoplast studies, **8** may potentially function as an allelochemical in some plant species at the cellular level.

5.1.3 Effect of Xanthine Alkaloids on Plant Metabolism

Traditionally, caffeine (**8**) and related alkaloids are utilized in cytological studies to induce the formation of binucleate cells [227] and as a consequence they are frequently used for a variety of purposes in plant cell research [228–232]. Nevertheless, few studies have investigated the effects of xanthine alkaloids on plant metabolism.

Caffeine (**8**) is a nucleotide derivative, so therefore exogenous **8** may directly influence nucleotide metabolism in plants. The short-term effect of 1 m*M* **8** on purine, pyrimidine and pyridine metabolism was examined in seedlings of rice (*Oryza sativa*) by Deng et al. [233]. Radiolabeled purine, pyrimidine and pyridine compounds were incubated with shoot-root axes of 4-day-old dark-grown seedlings for 6 h and the resultant in situ metabolism investigated. For purines, the incorporation of radioactivity from [8-^{14}C]adenine (**13**) and [8-^{14}C]guanine (**14**) into nucleotides was enhanced by **8**; in contrast, incorporation into CO_2 was reduced. The radioactivity incorporated into the ureides allantoin (**76**) and allantoic acid (**77**) from [8-^{14}C]guanine (**14**) and [8-^{14}C]inosine (**42**) was increased by **8**. For pyrimidines, **8** enhanced the incorporation of radioactivity from [2-^{14}C]uridine (**55**) into nucleotides, which was accompanied by a decrease in pyrimidine catabolism. Such differences were not observed with [2-^{14}C]uracil (**56**) metabolism and **8** did not influence the pyridine metabolism of [carbonyl-^{14}C]nicotinamide (**19**) and [2-^{14}C]nicotinic acid (**18**).

It was concluded that caffeine (**8**) increased the salvage pathways of adenine (**13**), guanine (**14**), and uridine (**55**) in rice seedlings. The increase seems to compensate for the reduction of catabolism as the fate of these purine and pyrimidine compounds shift to salvage reactions. Increased salvage activity of purine bases (**13** and **14**) and pyrimidine nucleosides (**55**) has been observed in the lag phase of proliferation of cultured plant cells [234–238]. Therefore, it can be speculated that **8** increased the nucleotide level for rapid proliferation of cells. However, reverse effects have been observed. Caffeine (**8**) at a concentration of 1 m*M* caused a slight inhibition of the growth of rice seedlings [233]. It can, therefore, be presumed that increments of salvage activity in caffeine-treated rice seedlings are not directly involved in the process of growth. In contrast, catabolites of purine and pyrimidine might be closely related to growth. For example, β-alanine (**57**), a catabolite of pyrimidine nucleotides, is a precursor for pantothenic acid (**58**) synthesis [239]. Allantoin (**77**), a catabolite of purine, plays an important role in the control of the metabolism of abscisic acid (**59**) [240]. Caffeine (**8**) displaces the balance of the conversion reactions of purine and pyrimidine metabolism in favor of nucleotide synthesis. This may initiate conditions for normal growth in rice seedlings [233].

5.1.4 Effect of Caffeine on Protein Expression Profiles

The effect of caffeine (**8**) on the expression profiles of water-soluble proteins was carried out in shoot-root axes of rice seedlings by Deng et al. [241]. Two-dimensional difference gel electrophoresis combined with matrix-assisted laser desorption/ionization time of flight/time mass spectrometry was employed for the separation and identification of proteins. The results indicated that the size of 65 protein spots was changed by treatment with 1 m*M* caffeine (**8**) and 12 proteins were identified from rice databases [241].

Caffeine reduced 51 protein spots and, arguably, growth inhibition might be related to expression of these proteins. Among them, caffeine markedly reduced the levels of five proteins, namely, β-tubulin, sucrose synthase, glyceraldehyde-3-phosphate dehydrogenase, reversibly glycosylated polypeptide/α-1,4-glucan protein synthase, and cytoplasmic malate dehydrogenase. Since microtubules, tubular polymers of tubulin, are a component of the cytoskeleton, a decrease of this protein could cause a reduction in cell division. Decrease of UDP-arabinopyranose mutase may also be closely related to the reduction of cell wall components, which may inhibit the growth of seedlings. Caffeine (**8**) also reduced the enzymes involved in carbohydrate metabolism. This may inhibit sucrose degradation and subsequent utilization of sucrose (**48**) for biosynthesis and energy metabolism.

The size of 14 protein spots was increased by caffeine (**8**) treatment, while quantities of seven proteins, alanyl-aminopeptidase, acetyl-CoA carboxylase, adenine phosphoribosyltransferase, NAD-malate dehydrogenase, ornithine carbamoyltransferase, glucose-6-phosphate isomerase, and nuclear RNA binding protein were substantially higher than in control samples. Caffeine-mediated enhancement of proteins is of interest. The data obtained in this study showed only accumulation of these proteins in the caffeine-treated seedlings and, therefore, further studies including fluctuations of enzymatic activities and amounts of these proteins during germination are necessary to elucidate the effects of **8** more fully.

5.1.5 Allelopathy in Natural Ecosystems

Experimental evidence from laboratory studies, discussed above, suggests the xanthine alkaloids have allelopathic activity. However, it is unclear to what extent caffeine (**8**) is involved in allelopathy in natural ecosystems because soil bacteria, such as *Pseudomonas putida*, degrade xanthine alkaloids (see Sect. 4.2.4).

The caffeine (**8**) concentration in soil around mature, old coffee trees has been reported; levels in Mexican canopy soil indicated that considerable amounts of the alkaloid are released from the tree litter (fallen leaves and fruits) and over the years accumulate in the vicinity of the roots. The annual amount of litter from mature coffee trees is estimated to be ~150–200 g dry matter/m^2/year, and this may release ~1–2 g caffeine/m^2/year [242]. Part of this may be degraded by soil bacteria, but Waller [243] argued that the antimicrobial activity of **8** may reduce catabolism of alkaloids in the soil and so prolong its retention and increase its accumulation.

5.2 *Chemical Defense Theory*

5.2.1 Chemical Defense Against Microorganisms and Animals

In 1984, James Nathanson [244] was the first to report that natural and synthetic xanthine alkaloids inhibit insect feeding and are pesticidal at concentrations known to occur in plants. These effects are due primarily to an inhibition of phosphodiesterase activity and an increase in intracellular cyclic AMP (**60**). At lower concentrations, methylxanthines are potent synergists of other pesticides known to activate adenylate cyclase in insects. These findings suggest that xanthine alkaloids may function as natural insecticides and that phosphodiesterase inhibitors, alone or in combination with other compounds, may be useful in insect control. Similar results are also reported by Hollingsworth et al. [245, 246], who reported that spraying tomato leaves with a 1% solution of **8** deters feeding by tobacco hornworms, while treatment of cabbage leaves and orchids with 0.01–0.1% solutions of **8** acted as a neurotoxin and killed or repelled slugs and snails.

Therefore, the high caffeine (**8**) content of young leaves, fruits and flower buds of species such as *Coffea arabica* and *Camellia sinensis* may act as a defense to protect young soft tissues from pathogens and herbivores [42, 128]. Caffeine (**8**) synthesized in response to infection of cacao leaves might inhibit the growth of fungal and bacterial pathogens and restrict their proliferation beyond the original site of infection [247].

5.2.2 Proof of the Chemical Defense Theory Demonstrated with Transgenic Plants

Recently, the chemical defense theory was verified experimentally by Sano and his co-workers using transgenic caffeine-forming plants. They reported that transgenic tobacco and chrysanthemum plants which produced caffeine (**8**) have, respectively, repellent effects against tobacco cutworm (*Spodoptera litura*) and fungal resistance [120, 248–250].

Kim and Sano [251] found that caffeine-producing transgenic plants constitutively expressed defense-related genes encoding pathogenesis related proteins 1a (PR1a) and proteinase inhibitor II (PI-II) under non-stressed conditions. Transgenic tobacco plants were highly resistant against tobacco mosaic virus and *Pseudomonas syringae*. Expression of *PR1a* and *PI-II* was higher in transgenic plants than in wild-type plants following infection. Exogenous caffeine (**8**) applied to wild-type tobacco leaves conferred similar resistance properties. These findings indicate that **8** acts as a signal molecule activating the defense system of the host plants by directly or indirectly initiating gene expression. Recently, it was shown that **8** induces the production of a mildly toxic secondary metabolite in planta that stimulates endogenous self-defense systems, thereby conferring tolerance or resistance against biotic stresses [14, 252]. These results, of what are essentially greenhouse-based studies, strongly suggest that **8** can have a key role in the chemical defense of plants. However, its ecological function in plants that produce caffeine (**8**) in a natural ecosystem remains to be determined.

6 Biotechnology of Xanthine Alkaloids

Experiments are often carried out on the production of useful phytochemicals by cell and tissue cultures. Caffeine (**8**) found in medicines and soft drinks, and caffeine required for these products is obtained by either large-scale chemical synthesis or as a by-product of procedures used to decaffeinate tea and coffee. As a consequence, caffeine (**8**) production by tissue and cell cultures of coffee and tea has had little economic impact. Early studies on tissue culture of caffeine-forming plants were summarized in a prior review [13].

Using the gene sequences of *N*-methyltransferases involved in caffeine biosynthesis, two types of transgenic plants have been made. One is the construction of genetically modified decaffeinated coffee and tea plants, in which **8** production is suppressed. The other is the introduction of caffeine biosynthesis into non-caffeine-producing crops in order to possess resistance against herbivores. Examples of transgenic plants using caffeine synthase-related genes are shown in Table 12.

6.1 Decaffeinated Coffee and Tea Plants

It is known that caffeine (**8**) often induces insomnia in humans. Furthermore, heavy coffee drinkers, especially the elderly, sometimes suffer unpleasant side effects from **8** including palpitations, gastrointestinal disturbances, anxiety, and tremor. A restricted daily coffee intake is also advised for those with liver disease and pregnant women [253]. It is against this background that there has been an increasing demand for decaffeinated coffee, with "decaf" now accounting for ~10% of coffee consumption worldwide [254]. The flavor and aroma of decaffeinated coffees have improved greatly in recent years as more sophisticated decaffeination methods have been introduced. The latest procedure involves the use of supercritical fluid extraction with liquid CO_2. Such methods require considerable investment in machinery and expenditure of energy, and, in the long-term, the demand for decaffeinated coffee could be better met by the use of a traditional breeding programme. Unfortunately, it would require a long time to establish new hybrid coffee plants on a commercial scale. The use of genetically modified plants may overcome this problem. Production of decaffeinated transgenic plants that were obtained by the introduction of antisense or double-stranded RNA interference (RNAi) constructs of the *N*-methyltransferases related genes has been reported for both coffee [255–258] and tea [259].

The studies on transgenic decaf coffee plants was initiated by Sano's group at the Nara Institute of Science and Technology in Japan. The first approach was to construct transgenic coffee plants with reduced caffeine content by the RNA interference method, in which mRNA of the target gene is selectively degraded by small double-stranded RNA species [255–257]. The 30-untranslated region and the coding region of *CaMXMT* cDNA were selected to design the RNAi constructs. Two different RNAi constructs, RNAi-S having a short insert with 150 bp, and RNAi-L with a long insert of 360 bp, were inserted into the pBIH1-IG vector, which was introduced into the EHA101 strain of *Agrobacterium tumefaciens* to transform *Coffea arabica* and *Coffea canephora* plants. The resulting transformed lines were assayed for expression of *N*-methyltransferase genes by RT-PCR, and it was found that the *CaMXMT*-RNAi construct suppressed transcripts for not only *CaMXMT* but also *CaXMT* and *CaDXMT*. The homogeneity between *CaMXMT*, *CaXMT* and *CaDXMT* is over 90% in the coding region [124], suggesting that the primary small double-stranded *CaMXMT*-RNA progressively produces many secondary small double-stranded RNAs spanning its coding region to the adjacent sequence of the initiator region, which, in turn, destroys mRNAs for *CaXMT* and *CaDXMT*. The reduced level of transcripts indicated decreased activities of the corresponding

Table 12 Transgenic plants using genes involved in caffeine biosynthesis

Plant species	Transferred gene	Method	Xanthine alkaloid content	Purpose	Reference
Coffea canephora *Coffea arabica*	*CaMXMT1*	Antisense, RNAi	3–10 mg/g f.w.	Decaffeinated coffee	[255]
Camellia sinensis	*Caffeine synthase* (*FJ554589*)	RNAi	15–40 mg/g d.w.	Decaffeinated tea	[259]
Nicotiana tabacum	*CaMXMT1*, *CaMXMTMT1/2*, *CaDXM1*	Sense	3–5 μg/g f.w.	Endogenous pesticides	[120]
Dendranthem grandiflorum cv. Shinba[a]	*CaMXMTCaDXM1*	Sense	3–4 μg/g d.w.	Fungal resistance	[249]

[a]Chrysanthemum

enzymes, and this was confirmed by directly measuring their products, theobromine (**5**) and caffeine (**8**). The caffeine (**8**) content in the controls was ~8.4 mg/g f.w., while that in both RNAi-S and RNAi-L was ~4.0 mg/g f.w., showing an average 50% reduction. However, the amount was variable depending on the line, with one notable example of *Coffea canephora* showing up to 70% reduction [255].

The caffeine biosynthesis activity in leaves of these low caffeine (**8**) transgenic plants is reduced. Metabolic studies indicate that substrates of caffeine biosynthesis are catabolized via the conventional purine catabolic pathway and, as a consequence, other than a reduced caffeine (**8**) content, these plants have the same profile of cellular metabolites as wild-type tissues [260]. Currently, caffeine-deficient beans from transgenic *Coffea arabica* plants have not been produced. When this is achieved with a >90% reduction in the content of **8**, because of the substantial market for decaffeinated coffee, it will at least have the potential to have major commercial implications.

Tea is a woody perennial crop with a long life cycle, self-incompatibility, and high inbreeding depression. These characteristics usually limit its genetic improvements by conventional breeding methods. To produce low-caffeine tea, conventional breeding may take 25 years or more [261]. Yadav's group at the Institute of Himalayan Bioresource Technology in India attempted to produce a selected cultivar of tea with a low-caffeine content using RNAi technology. To suppress the caffeine synthesis in the selected tea, *Camellia sinensis* cv. Kangra jat, they isolated a partial gene fragment of caffeine synthase (*CS*), which catalyzed the conversion of 7-methylxanthine (**2**) to caffeine (**8**) via theobromine (**5**), from the same cultivar and used it to design a RNAi construct (pFGC1008-CS). Somatic embryos were transformed with the developed construct using a biolistic method. Transformed somatic embryos showed reduction in the levels of *CS* transcript expression as well as in caffeine (**8**). Plants were regenerated from the transformed somatic embryos. Transgenic plants showed a significant suppression of *CS* transcript expression and also showed a reduction of 44–61% in **8** and 46–67% in **5** compared to the controls. These results suggest that the RNAi construct using a single partial fragment of the *CS* gene significantly reduced the expression of the targeted endogenous gene. The resultant transgenic plants were morphologically similar at maturity [259], but it is necessary to confirm that, except for the low xanthine alkaloid content, the chemical constituents are essentially same as a normal tea. The degree of reduction in the content of caffeine (**8**) and their effects on the tea production in terms of quality and yield need to be evaluated upon maturity. The stability and heritability pattern of this low caffeine (**8**) and theobromine (**5**) trait should also be studied to establish commercial decaf tea products.

6.2 Caffeine-Producing Transgenic Plants

As shown above, the ecological function of caffeine (**8**) may constitute a part of the chemical defense systems against pathogens and herbivores [125]. Sano's group successfully constructed caffeine-producing transgenic plants, with species that would not normally synthesize the purine alkaloid, and examined tolerance traits against biotic attackers [120]. A multi-gene transfer vector (pBINNMT777) was constructed using *CaXMT1*, *CaMXMT1*, and *CaDXMT1*, and all were driven by the cauliflower mosaic virus 35S RNA promoter and nopaline synthase (NOS) terminator. The T-DNA region of the plasmid also contained genes for neomycin phosphotransferase (NPT II) and hygromycin phosphotransferase (HPT) as the antibiotic resistance marker. The pBIN-NMT777 was introduced into tobacco [120] and chrysanthemum [249].

In transgenic tobacco leaves, the caffeine (**8**) concentration varied from 0 to 5 μg/g f.w. The highest concentration was found in aged leaves when plants entered the reproductive stage and formed flower buds. Immature fruits also contained **8** but at a lower level (<1.3 μg/g f.w.). In transgenic chrysanthemum, the caffeine (**8**) content of fully matured leaves was 3 μg/g f.w. Transgenic plants were tested for their tolerant traits against various biotic stresses including herbivore and pathogen attacks.

6.2.1 Antiherbivore Activity

Use was made of larvae of the tobacco cutworm (*Spodoptera litura*) and beet armyworm (*Spodoptera exigua*), which are severe pests for many crop plants. When presented with fresh tobacco leaves containing caffeine (**8**) at 5 μg/g and wild-type leaves, caterpillars positively avoided the transgenic leaves [120]. The repellent effect was also observed with low caffeine (**8**) content leaves (0.4 μg/g tissue), with larvae eating only 4% of the transgenic and up to 32% of the wild-type leaves [120]. A similar result was obtained with caterpillars of the beet armyworm feeding on transgenic *Chrysanthemum* producing **8** at 3 μg/g tissue [248].

6.2.2 Antipathogen Activity

More than 70% of known plant disease is caused by fungi, and 30% by bacteria, viruses and other pathogens [14]. Transgenic tobacco and chrysanthemum plants that produce caffeine (**8**) were examined for their response to pathogens. *Chrysanthemum* plants were infected with a necrotrophic fungus grey mold (*Botrytis cinerea*), which causes death of the flowers, leaves, buds, and fruits of many plant species [249]. In the wild-type plant, lesions appeared 72 h after inoculation and rapidly spread from the infected site to the rest of the leaf. The lesion size exceeded 16 mm in diameter five days after infection. In the transgenic lines, the lesions appeared 90 h after inoculation, and were smaller than the controls, ranging from 1 to 9 mm in diameter.

To investigate resistance against microbial pathogens, use was made of *Pseudomonas syringae* pv. *glycinea*, which causes wild-fire disease. In wild-type plants, distinct lesions were formed 24 h after infection and developed into severe necrosis after 48 h. In contrast, lesion development was strongly inhibited in the transgenic line even 48 h after inoculation [251]. The number of propagated bacteria was lower in the transgenic line than in the control.

Resistance against viral pathogens was also examined [251]. Tobacco mosaic virus (TMV) has a wide host range of over 120 plant species, and causes mottled patterns on leaves, ultimately leading to plant death. When healthy leaves of wild-type and transgenic tobacco plants were inoculated with TMV and kept at 30°C, they do not recognize infection and the virus particles propagated. Upon a temperature shift to 23°C, the hypersensitive response (HR) takes place, and a series of defense systems begin to operate. Physiologically, these responses can be assessed visibly by the formation and development of necrotic lesions. With wild-type plants, lesions appeared 48 h after the temperature shift, and develop further over a subsequent 48 h period. In transgenic lines, lesions were similarly formed 48 h after a temperature shift but did not further develop. In addition to slow lesion formation, the size of lesions on transgenic lines was much smaller. The total number of lesions was also reduced in transgenic lines, with only 15–30% of the control plants. These results point to the effectiveness of caffeine (**8**) in conferring tolerance against a wide range of pathogens.

Exogenously applied caffeine (**8**) is toxic for a diversity of organisms at an average concentration ranging from 100 μg to 3 mg/g f.w. (Table 11). For example, cabbage leaves sprayed with 0.1% caffeine (**8**) solution were toxic for snails and slugs. In contrast, the amount of **8** produced in transgenic plants was at most 5 μg/g f.w., which is 2–3 orders of magnitude lower. Arguably, these low internal levels of endogenous **8** activate the defense responses, possibly via salicylic acid (**51**)-induced expression of pathogenesis-related proteins. The higher concentrations of exogenous caffeine are themselves toxic to bacteria and fungi on the leaf surface while much smaller amounts may be absorbed and act in the same way as the endogenous **8**.

Despite the relatively low internal concentration of caffeine (**8**), the transgenic plants showed a tolerance against pests and pathogens. This apparent discrepancy in effective concentration raised a question as to whether or not **8** in transgenic plants is directly toxic for organisms. It is conceivable that endogenously produced **8** induced some chemical changes in the leaves, thereby indirectly affecting the plant defense responses by a different mechanism [14].

The concept of the activation of plant defense systems by caffeine (**8**) was proposed by Sano et al. [14]. Their simplest model for the action of **8**, the caffeine signal pathway, in enhancing the defense response, is summarized in Fig. 20. The molecular cascade of the caffeine signal may be described as follows: Step 1: Caffeine (**8**) directly blocks phosphodiesterase (PDE); Step 2: Degradation of cAMP (**60**) is inhibited and its level increases; Step 3: Increased **60** activates the cyclic nucleotide-gated channel (CNGC); Step 4: Activated CNGC increases cytosolic Ca^{2+} level; Step 5: Increased Ca^{2+} activates calcium-dependent protein kinase (CDPK) activity; Step 6: Activated CDPK directly phosphorylates phenylalanine

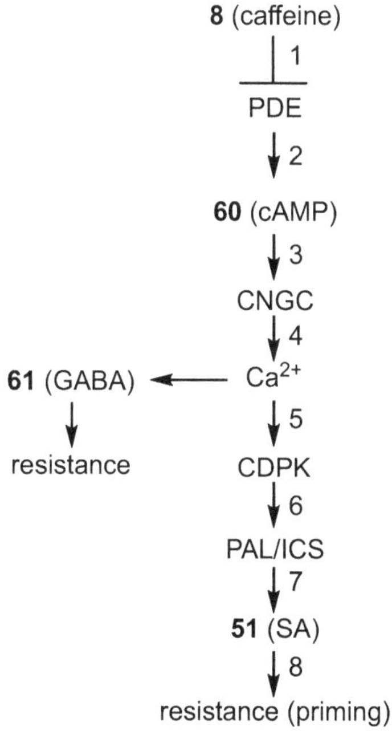

Fig. 20 Caffeine signal cascade hypothesis in plants proposed by Sano et al. [14]. The molecular cascade of caffeine signal is predicted based on available literature and some experimental data. See text. PDE, phosphodiesterase; CNGC, cyclic nucleotide-gated channel; CDPK, calcium-dependent protein kinase (CDPK) activity; PAL, phenylalanine ammonia lyase; ICS, isochorismic synthase; SA, salicylic acid

ammonia lyase (PAL) and/or isochorismic synthase (ICS); Step 7: Activities of phosphorylated PAL and ICS increase, resulting in acceleration of salicylic acid (SA) (**51**) production, and Step 8: The increased level of **51** primes the defense response. $GABA_A$ receptors and adenosine receptors have not been identified in plants. However, increased Ca^{2+} induces γ-aminobutyric acid (GABA) (**61**) accumulation, which enhances the defense reaction.

Although transgenic tobacco and chrysanthemum plants contain only a low amount of caffeine (**8**) they nonetheless exhibited a strong tolerance against herbivores and pathogens. The defense system was autonomously activated in the absence of external stresses. This can be regarded as the priming of defense responses, by which host plants go on standby to cope with a broad range of biotic stresses. This feature resembles mammalian immunization or vaccination, and it was proposed that plants can also be immunized by expressing a mildly toxic "antigenic" chemical, such as **8**, in planta. This hypothesis was proposed by Sano et al. [14] and while fascinating, direct evidence has yet to be obtained.

7 Summary and Perspectives

A summary of this contribution is illustrated in Fig. 21. Since highly purified caffeine synthase was obtained from tea leaves, a series of caffeine synthase-like *N*-methyltransferases belonging to the motif B′ family have been cloned and the properties of these recombinant enzymes in coffee plants characterized. Interestingly, these genes are not only related to the biosynthesis of caffeine (**8**) but also that of trigonelline (**20**), which also accumulates in coffee seeds at similar concentrations to **8**.

Disputes on the biosynthesis schemes with speculations based on complicated feeding experiments and the results from crude and labile enzyme preparations are now almost resolved. The main caffeine biosynthesis pathway is confirmed as a xanthosine (**16**) → 7-methylxanthosine (**23**), → 7-methylxanthine (**2**) → theobromine (**5**) → caffeine (**8**) pathway. However, the gene of the nucleosidase that catalyzes the conversion of 7-methylxanthosine (**23**) → 7-methylxanthine (**2**) has not yet been cloned. In contrast to their biosynthesis, the catabolism of xanthine alkaloids has not yet been elucidated at the molecular level, although feeding experiments indicate that demethylase reactions are involved and xanthine is temporally formed. Xanthine (**1**) is further catabolized by the conventional purine catabolic pathway via allantoin (**76**). For an ecological role of caffeine (**8**), the allelopathy and the chemical defense theories, has been postulated. The occurrence of chemical defense has been substantiated using caffeine-producing transgenic plants at laboratory level.

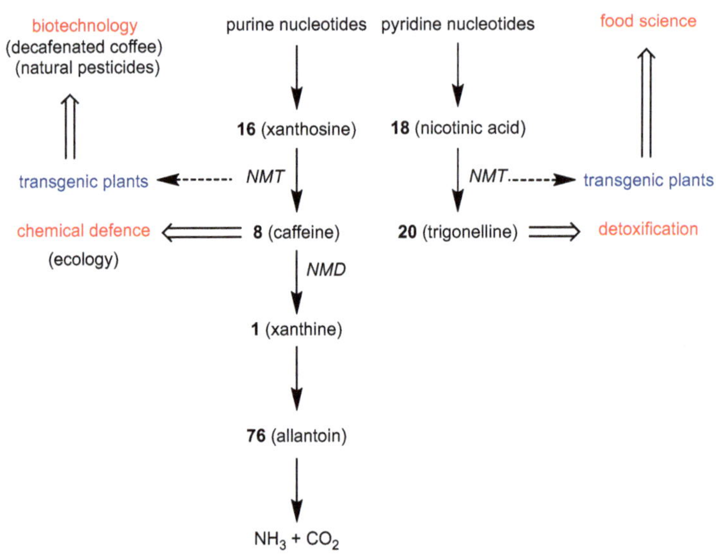

Fig. 21 Summary of this chapter

Since the genes encoding *N*-methyltransferases were elucidated, transgenic coffee plants with modified caffeine (**8**) contents have been produced. The use of genetic engineering to make fully flavored caffeine-free beverages will be of interest to the increasing numbers of consumers who are concerned about the potentially adverse effects of caffeine (**8**) consumption on their health. In addition, **8** acts as natural pesticides in caffeine-producing transgenic crops. The recent discovery of similar genes involved in the biosynthesis of trigonelline (**20**) could in the future be a useful means for creating trigonelline (**20**)-rich coffee beans, which could have potential beneficial effects on human health.

Acknowledgements We would like to thank Professor Andrew A. McCarthy, European Molecular Biology Laboratory, Grenoble, France for providing a personal communication on the current status of 7-methylxanthosine synthase. The authors also thank Professor Tatsuhito Fujimura, University of Tsukuba, Japan for drawing Fig. 2.

References

1. Ashihara H, Crozier A (2001) Caffeine: a well known but little mentioned compound in plant science. Trends Plant Sci 6:407
2. Runge FF (1820) Neueste phytochemische Entdeckungen zur Begründung einer wissenschaftlichen Phytochemie. In: Phytochemische Entdeckungen. Reimer, Berlin, p 204
3. Von Giese F (1820) Vermischte Notizen. 1. Kaffeestoff und Salzgehalt des Quassia Extrakts. Ann Chem 4:240
4. Oudry V (1927) Thein, eine organische Salzbase im Thee (*Thea chinesis*). Mag Pharm 19:49
5. Stenhouse J (1843) Über Thein und seine Darstellung. Liebigs Ann Chem 45:366
6. Daniell WF (1865) On the kola-nut of tropical West Africa (the guru nut of Soudan). Pharm J 6:450
7. Woskresensky A (1842) Über das Theobromin. Liebigs Ann Chem 41:125
8. Salomon G. Über Paraxanthin und Heteroxanthin. Ber Dtsch Chem Ges 18:3406
9. Chou CH, Waller GR (1980) Possible allelopathic constituents of *Coffea arabica*. J Chem Ecol 6:643
10. Fischer E, Ach L (1895) Synthese des Caffeins. Ber Dtsch Chem Ges 28:3135
11. Kato M, Mizuno K, Fujimura T, Iwama M, Irie M, Crozier A, Ashihara H (1999) Purification and characterization of caffeine synthase from tea leaves. Plant Physiol 120:579
12. Kato M, Mizuno K, Crozier A, Fujimura T, Ashihara H (2000) A gene encoding caffeine synthase from tea leaves. Nature 406:956
13. Ashihara H, Crozier A (1999) Biosynthesis and metabolism of caffeine and related purine alkaloids in plants. Adv Bot Res 30:117
14. Sano H, Kim Y-S, Choi Y-E (2013) Like cures like: caffeine immunizes plants against biotic stresses. Adv Bot Res 68:273
15. Tarka SM, Hurst WJ (1998) Introduction to the chemistry, isolation, and biosynthesis of methylxanthines. In: Spiller GA (ed) Caffeine. CRC Press, Boca Raton, FL, p 1
16. Kihlman BA (1977) Occurrence and biosynthesis of methylated oxypurines in plants. In: Caffeine and chromosome. Elsevier, Amsterdam, p 11
17. Willaman JJ, Schubert BG (1961) Alkaloid-bearing plants and their contained alkaloids, Technical Bulletin No. 1234. Agricultural Research Service, U.S. Department of Agriculture, Washington, DC

18. O'Connell FD (1969) Isolation of caffeine from *Banisteriopsis inebrians* (Malpighiaceae). Naturwissenschaften 56:139
19. Stewart I (1985) Identification of caffeine in citrus flowers and leaves. J Agric Food Chem 33:1163
20. Kretschmar JA, Baumann TW (1999) Caffeine in *Citrus* flowers. Phytochemistry 52:19
21. Smith AR, Pryer KM, Schuettpelz E, Korall P, Schneider H, Wolf PG (2006) A classification for extant ferns. Taxon 55:705
22. The Angiosperm Phylogeny Group (2009) An update of the angiosperm phylogeny group classification for the orders and families of flowering plants: APG III. Bot J Linn Soc 161:105
23. Nagata T, Sakai S (1984) Differences in caffeine, flavonols and amino acids contents in leaves of cultivated species of *Camellia*. Jpn J Breed 34:459
24. Nagata T, Sakai S (1985) Caffeine, flavanol and amino acid contents in leaves of hybrids and species of the section Dubiae in the genus *Camellia*. Jpn J Breed 35:1
25. Nagata T, Sakai S (1985) Purine base pattern of *Camellia irrawadiensis*. Phytochemistry 24:2271
26. Ye CX, Lin Y, Zhou H, Cheng F, Li X (1997) Isolation and analysis of purine alkaloids from *Camellia ptilophylla* Chang. Acta Scientiarum Naturalium Universitatis Sunyatseni 36:30
27. Johnson TB (1937) Purines in the plant kingdom: the discovery of a new purine in tea. J Am Chem Soc 59:1261
28. Ye C, Lin Y, Su J, Zhang H (1999) Purine alkaloids in *Camellia assamica* var. *kucha* Chang et Wang. Acta Scientiarum Naturalium Universitatis Sunyatseni 38:82
29. Zheng XQ, Ye CX, Kato M, Crozier A, Ashihara H (2002) Theacrine (1,3,7,9-tetramethyluric acid) synthesis in leaves of a Chinese tea, kucha (*Camellia assamica* var. *kucha*). Phytochemistry 60:129
30. Deng W-W (2011) Biosynthesis of secondary metabolites in tea plants. PhD thesis, Ochanomizu University, Tokyo
31. Ishida M, Kitao N, Mizuno K, Tanikawa N, Kato M (2009) Occurrence of theobromine synthase genes in purine alkaloid-free species of *Camellia* plants. Planta 229:559
32. Deng W-W, Jin Y, Yuan Y, Zhang Z-Z (2013) Profile of purine metabolism and purine alkaloid biosynthesis in *Schima* and *Eurya* plants. Bull Bot Res 33:410
33. Ashihara H, Kubota H (1986) Patterns of adenine metabolism and caffeine biosynthesis in different parts of tea seedlings. Physiol Plant 68:275
34. van Breda SV, Merwe CF, Robbertse H, Apostolides Z (2013) Immunohistochemical localization of caffeine in young *Camellia sinensis* (L.) O. Kuntze (tea) leaves. Planta 237:849
35. Petracco M (2005) Our everyday cup of coffee: the chemistry behind its magic. J Chem Educ 82:1161
36. Mazzafera P, Carvalho A (1992) Breeding for low seed caffeine content of coffee (*Coffea* L.) by interspecific hybridization. Euphytica 59:55
37. Anthony F, Clifford MN, Noirot M (1993) Biochemical diversity in the genus *Coffea* L.: chlorogenic acids, caffeine and mozambioside contents. Genet Resour Crop Evol 40:61
38. Silvarolla MB, Mazzafera P, Fazuoli LC (2004) Plant biochemistry: a naturally decaffeinated arabica coffee. Nature 429:826
39. Baumann TW, Oechslin M, Wanner H (1976) Caffeine and methylated uric acids: chemical patterns during vegetative development of *Coffea liberica*. Biochem Physiol Pflanzen 170:217
40. Petermann J, Baumann TW (1983) Metabolic relations between methylxanthines and methyluric acids in *Coffea*. Plant Physiol 73:961
41. Zheng XQ, Ashihara H (2004) Distribution, biosynthesis and function of purine and pyridine alkaloids in *Coffea arabica* seedlings. Plant Sci 166:807
42. Frischknecht PM, Ulmer-Dufek J, Baumann TW (1986) Purine alkaloid formation in buds and developing leaflets of *Coffea arabica*: expression of an optimal defence strategy? Phytochemistry 25:613

43. Fujimori N, Ashihara H (1994) Biosynthesis of theobromine and caffeine in developing leaves of *Coffea arabica*. Phytochemistry 36:1359
44. Keller H, Wanner H, Baumann TW (1972) Kaffeinsynthese in Fruchten und Gewebekulturen von *Coffea arabica*. Planta 108:339
45. Koshiro Y, Zheng XQ, Wang M, Nagai C, Ashihara H (2006) Changes in content and biosynthetic activity of caffeine and trigonelline during growth and ripening of *Coffea arabica* and *Coffea canephora* fruits. Plant Sci 171:242
46. Clifford MN, Ramirez-Martinez JR (1990) Chlorogenic acids and purine alkaloids contents of maté (*Ilex paraguariensis*) leaf and beverage. Food Chem 35:13
47. Alikaridis F (1987) Natural constituents of *Ilex* species. J Ethnopharmacol 20:121
48. Filip R, de Iglesias DIA, Rondina RVD, Coussio JD (1983) Análisis de las hojas y tallos de *Ilex argentina* Lillo. I. Xantinas. Acta Farm Bonaerense 2:87
49. Edwards AL, Bennett BC (2005) Diversity of methylxanthine content in *Ilex cassine* L. and *Ilex vomitoria* Ait.: assessing sources of the North American stimulant cassina. Econ Bot 59:275
50. Mazzafera P (1994) Caffeine, theobromine and theophylline distribution in *Ilex paraguariensis*. Rev Brasil Fisiol Veg 6:149
51. Crozier A, Ashihara H, Tomas-Barberan F (eds) (2012) Teas, cocoa and coffee: plant secondary metabolites and health. Wiley-Blackwell, Oxford, UK
52. Pereira-Caro G, Borges G, Nagai C, Jackson MC, Yokota T, Crozier A, Ashihara H (2013) Profiles of phenolic compounds and purine alkaloids during the development of seeds of *Theobroma cacao* cv. Trinitario. J Agric Food Chem 61:427
53. Timbie DJ, Sechrist L, Keeney PG (1978) Application of high-pressure liquid chromatography to the study of variables affecting theobromine and caffeine concentrations in cocoa beans. J Food Sci 43:560
54. Hammerstone JF Jr, Romanczyk LJ Jr, Aitkent WM (1994) Purine alkaloid distribution within *Herrania* and *Theobroma*. Phytochemistry 35:1237
55. Kufer J, McNeil CL (2009) The jaguar tree (*Theobroma bicolor* Bonpl.). In: McNeil CL (ed) Chocolate in Mesoamerica: a cultural history of cacao. Oxford University Press, Oxford, UK, p 542
56. Koyama Y, Tomoda Y, Kato M, Ashihara H (2003) Metabolism of purine bases, nucleosides and alkaloids in theobromine-forming *Theobroma cacao* leaves. Plant Physiol Biochem 41:977
57. Gurney KA, Evans LV, Robinson DS (1991) Extraction of purine alkaloids from cocoa tissues and determination by high-performance liquid chromatography. Phytochem Anal 2:15
58. Senanayake UM, Wijesekera ROB (1971) Theobromine and caffeine content of the cocoa bean during its growth. J Sci Food Agric 22:262
59. Sotelo A, Alvarez RG (1991) Chemical composition of wild-*Theobroma* species and their comparison to the cacao bean. J Agric Food Chem 39:1940
60. Bucheli P, Rousseau G, Alvarez M, Laloi M, McCarthy J (2001) Developmental variation of sugars, carboxylic acids, purine alkaloids, fatty acids, and endoproteinase activity during maturation of *Theobroma cacao* L. seeds. J Agric Food Chem 49:5046
61. Zheng X-Q, Koyama Y, Nagai C, Ashihara H (2004) Biosynthesis, accumulation and degradation of theobromine in developing *Theobroma cacao* fruits. J Plant Physiol 161:363
62. Belliardo F, Martelli A, Valle M (1985) HPLC determination of caffeine and theophylline in *Paullinia cupana* Kunth (Guarana) and *Cola* spp. samples. Z Lebensm Unters Forsch 180:398
63. Niemenak N, Onomo PE, Fotso, Lieberei R, Ndoumou DO (2008) Purine alkaloids and phenolic compounds in three *Cola* species and *Garcinia kola* grown in Cameroon. S Afr J Bot 74:629
64. Schimpl FC, Kiyota E, Mayer JLS, Gonçalves JFdC, da Silva JF, Mazzafera P (2014) Molecular and biochemical characterization of caffeine synthase and purine alkaloid concentration in guarana fruit. Phytochemistry 105:25
65. Weckerle CS, Stutz MA, Baumann TW (2003) Purine alkaloids in *Paullinia*. Phytochemistry 64:735

66. Baumann TW, Schulthess BH, Hanni K (1995) Guarana (*Paullinia cupana*) rewards seed dispersers without intoxicating them by caffeine. Phytochemistry 39:1063
67. Ashihara H, Yokota T, Crozier A (2013) Biosynthesis and catabolism of purine alkaloids. Adv Bot Res 68:111
68. Pierattini E, Francini A, Raffaelli A, Sebastiani L (2016) Degradation of exogenous caffeine by *Populus alba* and its effects on endogenous caffeine metabolism. Environ Sci Pollut Res 23:7289
69. Ashihara H, Ludwig IA, Katahira R, Yokota T, Fujimura T, Crozier A (2014) Trigonelline and related nicotinic acid metabolites: occurrence, biosynthesis, taxonomic considerations, and their roles in planta and in human health. Phytochem Rev 14:765
70. Bruton T, Alboloushi A, Garza Bdl, Kim BO, Halden RU (2010) Fate of caffeine in the environment and ecotoxicological considerations. In: Contaminants of emerging concern in the environment: ecological and human health considerations, ACS Symposium Series, vol 1048. American Chemical Society, Washington DC, p 257
71. Bonner J (1950) Plant biochemistry. Academic Press, New York
72. Bresler HW (1904) Uber die Bestimmung der Nucleinbasen im Safte von *Beta vulgaris*. Hoppe Seyler's Z Physiol Chem 4:535
73. Weevers T (1930) Die Funkion der Xanthinderivate im Pflanzenstoffwechsel. Arch Neerl Sci 3B(5):111
74. Anderson L, Gibbs M (1962) The biosynthesis of caffeine in the coffee plant. J Biol Chem 237:1941
75. Inoue T, Yamashita S, Kawamura Y, Sasaki G (1960) Studies on biogenesis of tea components. I. Absorption of ^{15}N to new leaves and old leaves. Yakugaku Zasshi 80:548
76. Inoue T, Kawamura Y (1961) Studies on biogenesis of tea components. II. Formation of caffeine in excised tea shoots. Chem Pharm Bull 9:236
77. Inoue T (1972) Studies on biogenesis of tea components. IV. Incorporation of glycine-2-^{14}C to caffeine. Proc Hoshi Coll Pharm (Tokyo) 13:60
78. Inoue T, Adachi F (1962) Studies on biogenesis of tea components. III. The origin of the methyl groups in caffeine. Chem. Pharm Bull 10:1212
79. Serenkov GP, Proiser E (1961) Biosynthesis of caffeine in tea leaves. Dokl Akad Nauk USSR 140:716
80. Proiser E, Serenkov GP (1963) Caffeine biosynthesis in tea leaves. Biokhimiya (Moscow) 28:857
81. Konishi S, Ozawa M, Takahashi E (1972) Metabolic conversion of *N*-methyl carbon of γ-glutamylmethylamide to caffeine in tea plants. Plant Cell Physiol 13:365
82. Konishi S, Inoue T, Takahashi E (1972) Localization of the carbon in caffeine biosynthesized from *N*-methyl carbon of γ-glutamylmethylamide in tea plants. Plant Cell Physiol 13:695
83. Suzuki T (1972) The participation of *S*-adenosylmethionine in the biosynthesis of caffeine in the tea plant. FEBS Lett 24:18
84. Suzuki T (1973) Metabolism of methylamine in the tea plant (*Thea sinensis* L.). Biochem J 132:753
85. Suzuki T, Takahashi E (1975) Metabolism of xanthine and hypoxanthine in the tea plant (*Thea sinensis* L.). Biochem J 146:79
86. Suzuki T, Takahashi E (1976) Caffeine biosynthesis in *Camellia sinensis*. Phytochemistry 15:1235
87. Looser E, Baumann TW, Wanner H (1974) The biosynthesis of caffeine in the coffee plant. Phytochemistry 13:2515
88. Ogutuga DBA, Northcote DH (1970) Caffeine formation in tea callus tissue. J Exp Bot 21:258
89. Ogutuga DBA, Northcote DH (1970) Biosynthesis of caffeine in tea callus tissue. Biochem J 117:715
90. Keller H, Wanner H, Baumann TW (1972) Kaffeinsynthese in Früchten und Gewebekulturen von *Coffea arabica*. Planta 108:339
91. Suzuki T, Takahashi E (1976) Metabolism of methionine and biosynthesis of caffeine in the tea plant (*Camellia sinensis* L.). Biochem J 160:171

92. Suzuki T, Takahashi E (1976) Further investigation of the biosynthesis of caffeine in tea plants (*Camellia sinensis* L.): methylation of transfer ribonucleic acid by tea leaf extracts. Biochem J 160:181
93. Loomis WD (1969) Removal of phenolic compounds during the isolation of plant enzymes. In: Methods in enzymology, vol 13. Academic Press, New York, p 555
94. Suzuki T, Takahashi E (1975) Biosynthesis of caffeine by tea-leaf extracts: enzymic formation of theobromine from 7-methylxanthine and of caffeine from theobromine. Biochem J 146:87
95. Roberts MF, Waller GR (1979) *N*-Methyltransferases and 7-methyl-*N*9-nucleoside hydrolase activity in *Coffea arabica* and the biosynthesis of caffeine. Phytochemistry 18:451
96. Baumann TW, Koetz R, Morath P (1983) *N*-Methyltransferase activities in suspension cultures of *Coffea arabica* L. Plant Cell Rep 2:33
97. Waller GR, MacVean CD, Suzuki T (1983) High production of caffeine and related enzyme activities in callus cultures of *Coffea arabica* L. Plant Cell Rep 2:109
98. Negishi O, Ozawa T, Imagawa H (1985) Methylation of xanthosine by tea-leaf extracts and caffeine biosynthesis. Agric Biol Chem 49:887
99. Negishi O, Ozawa T, Imagawa H (1988) *N*-Methylnucleosidase from tea leaves. Agric Biol Chem 52:169
100. Ashihara H, Nobusawa E (1981) Metabolic fate of [8-^{14}C]adenine and [8-^{14}C]hypoxanthine in higher plants. Z Pflanzenphysiol 104:443
101. Nobusawa E, Ashihara H (1983) Purine metabolism in cotyledons and embryonic axes of black gram (*Phaseolus mungo* L.) seedlings. Int J Biochem 15:1059
102. Ashihara H, Kubota H (1987) Biosynthesis of purine alkaloids in *Camellia* plants. Plant Cell Physiol 28:535
103. Ashihara H (1993) Purine metabolism and the biosynthesis of caffeine in maté leaves. Phytochemistry 33:1427
104. Negishi O, Ozawa T, Imagawa H (1992) Biosynthesis of caffeine from purine nucleotides in tea plant. Biosci Biotechnol Biochem 56:499
105. Suzuki T, Ashihara H, Waller GR (1992) Purine and purine alkaloid metabolism in *Camellia* and *Coffea* plants. Phytochemistry 31:2575
106. Koshiishi C, Kato A, Yama S, Crozier A, Ashihara H (2001) A new caffeine biosynthetic pathway in tea leaves: utilisation of adenosine released from the *S*-adenosyl-L-methionine cycle. FEBS Lett 499:50
107. Nazario GM, Lovatt CJ (1993) Separate de novo and salvage purine pools are involved in the biosynthesis of theobromine but not caffeine in leaves of *Coffea arabica* L. Plant Physiol 103:1203
108. Ashihara H, Monteiro AM, Gillies FM, Crozier A (1996) Biosynthesis of caffeine in leaves of coffee. Plant Physiol 111:747
109. Schulthess BH, Morath P, Baumann TW (1996) Caffeine biosynthesis starts with the metabolically channelled formation of 7-methyl-XMP. A new hypothesis. Phytochemistry 41:169
110. Schulthess BH, Baumann TW (1995) Are xanthosine and 7-methylxanthosine caffeine precursors? Phytochemistry 39:1363
111. Mizuno K, Kato M, Irino F, Yoneyama N, Fujimura T, Ashihara H (2003) The first committed step reaction of caffeine biosynthesis: 7-methylxanthosine synthase is closely homologous to caffeine synthases in coffee (*Coffea arabica* L.). FEBS Lett 547:56
112. Gillies FM, Jenkins GI, Ashihara H, Crozier A (1995) In vitro-biosynthesis of caffeine: the stability of *N*-methyltransferase activity in cell-free preparations from liquid endosperm of *Coffea arabica*. In: Proceedings of the 16th international conference on coffee science, ASIC, Paris, p 599
113. Mazzafera P, Wingsle G, Olsson O, Sandberg G (1994) *S*-Adenosyl-L-methionine:theobromine 1-*N*-methyltransferase, an enzyme catalysing the synthesis of caffeine in coffee. Phytochemistry 17:1577

114. Ashihara H, Kato M, Crozier A (1996) Caffeine biosynthesis in leaves of *Camellia sinensis*: substrate specificity of *N*-methyltransferase. In: Principles regulating biosynthesis and storage of secondary products. Phytochemical Society of Europe, Halle-Wittenberg, p 40
115. Kato M, Kanehara T, Shimizu H, Suzuki T, Gillies FM, Crozier A, Ashihara H (1996) Caffeine biosynthesis in young leaves of *Camellia sinensis*: in vitro studies on *N*-methyltransferase activity involved in the conversion of xanthosine to caffeine. Physiol Plant 98:629
116. Mosli Waldhauser SS, Gillies FM, Crozier A, Baumann TW (1997) Separation of the *N*-7 methyltransferase, the key enzyme in caffeine biosynthesis. Phytochemistry 45:1407
117. Kato M, Mizuno K, Crozier A, Fujimura T, Ashihara H (2000) Plant biotechnology: caffeine synthase gene from tea leaves. Nature 406:956
118. Moisyadi S, Neupane KR, Stiles JI (1998) Cloning and characterization of a cDNA encoding xanthosine-N7-methyltransferase from coffee (*Coffea arabica*). Acta Hort 461:367
119. Moisyadi S, Neupane KR, Stiles JI (1999) Cloning and characterization of xanthosine-*N*7-methyltransferase, the first enzyme of the caffeine biosynthetic pathway. In: Proceedings of 18th international conference on coffee science, ASIC, Paris, p 327
120. Uefuji H, Tatsumi Y, Morimoto M, Kaothien-Nakayama P, Ogita S, Sano H (2005) Caffeine production in tobacco plants by simultaneous expression of three coffee *N*-methyltransferases and its potential as a pest repellent. Plant Mol Biol 59:221
121. Mizuno K, Okuda A, Kato M, Yoneyama N, Tanaka H, Ashihara H, Fujimura T (2003) Isolation of a new dual-functional caffeine synthase gene encoding an enzyme for the conversion of 7-methylxanthine to caffeine from coffee (*Coffea arabica* L.). FEBS Lett 534:75
122. Mizuno K, Tanaka H, Kato M, Ashihara H, Fujimura T (2001) cDNA cloning of caffeine (theobromine) synthase from coffee (*Coffea arabica* L.). In: Proceedings of 18th international conference on coffee science, ASIC, Paris, p 815
123. Ogawa M, Herai Y, Koizumi N, Kusano T, Sano H (2001) 7-Methylxanthine methyltransferase of coffee plants. Gene isolation and enzymatic properties. J Biol Chem 276:8213
124. Uefuji H, Ogita S, Yamaguchi Y, Koizumi N, Sano H (2003) Molecular cloning and functional characterization of three distinct *N*-methyltransferases involved in the caffeine biosynthetic pathway in coffee plants. Plant Physiol 132:372
125. Ashihara H, Sano H, Crozier A (2008) Caffeine and related purine alkaloids: biosynthesis, catabolism, function and genetic engineering. Phytochemistry 69:841
126. Negishi O, Ozawa T, Imagawa H (1985) Conversion of xanthosine into caffeine in tea plants. Agric Biol Chem 49:251
127. Yue Y, Guo H (2014) Quantum mechanical/molecular mechanical study of catalytic mechanism and role of key residues in methylation reactions catalyzed by dimethylxanthine methyltransferase in caffeine biosynthesis. J Chem Inf Model 54:593
128. Fujimori N, Suzuki T, Ashihara H (1991) Seasonal variations in biosynthetic capacity for the synthesis of caffeine in tea leaves. Phytochemistry 30:2245
129. McCarthy AA, McCarthy JG (2007) The structure of two *N*-methyltransferases from the caffeine biosynthetic pathway. Plant Physiol 144:879
130. Ashihara H, Gillies FM, Crozier A (1997) Metabolism of caffeine and related purine alkaloids in leaves of tea (*Camellia sinensis* L.). Plant Cell Physiol 38:413
131. Yoneyama N, Morimoto H, Ye CX, Ashihara H, Mizuno K, Kato M (2006) Substrate specificity of *N*-methyltransferase involved in purine alkaloids synthesis is dependent upon one amino acid residue of the enzyme. Mol Genet Genomics 275:125
132. Kato M, Mizuno K (2004) Caffeine synthase and related methyltransferases in plants. Front Biosci 9:1833
133. Joshi CP, Chiang VL (1998) Conserved sequence motifs in plant *S*-adenosyl-L-methionine-dependent methyltransferases. Plant Mol Biol 37:663
134. Ross JR, Nam KH, D'Auria JC, Pichersky E (1999) *S*-Adenosyl-methionine:salicylic acid carboxyl methyltransferase, an enzyme involved in floral scent production and plant defense, represents a new class of plant methyltransferases. Arch Biochem Biophys 367:9

135. Dudareva N, Murfitt LM, Mann CJ, Gorenstein N, Kolosova N, Kish CM, Bonham C, Wood K (2000) Developmental regulation of methyl benzoate biosynthesis and emission in snapdragon flowers. Plant Cell 12:949
136. Seo HS, Song JT, Cheong JJ, Lee YH, Lee YW, Hwang I, Lee JS, Choi YD (2001) Jasmonic acid carboxyl methyltransferase: a key enzyme for jasmonate-regulated plant responses. Proc Natl Acad Sci USA 98:4788
137. Yang Y, Yuan JS, Ross J, Noel JP, Pichersky E, Chen F (2006) An *Arabidopsis thaliana* methyltransferase capable of methylating farnesoic acid. Arch Biochem Biophys 448:123
138. Zhao N, Ferrer J-L, Ross J, Guan J, Yang Y, Pichersky E, Noel JP, Chen F (2008) Structural, biochemical, and phylogenetic analyses suggest that indole-3-acetic acid methyltransferase is an evolutionarily ancient member of the SABATH family. Plant Physiol 146:455
139. Varbanova M, Yamaguchi S, Yang Y, McKelvey K, Hanada A, Borochov R, Yu F, Jikumaru Y, Ross J, Cortes D, Ma CJ, Noel JP, Mander L, Shulaev V, Kamiya Y, Rodermel S, Weiss D, Pichersky E (2007) Methylation of gibberellins by *Arabidopsis* GAMT1 and GAMT2. Plant Cell 19:32
140. Murata J, Roepke J, Gordon H, De Luca V (2008) The leaf epidermome of *Catharanthus roseus* reveals its biochemical specialization. Plant Cell 20:524
141. D'Auria JC, Chen F, Pichersky E (2003) The SABATH family of MTS in *Arabidopsis thaliana* and other plant species. In: John TR (ed) Recent advances in phytochemistry, vol 37. Elsevier, Oxford, UK, p 253
142. Zubieta C, Ross JR, Koscheski P, Yang Y, Pichersky E, Noel JP (2003) Structural basis for substrate recognition in the salicylic acid carboxyl methyltransferase family. Plant Cell 15:1704
143. Denoeud F, Carretero-Paulet L, Dereeper A, Droc G, Guyot R, Pietrella M, Zheng C, Alberti A, Anthony F, Aprea G, Aury J-M, Bento P, Bernard M, Bocs S, Campa C, Cenci A, Combes M-C, Crouzillat D, Da Silva C, Daddiego L, De Bellis F, Dussert S, Garsmeur O, Gayraud T, Guignon V, Jahn K, Jamilloux V, Joët T, Labadie K, Lan T, Leclercq J, Lepelley M, Leroy T, Li L-T, Librado P, Lopez L, Muñoz A, Noel B, Pallavicini A, Perrotta G, Poncet V, Pot D, Priyono, Rigoreau M, Rouard M, Rozas J, Tranchant-Dubreuil C, VanBuren R, Zhang Q, Andrade AC, Argout X, Bertrand B, de Kochko A, Graziosi G, Henry RJ, Jayarama MR, Nagai C, Rounsley S, Sankoff D, Giuliano G, Albert VA, Wincker P, Lashermes P (2014) The coffee genome provides insight into the convergent evolution of caffeine biosynthesis. Science 345:1181
144. Pichersky E, Lewinsohn E (2011) Convergent evolution in plant specialized metabolism. Annu Rev Plant Biol 62:549
145. Perrois C, Strickler S, Mathieu G, Lepelley M, Bedon L, Michaux S, Husson J, Mueller L, Privat I (2015) Differential regulation of caffeine metabolism in *Coffea arabica* (Arabica) and *Coffea canephora* (Robusta). Planta 241:179
146. Mizuno K, Matsuzaki M, Kanazawa S, Tokiwano T, Yoshizawa Y, Kato M (2014) Conversion of nicotinic acid to trigonelline is catalyzed by *N*-methyltransferase belonged to motif B′ methyltransferase family in *Coffea arabica*. Biochem Biophys Res Commun 452:1060
147. Ashihara H, Ludwig I, Katahira R, Yokota T, Fujimura T, Crozier A (2015) Trigonelline and related nicotinic acid metabolites: occurrence, biosynthesis, taxonomic considerations, and their roles in planta and in human health. Phytochem Rev 14:765
148. Ashihara H (2015) Plant biochemistry: trigonelline biosynthesis in *Coffea arabica* and *Coffea canephora*. In: Preedy VR (ed) Coffee in health and disease prevention. Academic Press, San Diego, p 19
149. Arabidopsis-Genome-Initiative (2000) Analysis of the genome sequence of the flowering plant *Arabidopsis thaliana*. Nature 408:796
150. International-Rice-Genome-Sequencing-Project (2005) The map-based sequence of the rice genome. Nature 436:793
151. Zrenner R, Ashihara H (2011) Nucleotide metabolism. In: Ashihara H, Crozier A, Komamine A (eds) Plant metabolism and biotechnology. Wiley, Chichester, UK, p 135

152. Ito E, Ashihara H (1999) Contribution of purine nucleotide biosynthesis de novo to the formation of caffeine in young tea (*Camellia sinensis*) leaves. J Plant Physiol 254:145
153. Yabuki N, Ashihara H (1991) Catabolism of adenine nucleotides in suspension-cultured plant cells. Biochim Biophys Acta 1073:474
154. Fujimori N, Ashihara H (1993) Biosynthesis of caffeine in flower buds of *Camellia sinensis*. Ann Bot 71:279
155. Baumann TW (2015) Revisiting caffeine biosynthesis — speculations about the proximate source of its purine ring. Nat Prod Commun 10:793
156. Kremers RE (1954) Speculation on DPN as a biochemical precursor of caffeine and trigonelline in coffee. J Am Pharm Assoc 43:423
157. Negishi O, Ozawa T, Imagawa H (1994) Guanosine deaminase and guanine deaminase from tea leaves. Biosci Biotechnol Biochem 58:1277
158. Ashihara H, Takasawa Y, Suzuki T (1997) Metabolic fate of guanosine in higher plants. Physiol Plant 100:909
159. Mager J, Magasanik B (1960) Guanosine-5′-phosphate reductase and its role in the interconversion of purine nucleotides. J Biol Chem 235:1474
160. Stephens RW, Whittaker VK (1973) Calf thymus GMP reductase: control by XMP. Biochem Biophys Res Commun 53:975
161. Renart MF, Sillero A (1974) GMP reductase in *Artemia salina*. Biochim Biophys Acta 341:178
162. Le Floc'h F, Lafleuriel J, Guillot A (1982) Interconversion of purine nucleotides in Jerusalem artichoke shoots. Plant Sci Lett 27:309
163. Ashihara H, Mitsui K, Yabuki N, Nygaard P (1991) Adenosine metabolism and growth of adenosine-requiring mutant cells of *Datura innoxia*. Int J Purine Pyrimidine Res 2:129
164. Shoji T, Hashimoto T (2011) Nicotine biosynthesis. In: Ashihara H, Crozier A, Komamine A (eds) Plant metabolism and biotechnology. Wiley, Chichester UK, p 191
165. Yazaki K, Sugiyama A, Morita M, Shitan N (2008) Secondary transport as an efficient membrane transport mechanism for plant secondary metabolites. Phytochem Rev 7:513
166. Fujimori N, Ashihara H (1990) Adenine metabolism and the synthesis of purine alkaloids in flowers of *Camellia* plants. Phytochemistry 29:3513
167. Terrasaki Y, Suzuki T, Ashihara H (1994) Purine metabolism and the biosynthesis of purine alkaloids in tea fruits during development. Plant Physiol (Life Sci Adv) 13:135
168. Li Y, Ogita S, Keya CA, Ashihara H (2008) Expression of caffeine biosynthesis genes in tea (*Camellia sinensis*). Z Naturforsch 63c:267
169. Li YH, Gu W, Ye S (2007) Expression and location of caffeine synthase in tea plants. Russ J Plant Physiol 54:698
170. Baumann TW, Wanner H (1972) Untersuchungen Aber den Transport von Kaffein in der Kaffeepflanze (*Coffea arabica*). Planta 108:11
171. Mosli Waldhauser SS, Baumann TW (1996) Compartmentation of caffeine and related purine alkaloids depends exclusively on the physical chemistry of their vacuolar complex formation with chlorogenic acids. Phytochemistry 42:985
172. Kato A, Crozier A, Ashihara H (1998) Subcellular localization of the *N*-3 methyltransferase involved in caffeine biosynthesis in tea. Phytochemistry 48:777
173. Kumar V, Satyanarayana KV, Ramakrishna A, Chandrashekar A, Ravishankar GA (2007) Evidence for localization of *N*-methyltransferase (MMT) of caffeine biosynthetic pathway in vacuolar surface of *Coffea canephora* endosperm elucidated through localization of GUS reporter gene driven by NMT promoter. Curr Sci 93:383
174. Foyer CH (1984) Photosynthesis. Cell biology: a series of monographs, vol. 1. Wiley, New York
175. Kodama Y, Shinya T, Sano H (2008) Dimerization of *N*-methyltransferases involved in caffeine biosynthesis. Biochimie 90:547
176. Zrenner R, Stitt M, Sonnewald U, Boldt R (2006) Pyrimidine and purine biosynthesis and degradation in plants. Annu Rev Plant Biol 57:805

177. Hung W-F, Chen L-J, Boldt R, Sun C-W, Li H-M (2004) Characterization of *Arabidopsis* glutamine phosphoribosyl pyrophosphate amidotransferase-deficient mutants. Plant Physiol 135:1314
178. Smith PC, Mann A, Goggin D, Atkins C (1998) Air synthetase in cowpea nodules: a single gene product targeted to two organelles? Plant Mol Biol 36:811
179. Atkins CA, Smith P, Storer PJ (1997) Reexamination of the intracellular localization of de novo purine synthesis in cowpea nodules. Plant Physiol 113:127
180. van der Graaff E, Hooykaas P, Lein W, Lerchl J, Kunze G, Sonnewald U, Boldt R (2004) Molecular analysis of "de novo" purine biosynthesis in solanaceous species and in *Arabidopsis thaliana*. Front Biosci 9:1803
181. Schoor S, Farrow S, Blaschke H, Lee S, Perry G, von Schwartzenberg K, Emery N, Moffatt B (2011) Adenosine kinase contributes to cytokinin interconversion in *Arabidopsis*. Plant Physiol 157:659
182. Jorgensen K, Rasmussen AV, Morant M, Nielsen AH, Bjarnholt N, Zagrobelny M, Bak S, Moller BL (2005) Metabolon formation and metabolic channeling in the biosynthesis of plant natural products. Curr Opin Plant Biol 8:280
183. Deng W, Li M, Gu C, Li D, Ma L, Jin Y, Wan X (2015) Low caffeine content in novel grafted tea with *Camellia sinensis* as scions and *Camellia oleifera* as stocks. Nat Prod Commun 10:789
184. Kato M, Kitao N, Ishida M, Morimoto H, Irino F, Mizuno K (2010) Expression for caffeine biosynthesis and related enzymes in *Camellia sinensis*. Z Naturforsch 65c:245
185. Mohanpuria P, Kumar V, Joshi R, Gulati A, Ahuja P, Yadav S (2009) Caffeine biosynthesis and degradation in tea [*Camellia sinensis* (L.) O. Kuntze] is under developmental and seasonal regulation. Mol Biotechnol 43:104
186. Bailey BA, Bae H, Strem MD, Antunez de Mayolo G, Guiltinan MJ, Verica JA, Maximova SN, Bowers JH (2005) Developmental expression of stress response genes in *Theobroma cacao* leaves and their response to Nep1 treatment and a compatible infection by *Phytophthora megakarya*. Plant Physiol Biochem 43:611
187. Bailey BA, Strem MD, Bae H, de Mayolo GA, Guiltinan MJ (2005) Gene expression in leaves of *Theobroma cacao* in response to mechanical wounding, ethylene, and/or methyl jasmonate. Plant Sci 168:1247
188. Kumar A, Giridhar P (2015) Salicylic acid and methyljasmonate restore the transcription of caffeine biosynthetic *N*-methyltransferases from a transcription inhibition noticed during late endosperm maturation in coffee. Plant Gene 4:38
189. Kumar A, Simmi PS, Naik GK, Giridhar P (2015) RP-HPLC and transcript profile indicate increased leaf caffeine in *Coffea canephora* plants by light. J Biol Earth Sci 5:1
190. Kumar A, Naik GK, Simmi PS, Giridhar P (2015) Salinity and drought response alleviate caffeine content of young leaves of *Coffea canephora* var. Robusta cv. S274. J Appl Biol Biotechnol 3:50
191. Deng WW, Li Y, Ogita S, Ashihara H (2008) Fine control of caffeine biosynthesis in tissue cultures of *Camellia sinensis*. Phytochem Lett 1:195
192. Yu CL, Kale Y, Gopishetty S, Louie TM, Subramanian M (2008) A novel caffeine dehydrogenase in *Pseudomonas* sp. strain CBB1 oxidizes caffeine to trimethyluric acid. J Bacteriol 190:772
193. Kalberer P (1964) Untersuchungen zum Abbau des Kaffeins in den Blattern von *Coffea arabica*. Beril Schw Bot Geselshaft 74:62
194. Kalberer P (1965) Breakdown of caffeine in the leaves of *Coffea arabica* L. Nature 205:597
195. Suzuki T, Waller GR (1984) Biosynthesis and biodegradation of caffeine, theobromine, and theophylline in *Coffea arabica* L. fruits. J Agric Food Chem 32:845
196. Suzuki T, Waller GR (1984) Biodegradation of caffeine: formation of theophylline and caffeine in mature *Coffea arabica* fruits. J Sci Food Agric 35:66
197. Ashihara H, Monteiro AM, Moritz T, Gillies FM, Crozier A (1996) Catabolism of caffeine and related purine alkaloids in leaves of *Coffea arabica* L. Planta 198:334

198. Yin Y, Katahira R, Ashihara H (2015) Metabolism of purine alkaloids and xanthine in leaves of maté (*Ilex paraguariensis*). Nat Prod Commun 10:707
199. Moffatt BA, Ashihara H (2002) Purine and pyrimidine nucleotide synthesis and metabolism. In: The Arabidopsis book, vol 1. American Society of Plant Biologists, Rockville, MD, p 1
200. Munoz A, Raso M, Pineda M, Piedras P (2006) Degradation of ureidoglycolate in French bean (*Phaseolus vulgaris*) is catalysed by a ubiquitous ureidoglycolate urea-lyase. Planta 224:175
201. Winkler RG, Blevins DG, Randall DD (1988) Ureide catabolism in soybeans. Plant Physiol 86:1084
202. Deng W-W, Ashihara H (2010) Profiles of purine metabolism in leaves and roots of *Camellia sinensis* seedlings. Plant Cell Physiol 51:2105
203. Nazario GM, Lovatt CJ (1993) Regulation of purine metabolism in intact leaves of *Coffea arabica*. Plant Physiol 103:1195
204. Ashihara H, Crozier A (1999) Biosynthesis and catabolism of caffeine in low-caffeine-containing species of *Coffea*. J Agric Food Chem 47:3425
205. Ashihara H, Shimizu H, Takeda Y, Suzuki T, Gillies FM, Crozier A (1995) Caffeine metabolism in high and low caffeine containing cultivars of *Camellia sinensis*. Z Naturforsch 50c:602
206. Ashihara H, Kato M, Ye CX (1998) Biosynthesis and metabolism of purine alkaloids in leaves of cocoa tea (*Camellia ptilophylla*). J Plant Res 111:599
207. Ito E, Crozier A, Ashihara H (1997) Theophylline metabolism in higher plants. Biochim Biophys Acta 1336:323
208. Mazzafera P (2004) Catabolism of caffeine in plants and microorganisms. Front Biosci 9:1348
209. Madyastha KM, Sridhar GR (1998) A novel pathway for the metabolism of caffeine by a mixed culture consortium. Biochem Biophys Res Commun 249:178
210. Yu CL, Louie TM, Summers R, Kale Y, Gopishetty S, Subramanian M (2009) Two distinct pathways for metabolism of theophylline and caffeine are coexpressed in *Pseudomonas putida* CBB5. J Bacteriol 191:4624
211. Summers RM, Louie TM, Yu CL, Subramanian M (2011) Characterization of a broad-specificity non-haem iron *N*-demethylase from *Pseudomonas putida* CBB5 capable of utilizing several purine alkaloids as sole carbon and nitrogen source. Microbiology 157:583
212. Summers RM, Louie TM, Yu C-L, Gakhar L, Louie KC, Subramanian M (2012) Novel, highly specific *N*-demethylases enable bacteria to live on caffeine and related purine alkaloids. J Bacteriol 194:2041
213. Scheline RR (1991) CRC handbook of mammalian metabolism of plant compounds. CRC Press, Boca Raton, FL
214. Arnaud MJ (2011) Pharmacokinetics and metabolism of natural methylxanthines in animal and man. In: Arnaud MJ (ed) Methylxanthines, vol 200, Handbook of experimental pharmacology. Springer, Heidelberg, p 33
215. Molisch H (1937) Der Einfluss einer Pflanze auf die andere-Allelopathie. Gustav Fischer, Jena
216. Anaya AL, Cruz-Ortega R, Waller GR (2006) Metabolism and ecology of purine alkaloids. Front Biosci 11(Suppl 1):2354
217. Lovett JV, Hoult AHC (1998) Allelopathy in plants. In: Roberts MF, Wink MP (eds) Alkaloids: biochemistry, ecology, and medicinal applications. Plenum, London, p 337
218. Anaya AL, Ramos L, Hernandez JG, Cruz-Ortega R (1987) Allelopathy in Mexico. In: Waller GR (ed) Allelochemicals: role in agriculture and forestry, Symposium series 330. American Chemical Society, Washington, DC, p 89
219. Waller GR, Kumari D, Friedman J, Friedman N, Chou CH (1986) Caffeine autotoxicity in *Coffea arabica* L. In: Putnam AR, Tang C-S (eds) The science of allelopathy. Wiley, New York, p 243

220. Friedman J, Waller G (1983) Caffeine hazards and their prevention in germinating seeds of coffee (*Coffea arabica* L.). J Chem Ecol 9:1099
221. Baumann TW, Gabriel H (1984) Metabolism and excretion of caffeine during germination of *Coffea arabica* L. Plant Cell Physiol 25:1431
222. Chou C-H, Waller G (1980) Possible allelopathic constituents of *Coffea arabica*. J Chem Ecol 6:643
223. Rizvi SJH, Mukerji D, Mathur SN (1981) Selective phyto-toxicity of 1,3,7-trimethylxanthine between *Phaseolus mungo* and some weeds. Agric Biol Chem 45:1255
224. Smyth D (1992) Effect of methylxanthine treatment on rice seedling growth. J Plant Growth Regul 11:125
225. Sasamoto H, Fujii Y, Ashihara H (2015) Effect of purine alkaloids on the proliferation of lettuce cells derived from protoplasts. Nat Prod Commun 10:751
226. Sasamoto H, Murashige-Baba T, Inoue A, Sato T, Hayashi S, Hasegawa A (2013) Development of a new method for bioassay of allelopathy using protoplasts of a leguminous plant *Mucuna pruriens* with a high content of the allelochemical L-DOPA. J Plant Stud 2:71
227. Kihlman BA (1977) Caffeine and chromosomes. Elsevier, Amsterdam
228. Kihlman B (1949) The effect of purine derivatives on chromosomes. Hereditas 35:393
229. Kihlman B, Leven A (1949) The cytological effect of caffeine. Hereditas 35:109
230. Mineyuki Y, Letham DS, Hocart CH (1989) New 3-substituted xanthines: potent inhibitors of cell plate formation. Cell Biol Int Rep 13:129
231. Amino S, Nagata T (1996) Caffeine-induced uncoupling of mitosis from DNA replication in tobacco BY-2 cells. J Plant Res 109:219
232. Manandhar G, Apostolakos P, Galatis B (1996) Cell division of binuclear cells induced by caffeine: spindle organization and determination of division plane. J Plant Res 109:265
233. Deng WW, Katahira R, Ashihara H (2015) Short-term effect of caffeine on purine, pyrimidine and pyridine metabolism in rice (*Oryza sativa*) seedlings. Nat Prod Commun 10:737
234. Shimazaki A, Ashihara H (1982) Adenine and guanine salvage in cultured cells of *Catharanthus roseus*. Ann Bot 50:531
235. Shimazaki A, Hirose F, Ashihara H (1982) Changes in adenine nucleotide levels and adenine salvage during the growth of *Vinca rosea* cells in suspension culture. Z Pflanzenphysiol 106:191
236. Yin Y, Katahira R, Ashihara H (2014) Metabolism of purine nucleosides and bases in suspension-cultured *Arabidopsis thaliana* cells. Eur Chem Bull 3:925
237. Yin Y, Shimano F, Ashihara H (2007) Involvement of rapid nucleotide synthesis in recovery from phosphate starvation of *Catharanthus roseus* cells. J Exp Bot 58:1025
238. Kanamori-Fukuda I, Ashihara H, Komamine A (1981) Pyrimidine nucleotide biosynthesis in *Vinca rosea* cells: changes in the activity of de novo and salvage pathways during growth in a suspension culture. J Exp Bot 32:69
239. Katahira R, Ashihara H (2006) Dual function of pyrimidine metabolism in potato (*Solanum tuberosum*) plants: pyrimidine salvage and supply of β-alanine to pantothenic acid synthesis. Physiol Plant 127:38
240. Watanabe S, Matsumoto M, Hakomori Y, Takagi H, Shimada H, Sakamoto A (2014) The purine metabolite allantoin enhances abiotic stress tolerance through synergistic activation of abscisic acid metabolism. Plant Cell Environ 37:1022
241. Deng WW, Sasamoto H, Ashihara H (2015) Effect of caffeine on the expression pattern of water-soluble proteins in rice (*Oryza sativa*) seedlings. Nat Prod Commun 10:733
242. Epifanio JA (1981) Ecology of the cafetalero agroecosystem. Universidad Nacional Autonoma de Mexico, Mexico, DF
243. Waller GR (1989) Biochemical frontiers of allelopathy. Biol Plant 31:418
244. Nathanson JA (1984) Caffeine and related methylxanthines: possible naturally occurring pesticides. Science 226:184
245. Hollingsworth RG, Armstrong JW, Campbell E (2002) Pest control caffeine as a repellent for slugs and snails. Nature 417:915

246. Hollingsworth RG, Armstrong JW, Campbell E (2003) Caffeine as a novel toxicant for slugs and snails. Ann Appl Biol 142:91
247. Aneja M, Gianfagna T (2001) Induction and accumulation of caffeine in young, actively growing leaves of cocoa (*Theobroma cacao* L.) by wounding or infection with *Crinipellis perniciosa*. Physiol Mol Plant Pathol 59:13
248. Kim Y-S, Lim S, Kang K-K, Jung Y-J, Lee Y-H, Choi Y-E, Sano H (2011) Resistance against beet armyworms and cotton aphids in caffeine-producing transgenic chrysanthemum. Plant Biotechnol 28:393
249. Kim YS, Lim S, Yoda H, Choi CS, Choi YE, Sano H (2011) Simultaneous activation of salicylate production and fungal resistance in transgenic chrysanthemum producing caffeine. Plant Signal Behav 6:409
250. Kim YS, Uefuji H, Ogita S, Sano H (2006) Transgenic tobacco plants producing caffeine: a potential new strategy for insect pest control. Transgenic Res 15:667
251. Kim YS, Sano H (2008) Pathogen resistance of transgenic tobacco plants producing caffeine. Phytochemistry 69:882
252. Kim YS, Choi YE, Sano H (2010) Plant vaccination: stimulation of defense system by caffeine production in planta. Plant Signal Behav 5:489
253. Crozier TWM, Stalmach A, Lean MEJ, Crozier A (2012) Espresso coffees, caffeine and chlorogenic acid intake: potential health implications. Food Funct 3:30
254. Crozier A, Ashihara H (2006) The cup that cheers. Caffeine biosynthesis: biochemistry and molecular biology. Biochemist 28(5):23
255. Ogita S, Uefuji H, Yamaguchi Y, Nozomu K, Sano H (2003) Producing decaffeinated coffee plants. Nature 423:823
256. Ogita S, Uefuji H, Morimoto M, Sano H (2004) Application of RNAi to confirm theobromine as the major intermediate for caffeine biosynthesis in coffee plants with potential for construction of decaffeinated varieties. Plant Mol Biol 54:931
257. Ogita S, Uefuji H, Morimoto M, Sano H (2005) Metabolic engineering of caffeine production. Plant Biotechnol 22:461
258. Ashihara H, Ogita S, Crozier A (2011) Purine alkaloid metabolism. In: Ashihara H, Crozier A, Komamine A (eds) Plant metabolism and biotechnology. Wiley, Chichester, UK, p 163
259. Mohanpuria P, Kumar V, Ahuja P, Yadav S (2011) Producing low-caffeine tea through post-transcriptional silencing of caffeine synthase mRNA. Plant Mol Biol 76:523
260. Ashihara H, Zheng XQ, Katahira R, Morimoto M, Ogita S, Sano H (2006) Caffeine biosynthesis and adenine metabolism in transgenic *Coffea canephora* plants with reduced expression of *N*-methyltransferase genes. Phytochemistry 67:882
261. Mohanpuria P, Kumar V, Yadav S (2010) Tea caffeine: metabolism, functions, and reduction strategies. Food Sci Biotechnol 19:275
262. Nakayama F, Mizuno K, Kato M (2015) Biosynthesis of caffeine underlying the diversity of motif B′ methyltransferase. Nat Prod Commun 10:799

Hiroshi Ashihara is Professor Emeritus at Ochanomizu University in Tokyo. He obtained his Ph.D. from the University of Tokyo in 1975, and did postdoctoral research at the University of Sheffield, UK from 1977 to 1979. He began research on the regulatory mechanism and in planta function of the pentose phosphate pathway. His research was expanded to the biosynthesis of nucleotides of which the precursor, 5-phosphoribosyl-1-phosphate, is provided by the pentose phosphate pathway. He has studied de novo and salvage pathways of pyrimidine, purine, and pyridine nucleotide biosynthesis and their catabolic pathways for nearly 40 years at Ochanomizu University. He further developed his research interests to include the alkaloids that are derived from these nucleotides. He has published more than 200 original papers, about 20 reviews, four Japanese university textbooks, and has co-edited three English books with Alan Crozier. He has been a visiting professor at the University of Copenhagen, Denmark and the University of Ryukyus, Okinawa, Japan, and a visiting scientist at the University of Calgary, Canada. He has collaborated with many researchers in Japan and in overseas countries including the UK, USA, Canada, Denmark, Switzerland, and China. He is currently interested in metabolic function in plants and the application of metabolic engineering to nucleotides and related compounds.

Kouichi Mizuno obtained his Ph.D. at the University of Tsukuba in Japan, after which he did postdoctoral research at the National Food Research Institute in Tsukuba, studying starch biosynthesis using biochemical and molecular biological approaches. He initially isolated and characterized three types of starch branching enzymes from rice. In 1998, he joined the Fujimura Laboratory at the University of Tsukuba, as an Assistant Professor, and began collaborating with Hiroshi Ashihara at Ochanomizu University. He worked on caffeine biosynthesis and isolated and characterized the genes of caffeine synthases from coffee. In 2002, he was appointed Associate Professor at Akita Prefectural University. His current research focuses on the biosynthesis of secondary metabolites, including caffeine and related alkaloids.

Takao Yokota is Professor Emeritus at Teikyo University in Utsunomiya. He was appointed as Assistant Professor at the University of Tokyo in 1970 and one year later obtained a Ph.D. from the same university. He started his career as a natural products chemist in the area of plant hormones. Initially, his research was on the structural determination of gibberellins and their glucosides. Subsequently, he developed an interest in the biochemistry and bioactivity of plant steroid hormones, the brassinosteroids. He isolated a number of brassinosteroids, such as castasterone, from various plant sources and determined their structures. In 1990, he moved to a Professorship in the Department of Biosciences, Teikyo University, where he discovered pea mutants that have lesions in either the biosynthesis or the receptors of brassinosteroids. Furthermore, he succeeded in the isolation of a new strigolactone named orobanchol from red clover, which is now known to be a widely distributed plant hormone. He is currently continuing research on jasmonates and plant progesterone.

Alan Crozier obtained a Ph.D. in plant biochemistry from the University of London in 1967. He then moved to a postdoctoral position at the University of Calgary in Canada for a three-year period before taking up an appointment as a Lecturer in Botany at the University of Canterbury in Christchurch, New Zealand. In 1973, he moved to the University of Glasgow in the UK where he continued research on the biosynthesis of the plant hormones, gibberellins and indole-3-acetic acid. In the early 1990s, he began a long and fruitful collaboration on purine alkaloids with Hiroshi Ashihara and in 1996 began research on the bioavailability of dietary flavonoids and related compounds. In 2014 he moved to the University of California to continue research on the absorption and metabolism of flavan-3-ols in cocoa. He has published over 300 papers and edited nine books. He was a Thomson Reuters Highly Cited Researcher in 2014 and 2015 and is also an Eminent Scientist of RIKEN for internationally distinguished achievements in the field of plant hormones and secondary metabolites.

The Iboga Alkaloids

Catherine Lavaud and Georges Massiot

Contents

Dedicated to the memory of Howard J Lotsof

C. Lavaud (✉)
Faculty of Pharmacy, Université de Reims Champagne-Ardenne, Institut de Chimie Moléculaire de Reims, UMR CNRS 7312, Case postale 44, UFR des Sciences Exactes et Naturelles, BP 1039, 51687 Reims, Cedex 2, France
e-mail: catherine.lavaud@univ-reims.fr

G. Massiot
Université de Reims Champagne-Ardenne, Institut de Chimie Moléculaire de Reims, UMR CNRS 7312, Case postale 44, UFR des Sciences Exactes et Naturelles, BP 1039, 51687 Reims, Cedex 2, France
e-mail: georges.massiot@univ-reims.fr

A.D. Kinghorn, H. Falk, S. Gibbons, J. Kobayashi (eds.), *Progress in the Chemistry of Organic Natural Products 105*, DOI 10.1007/978-3-319-49712-9_2

1 Introduction

There are four levels of definition for the term "Iboga Alkaloids". The most basic definition relates to the alkaloids isolated from a plant named "iboga", *Tabernanthe iboga* (Plate 1), in botanical words. The second level refers to those psychoactive alkaloids from plants used in ceremonies and cults in Central Africa, and which are the object of interesting pharmacological developments. The third definition belongs to chemists, who see behind the word "iboga" a specific arrangement of atoms containing an indole nucleus and an isoquinuclidine system. In turn, the fourth definition relies on biosynthesis, and characterizes those alkaloids in which two carbon atoms of the original secologanoside have departed from their original positions. In this series more than in others, nomenclatural problems are associated with the last-mentioned definition, so the present authors have decided to follow herein the so-called biogenetic nomenclature [1] and not that of IUPAC. For a complete overview of the monoterpene indole alkaloids classification, the reader is invited to read a survey by L. Szabó, in which these compounds are classified in three types and nine main skeletons [2]. Of

Plate 1 Iboga (*Tabernanthe iboga*) at the Centre National Floristique de l'Université Félix Houphouët-Boigny de Cocody (Abidjan); photograph courtesy of Philomène Akoua Kouassi, University of Cocody (Ivory Coast)

this nomenclature, the present chapter is concerned only with alkaloids based on the ibogan, isoplumeran, and isoeburnan skeletons.

Iboga alkaloids are produced by a small number of plants of the family Apocynaceae, inclusive of the genera *Catharanthus*, *Tabernaemontana*, *Corynanthe*, *Voacanga*, and *Aspidosperma*. There are about 100 alkaloids of this type, and among these, two have emerged for the chemistry or biology they have inspired. Included is catharanthine (**1**) (Fig. 1), one of the major alkaloids from *Catharanthus roseus* (Plate 2), the tropical periwinkle (also known as "Madagascar periwinkle"), and this is used in the partial synthesis of the anticancer drug Navelbine®. The second alkaloid is ibogaine (**2**), which has been a subject of intense attention for its putative anti-addiction properties and in general for its action on the central nervous system. It is worth noting at this point that these alkaloids belong to opposite optical series. Despite years of investigation, there still is fascinating chemistry being developed around this scaffold in order to provide a better access to these compounds and to better understand the coupling reactions used to make "dimers". The biology of ibogaine was the object of an entire volume in *The Alkaloids: Chemistry and Pharmacology* in 2001 [3]. The purpose of this contribution is to provide an update on these aspects in the hope of promoting innovative research in the field. The literature covered spans the years 2000–2016, and the interested reader is invited to consult references [3–5] for earlier contributions.

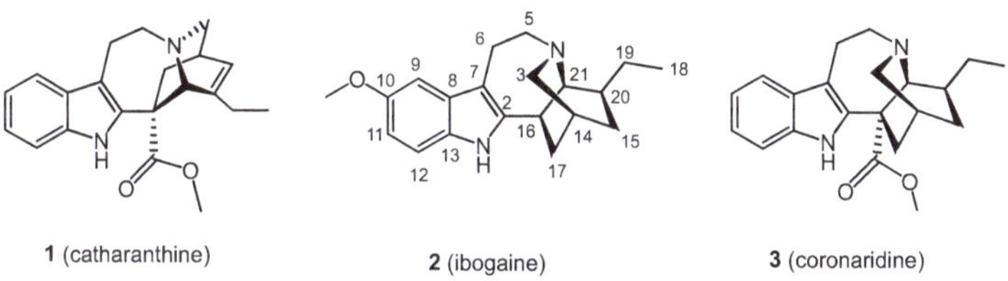

Fig. 1 Catharanthine (**1**), ibogaine (**2**), coronaridine (**3**)

Plate 2 Madagascar periwinkle (*Catharanthus roseus*); photograph courtesy of Bruno David, Pierre Fabre Laboratories, Toulouse (France)

2 Biosynthesis

Despite intensive work on *Catharanthus roseus*, which is a high-yield producer of the two iboga alkaloids catharanthine and coronaridine (**3**), there is no recent publication on the biosynthesis of these particular molecules [6–8]. There are no "omics" studies available on this part of the biosynthesis scheme and most details known date back to the 1970s with the incorporation of the strictosidine derivative preakuammicine (**8**) and more recently of stemmadenine (**5**) into catharanthine (**1**) [9]. The postulated intermediate dehydrosecodine (**6**), which ought to be highly unstable, remains undetected. While preakuammicine (**8**) is the well-accepted link in the sequence leading to the ibogans, there is no reason why its infamous and isomeric cousin precondylocarpine is not. The temptation is great to propose that **8** is the precursor of the coronaridine series while precondylocarpine (**9**) would lead to the catharanthine series (Figs. 2 and 3).

The formation of the azabicyclo ring system is thought to proceed via a Diels-Alder-like reaction and the question of its catalyst by means of an enzyme is of high relevance since examples of Diels-Alderases remain extremely rare. Only recently, one such enzyme has been isolated and its structure elucidated, but its genuine mechanism of action (catalyst or chiral template) is still an object of investigation [10,11]. It is to be noted that although dehydrosecodine (**6**) is achiral, coronaridine (**3**) and catharanthine (**1**) are chiral and belong to opposite series. While catharanthine (**1**) is predominant in *C. roseus*, the coronaridine series is generally prevalent in Nature, with ibogaine being the most representative alkaloid of the series. There is no reason, however, not to consider the two alternatives when working on the structural elucidation of an iboga-type alkaloid. The genus *Catharanthus* is of high relevance since it presents the unique feature of accumulating alkaloids in the two series, although at a different oxidation level. These are some of the reasons why the unraveling of the

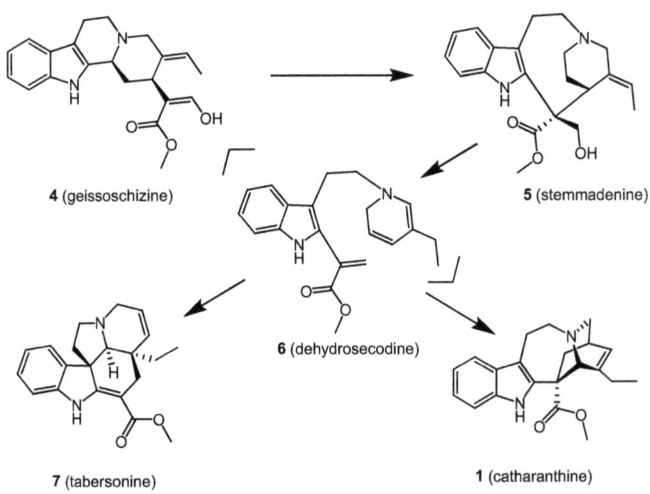

Fig. 2 Intermediates in the biosynthesis of catharanthine (**1**)

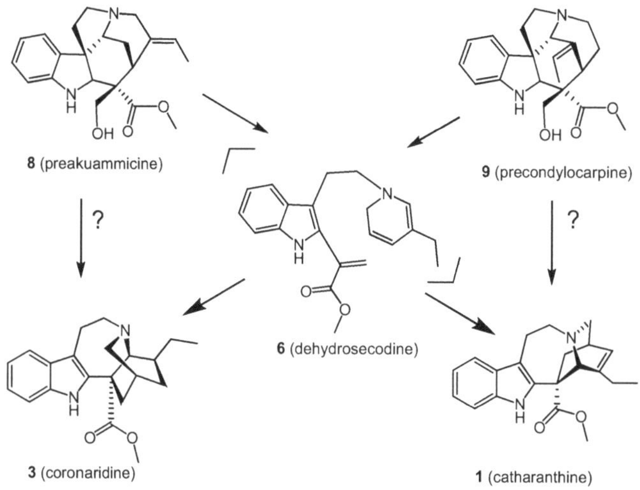

Fig. 3 The involvement of preakummicine (**8**) and of precondylocarpine (**9**) in the biosynthesis of iboga alkaloids

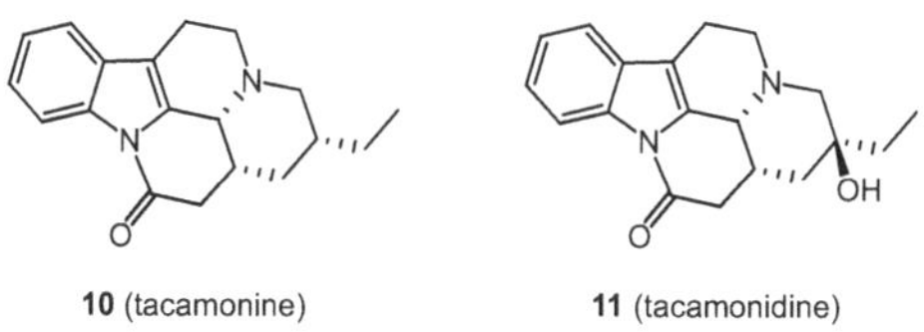

Fig. 4 Iboga alkaloids of the pseudo-vinca type

biosynthesis of the iboga alkaloids is important, not only for this particular field but for the whole of bio-organic chemistry.

Not all iboga alkaloids possess the isoquinuclidine skeleton, and there are a few examples of known alkaloids with the pseudo-aspidosperma (i.e. cleavamine) or pseudo vinca (i.e. tacamonine (**10**), Fig. 4) arrangement. For the moment, these alkaloids are but chemical curiosities and no experimental work has been performed to explain their biosynthesis. Soon after the isolation of tacamonine (**10**) from the well-known plant, *Tabernaemontana eglandulosa*, it has been suggested that its formation was due to an evolution of the plant when cultivated in greenhouses in The Netherlands [12]. This hypothesis was later seen not to be true since another of these alkaloids, tacamonidine (**11**) was isolated from wild *Tabernaemontana corymbosa* [13].

3 Structural Elucidation and Reactivity

The widespread availability of high-field NMR spectrometers and of sophisticated pulse sequences now enables chemists to determine structures of moderate complexity, such as the iboga alkaloids and even the "dimers", leading to full

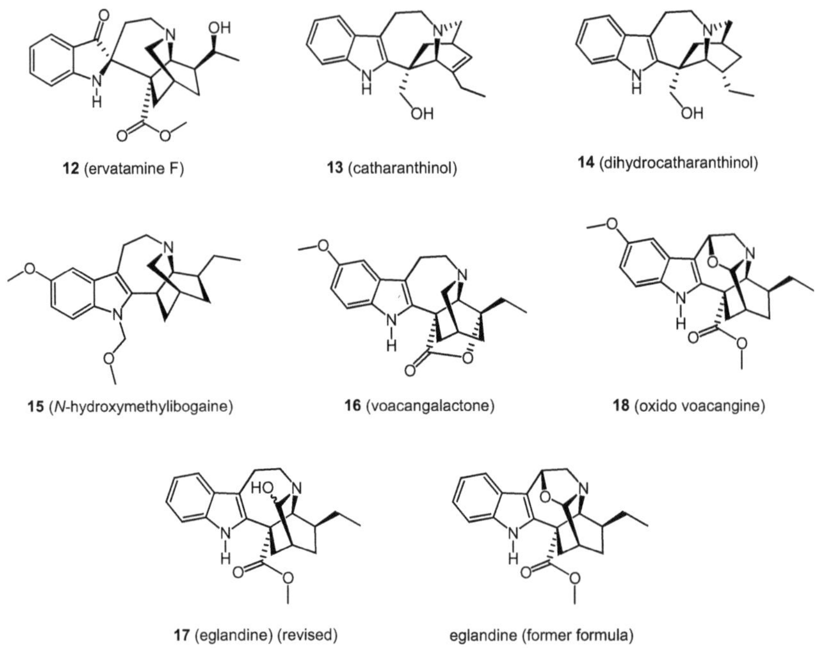

Fig. 5 Examples of structures of iboga alkaloids resolved by the Mosher method, X-ray crystallography, or total synthesis

unambiguous proton and carbon NMR assignments. Relative configurations in rings and ring junctions may be dealt with NOEs and coupling constant measurements. In these series, the only open chain fragment is the ethyl side chain, where chirality is present at C-19. The problem of its determination is best solved with one of the Mosher methods, an example of which is found in the structure elucidation of ervatamine F (**12**) (Fig. 5). In addition, X-ray crystallography allows the solving of structural problems whenever a single crystal is available and Sect. 4 lists several recent examples of such structural determinations. Catharanthinol (**13**), dihydrocatharanthinol (**14**), and *N*-hydroxymethylibogaine (**15**) have had their structure solved by X-ray crystallography, but unfortunately without absolute configuration determinations [14,15].

The problem of the absolute configuration of these alkaloids remains a difficulty and there are currently four methods available for this purpose. NMR spectroscopy is one of these, provided one stereogenic center is linked to the others and determined using Mosher derivatives. This procedure has often been used to determine the configuration of C-19, which unfortunately cannot be easily linked with the remaining stereogenic centers [16]. Circular dichroism is a general and sensitive method, which has been used to differentiate compounds in the (+)-catharanthine and (−)-ibogaine series [17]. It is now used to determine the absolute configurations of the linkages in dimers by application of the exciton chirality rule, which was first proposed by Harada and Nakanishi [18], and then later used by others [19]. Ab initio calculation of electronic CD spectra and comparison with the experimentally obtained values also allows the determination of absolute configurations, but so far most examples have dealt with monomers [16]. X-ray crystallography using anomalous scattering is the most unambiguous method. It is more and more frequently used to solve structures. Finally, partial or total synthesis has often been used to link a new alkaloid to another of known absolute configuration.

Chemical reactivity, and partial and total synthesis, were until recently integral parts of the structural elucidation process. Simple reactions were used to determine functionalities or to prepare crystalline derivatives. As a bonus, it often happened that new and unexpected reactions were discovered during the process. With the advent of highly sensitive spectroscopic techniques, this approach is almost no longer utilized, and only one example could be found in the more recent literature of a total synthesis, aimed at establishing the structure of voacangalactone (**16**) [20].

The most intriguing aspects of the iboga alkaloids are their unusual oxidation reactions, which arise from the particular geometry of the isoquinuclidine ring system. The original experiments were described by Bartlett and Taylor, who transformed coronaridine (**3**) (Fig. 6) with iodine in aqueous methanol into the 3-hydroxy- and 3-oxo-coronaridine derivatives [21]. These reactions went unnoticed for a long time, even though their products were later shown to be naturally occurring substances. Eglandine (**17**) (3-OH coronaridine) is one example, but its structure was assigned initially as containing a tetrahydro-oxazole ring and later had to be corrected [22,23]. The present chapter authors long questioned the natural occurrence of these tetrahydro-oxazoles until definitive proof of their existence came from the X-ray structure of oxindole **57** (ervaoffine A) [24]. While working on this chapter, the authors corresponded with Professor Hiromitsu Takayama, whose group isolated the 3,6-oxido voacangine (**18**), and he provided good NMR spectroscopic evidence for the cyclic structure of oxidovoacangine [25]. However, the probability is high that derivatives with substitutions at C-3 (OMe, OEt, or acetonyl) originate from the intermediate carbinolamines. Interestingly, these do not lend themselves to reduction or Pictet-Spengler-like reactions although they can be considered as masked aldehydes. One of the reasons why these particular structures are not easily identified is that they "misbehave" in mass spectrometers and often show [M-2] fragments (a result of a dismutation). As a consequence, there is a considerable tendency to make an extra bond as in ervaoffine A (**57**) or as in ervataine (**19**) [26]. Ervataine (**19**) clearly violates Bredt's

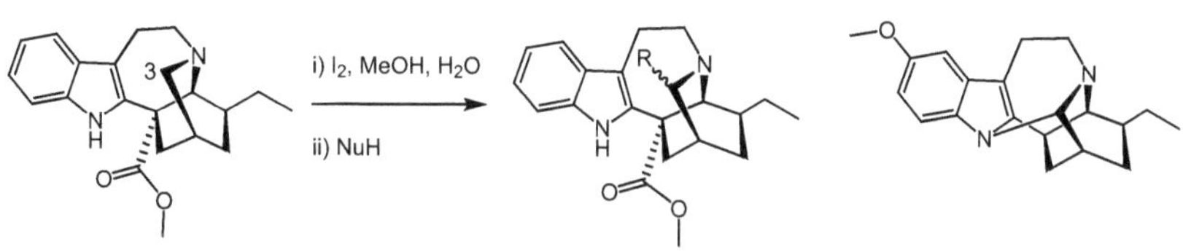

3 (coronaridine) R = H, then Nu **19** (ervataine)

Fig. 6 Oxidation of coronaridine at position 3 and "structure" of ervataine (**19**)

rule [27] and its most probable structure is 3-OH ibogaine (C-3 showed a ^{13}C NMR resonance at 93.2 ppm).

More classical oxidations arise at C-7, yielding the pseudo-indoxyl chromophore after rearrangement or at C-2 as an intermediate to oxindole. The addition of dioxygen on the double bond of indole and fragmentation produces *N*-acyl derivatives as observed in the degradation of tryptophan into kynurenine. All these chromophores are detected in natural products (vide infra) and these products, if they are not extraction artefacts, may be considered as end products in biosynthesis processes (Fig. 7).

The most studied oxidation in the ibogan series is related to the *N*-oxides, which are easily prepared under acidic peroxide conditions. While in the ibogaine/coronaridine series the *N*-oxide is relatively stable, the *N*-oxide of catharanthine (**20**) (Fig. 8) lends itself to a Meisenheimer rearrangement, bringing about a N-4—C-3 bond cleavage. Such products have been observed but knowledge of their biosynthesis remains at the hypothetical stage. Fragmentation of the *N*-oxide of catharanthine is the key reaction in the synthesis of anhydrovinblastine and the related antitumor alkaloids vinblastine, vincristine, vinorelbine, and vinflunine. The subject has recently been reviewed and the interested reader will find an account of this chemistry involved in books by Langlois and Langlois [28] and by Guéritte and Le Roux [29].

The formal addition of HCN across the C-16—C-21 bond of catharanthine (**1**) provides an intermediate aminonitrile (**21**), which allowed Beatty and Stephenson to bridge the gap between this series and the pseudo-tabersonine and -vincadifformine alkaloids [30]. The initial reaction was brought about by visible light irradiation of **1** in the presence of the polyfluorinated catalyst Ir(dF(CF_3)ppy)$_2$(dtbbpy)PF_6 and with trimethylsilyl cyanide as reactant (Fig. 9). Remarkably, this reaction was run in a flow reactor in order to better control reaction parameters and confine the hazards linked to HCN generation. Although the conversion to pseudotabersonine (**24**) only required iminium generation and its "through nitrogen" isomerization, it took carefully controlled acidic conditions to achieve the transformation. One of the unexpected difficulties was a partial racemization linked

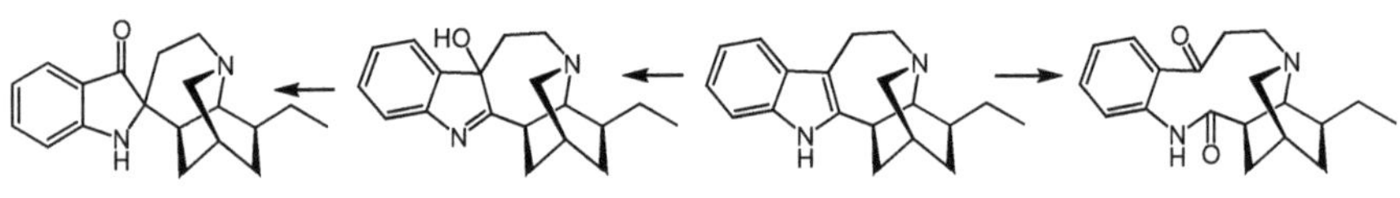

Fig. 7 Oxidation pathways for iboga alkaloids

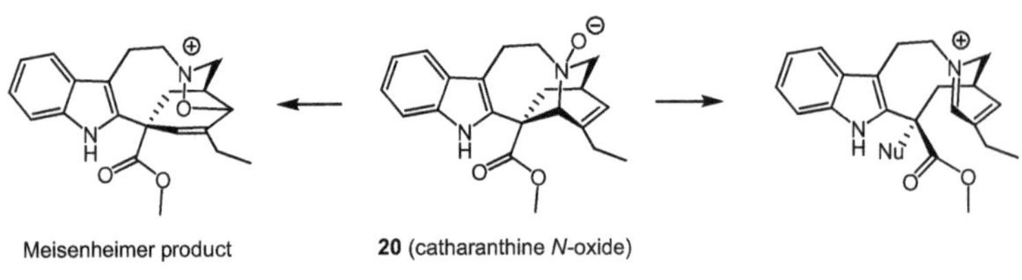

Fig. 8 The chemistry of catharanthine *N*-oxide (Meisenheimer product, left; reactivity under Polonovski conditions, right; Nu may be vindoline)

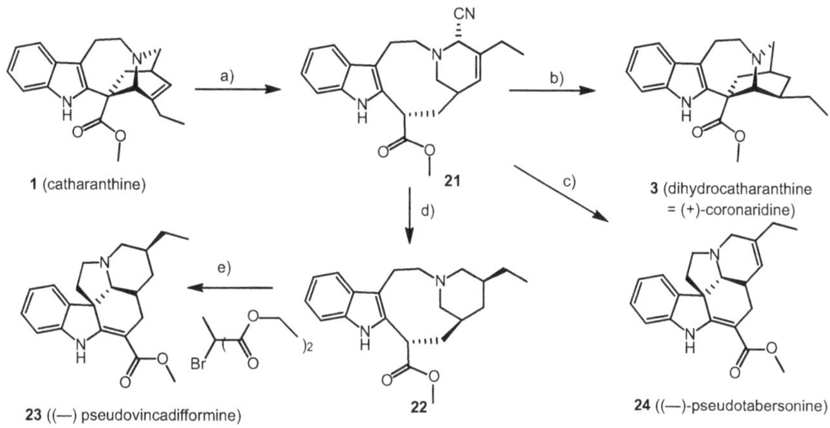

a) $Ir(dF(CF_3)ppy)_2(dtbbpy)PF_6$, TMSCN, $h\nu$, b) i) H_2, Pd/C, ii) TFA, reflux, c) TFA, degassed toluene
d) H_2, Pd/C then $NaBH_4$, e) $Ru(bpy)_3Cl_2$, $h\nu$

Fig. 9 Interconversion of catharanthine (**1**), pseudotabersonine (**22**), and pseudovincadifformine (**23**) via cyano compound **21**

to C-14 equilibration, and treatment of **21** with TFA led to an almost complete conversion into **24**. The optical rotation value and chiral HPLC parameters of the reaction product showed it to be a 2:1 mixture of enantiomers. While it was initially thought that these difficulties would be absent in the dihydro series, pseudovincadifformine (**23**) was not obtained but (+)-dihydrocatharanthine (**3**) occurred instead. Compound **23** was eventually reached in good yield through oxidative photoirradiation of **22**, the over reduction product of compound **21**. Among other things, this is to our knowledge, one of the rare articles daring to state that catharanthine (**1**) is a good starting material for partial synthesis due to its natural availability, a fact which is largely overlooked.

4 New Molecules

The list of compounds biogenetically linked to the iboga unit continues to be fed by new alkaloids but among the three skeletons of this group, the isoplumeran and isoeburnan variants remain scarce. This Section is divided into two parts: the monomers and the dimers, and, in each, molecules are grouped according to oxidation levels. The Apocynaceae family has continued to provide new occurrences, but at variance with other much more common indolomonoterpenes, the isolation of alkaloids of the iboga class is limited to only a few species in the genera *Ervatamia*, *Tabernaemontana*, *Voacanga*, and *Catharanthus*, with *Ervatamia* being considered as a synonym of *Tabernaemontana* by many authors (Tables 1 and 2). "The Plant List" [31] was chosen as the botanical reference. The total number of

Table 1 Sources of new monomeric iboga alkaloids

Plant name	Synonym	New monomeric alkaloids	Ref.
Catharanthus roseus (L.) G. Don		**63a** or **63b**	[45]
Ervatamia hainanensis Tsiang	*Tabernaemontana bufalina* Lour.	**39**, **41**, **42**, **48**, **49**, **53**, **54**	[39]
		12, **27**, **28**, **33**, **46**	[16]
Ervatamia officinalis Tsiang	*Tabernaemontana bovina* Lour.	**38**, **50**, **52**, **56**, **57**, **59**, **60**	[24]
Tabernaemontana corymbosa Roxb. ex Wall.		**30**, **31**, **32**, **33**, **40**	[36]
		11, **34**, **55**, **59**	[13]
		61, **62**	[44]
		35	[37]
		36, **37**	[38]
		25, **26**, **29**, **44**, **45**	[32]
		65, **66**, **67**, **68**	[46]
Tabernaemontana divaricata (L). R.Br. ex Roem. & Schult.		**43**, **47**	[40]
Tabernaemontana hystrix Steud.		**51**	[42]
Voacanga africana Stapf ex Scott-Elliot		**16**	[20]

Table 2 Sources of new bisindole iboga alkaloids

Plant	Synonym	New bisindole alkaloids	Ref.
Tabernaemontana corymbosa Roxb. ex Wall.		**77**, **78**, **80**	[50]
		88, **89**, **92**, **95**, **96**, **97**, **98**, **99**, **100**, **101**	[51]
		102, **103**	[54]
		83, **86**	[13]
		94, **108**, **120**, **122**	[53]
		79, **80**	[36]
		76, **81**, **82**, **87**, **123**, **124**	[49]
		125, **126**	[46]
Tabernaemontana divaricata (L). R. Br. ex Roem. & Schult.		**90**	[52]
		104, **105**, **106**, **107**, **121**	[40]
Tabernaemontana sphaerocarpa Blume		**75**	[48]
Muntafara sessilifolia Baker	*Tabernaemontana sessifolia* Baker	**111**, **112**, **116**, **117**, **118**, **119**	[56]
		109, **110**, **111**, **113**, **114**	[55]
Ervatamia chinensis (Merr.) Tsiang	*Tabernaemontana corymbosa*	**83**, **84**, **85**, **93**	[19]

species mentioned below does not exceed ten for a total of a little more than a hundred compounds. This may be the tip of the iceberg since the number of *Tabernaemontana*, *Ervatamia*, and *Voacanga* species is close to 700. The limited presence of iboga alkaloids in these particular genera is probably a consequence of their elaborate biosynthesis, pertaining to well-evolved plants. Alkaloids are considered here as new when not mentioned in previous reviews.

4.1 Monomers

4.1.1 Ibogamine and Coronaridine Derivatives

Many of the newly isolated structures may be viewed as simple derivatives of ibogamine or coronaridine, with a limited range of oxidations. Typically, these variations concern oxidations at C-19 in the side chain, C-10 and C-11 of the indole nucleus, and more rarely C-15 and C-20 in the isoquinuclidine moiety. Molecules with substituents at C-11 remain scarce.

Four new 19-hydroxy derivatives have been reported: (19*S*)-hydroxyibogamine (**25**) (Fig. 10), 19-*epi*-isovoacristine (**26**) [32], and ervatamines A (**28**) and H (**27**) [16]. The relative configuration of C-19 was determined by comparison of the ^{13}C NMR chemical shifts of C-15 and C-21, as proposed by Wenkert in 1976 [33,34]. The problem of absolute configuration was ignored in Ref. [32], while ECD provided a solution with ab initio calculations for compound **28**, and a comparison with ECD of coronaridine for **27**. C-ring contraction is a very rare occurrence in the biosynthesis of indole alkaloids, for which the origin is a C-5—C-6 Polonosvski fragmentation. Ervatamine A (**28**) adds to this very short list and may be considered as a collateral pathway in the reaction of the corresponding ibogan *N*-oxide. This is the second ibogan-type alkaloid isolated where ring C is contracted to form an unusual six-membered ring as in flabelliformidine [35]. Ervatamine H (**27**)

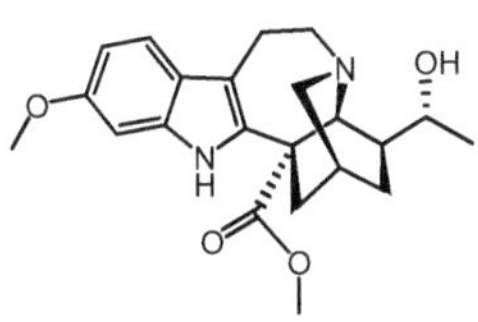

26 (19-epiisovoacristine)

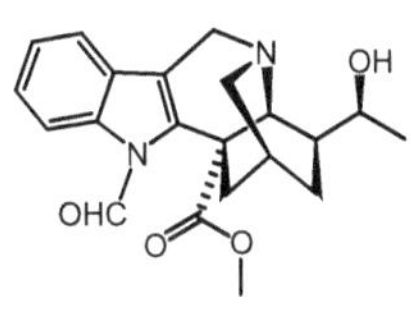

28 (ervatamine A)

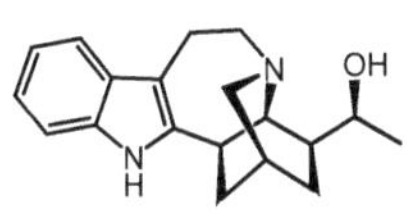

25 ((19*S*)-hydroxyibogaine)

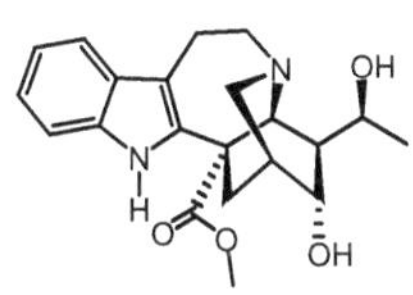

27 (ervatamine H)

Fig. 10 The 19-hydroxy derivatives **25–28**

has a very unusual oxidation of C-15, which could also be envisioned as a hydration of a Δ^{14} double bond. Observation of a NOE correlation between H-18 and H-15, and ECD comparison with coronaridine allowed the (14*R*,15*R*,16*S*,20*S*,21*S*)-configuration to be proposed.

Five new alkaloids were found to contain an additional degree of oxidation and a ketone at C-19: isovoacryptine (**29**) [32], conodusines A-C (**30–32**) [36], and (−)-ervatamine I (**33**) [16], for which an isomer was named (+)-conodusine E (Fig. 11) [36]. The report on conodusine E was published subsequent to that on ervatamine I (**33**), and, given the similarities between the data for the two compounds, except for their optical rotations, great care was accorded to the determination of the relative and absolute configurations. First, a chemical correlation was carried out with (−)-heyneanine (Dess-Martin oxidation), then the ECD was measured and compared with the calculated ECD values for the two enantiomers, and finally an X-ray analysis was performed. All these experiments converged to the structure depicted as **33**. The Malaysian authors proposed that the so-called ervatamine I is an enantiomer of their own compound and therefore an alkaloid of the catharanthine series. However, the present authors feel that this hypothesis creates a precedent: the isolation of a catharanthine-like compound in a *Tabernaemontana* species. The sole rationale given for the absolute configuration of ervatamine I is the sign of the optical rotation (the same as coronaridine), which is an argument to be considered of little value at the present time. Conodusines A and B are isomeric at C-20, a position α to a carbonyl, and hence prone to isomerization. The possibility of conodusine B being an artefact was discussed but could not be established definitively. The same possibility holds for conodusine C (**32**), the *N*-oxide of conodusine B (**31**).

Voacanga africana is one of the most highly investigated West African medicinal plants. From this species, a Japanese group isolated a new bioactive alkaloid, voacangalactone (**16**) (Fig. 12), for which the structure was determined by spectroscopic means and by total synthesis [20]. All alkaloids with hydroxy groups at positions C-15, C-19, and C-20 may stem from precursors with a double bond between C-19 and C-20 or C-15 and C-20, isomeric with catharanthine (**1**), and until now never detected in this series.

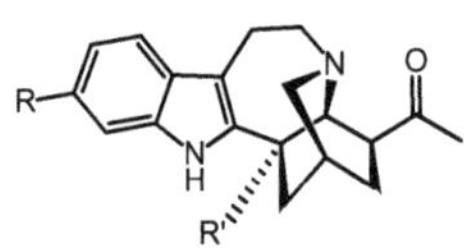

29 R = OMe, R' = CO_2Me
(isovoacryptine)
30 R = R' = H
(conodusine A)
33 R = H, R' = CO_2Me
(conodusine E)

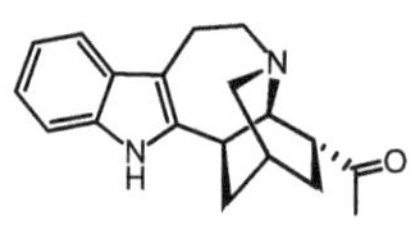

31 (conodusine B)

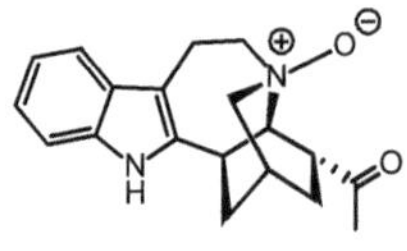

32 (conodusine C)

Fig. 11 The 19-keto derivatives **29–32**

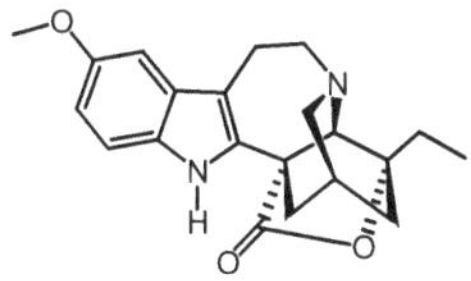

16 (voacangalactone)

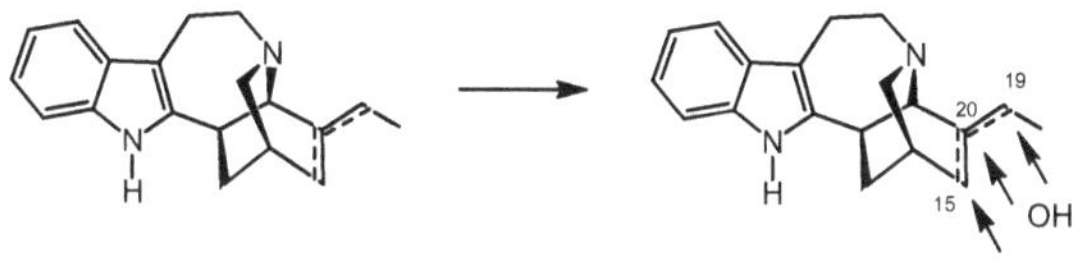

Fig. 12 Voacangalactone (**16**) and putative formation of derivatives with C-15, C-19, and C-20 OH

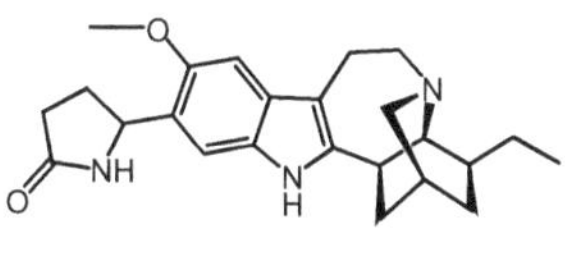

34 (cononusine)

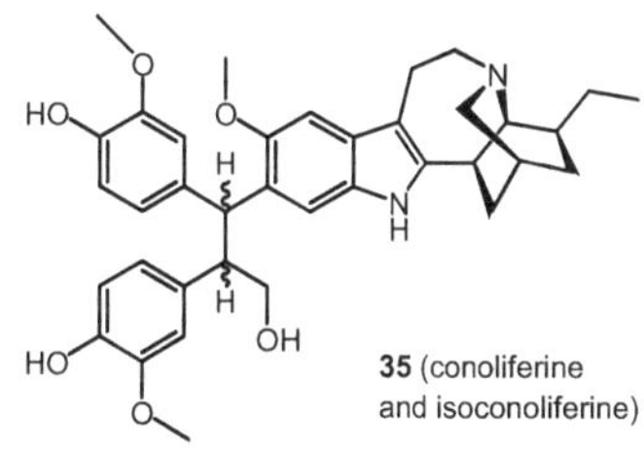

35 (conoliferine and isoconoliferine)

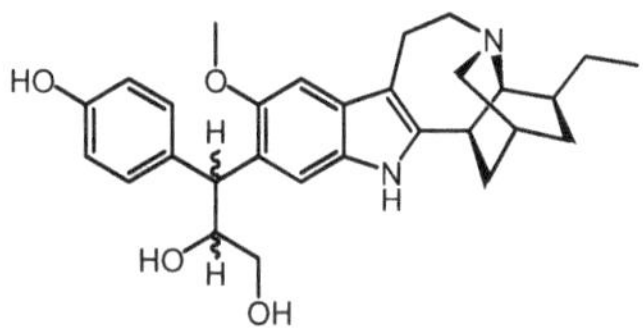

36 (conomicidine A and isoconomicidine A)

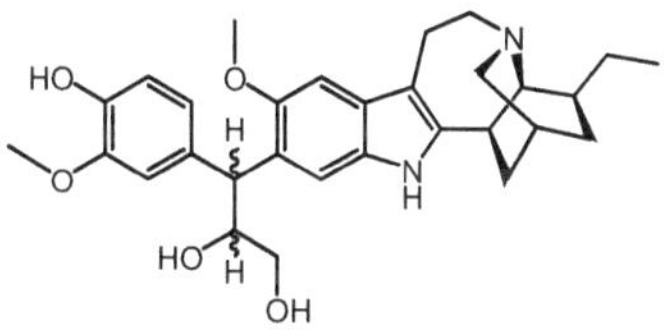

37 (conomicidine B and isoconomicidine B)

Fig. 13 Pyrrolidone and lignan conjugates of ibogaine

Nucleophilic attack on C-11 is favored when C-10 is substituted by a methoxy group, which is the situation for ibogaine, and leads to a group of iboga alkaloids diversely substituted at C-11, all found in *Tabernaemontana corymbosa* from Malaysia. These were: cononusine (**34**) with a pyrrolidone unit [13], and the lignan conjugates conoliferine and isoconoliferine (**35**) [37], conomicidines A and B (**36**), and isoconomicidines A and B (**37**) (Fig. 13) [38]. These included pairs of diastereomers that were not separated even though the authors discussed at length the relative configurations of the new chiral centers. The question of the absolute configuration of the iboga moiety was not a subject of debate, assuming that the configuration is prevalent in *Tabernaemontana*. In cononusine (**34**), the relative and absolute configurations of the chiral center in the pyrrolidone ring could not be determined. The lignan conjugates were assumed to arise from a nucleophilic attack of ibogaine on a quinone methide derived from one or two coniferyl units.

4.1.2 3-Alkyl- or 3-Oxo-ibogamine/-coronaridine Derivatives

These alkaloids are, for most of them, oxidized forms of known compounds. Alkaloids **38–40** (Fig. 14) are all 7-hydroxylated indolenines, further oxidized as lactams at position C-3. They derive, respectively, from ibogaine, coronaridine, and conodusine A (20-oxo-ibogamine), although no chemical correlations were attempted as structural proof. The reactions leading to these compounds are well established in the literature on indolomonoterpenes and their biosynthesis follows the usual oxidation pathways. The most well-documented structure is 3-oxo-(7*R*)-coronaridinehydroxyindolenine (**39**), which was determined by X-ray diffraction, with Flack parameter calculations, thus making the absolute configuration definite [39]. The two other compounds have had their configuration determined by comparison of ECD spectra (**38**: λ_{max} ($\Delta\varepsilon$) 230 (−9.0), 258 (+18.6), 289 (−4.4) nm [24]; **39**: λ_{max} ($\Delta\varepsilon$) 221 (+26.3), 260 (−10.9), 294 (0) nm; **40**: λ_{max} ($\Delta\varepsilon$) 221 (+11.6), 257 (−5.7), 289 (+1.7) nm [36]). It would have been advisable to have the ECD also recorded for tabernaricatine F (**43**), for which the configuration of C-7 was proposed after observation of a NOE between OH and H-16.

Following the same biosynthesis, three other hydroxy indolenines (**41–43**) (Fig. 15) have been encountered in the coronaridine/ibogaine series, displaying

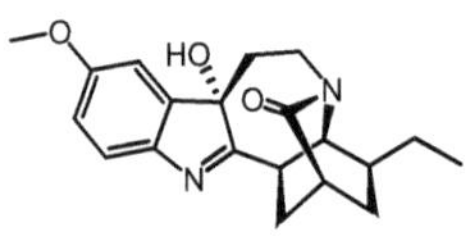

38 (3-oxo-(7*S*)-ibogaine hydroxyindolenine)

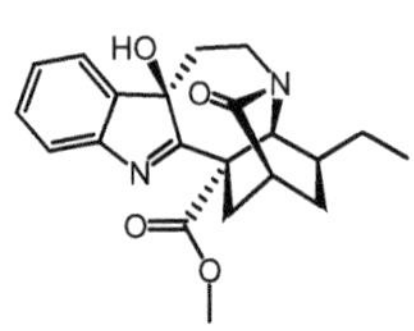

39 (3-oxo-(7*R*)-coronaridine hydroxyindolenine)

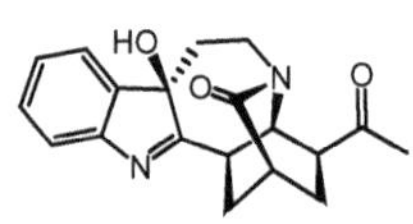

40 (conodusine D)

Fig. 14 The hydroxy indolenines (**38–40**)

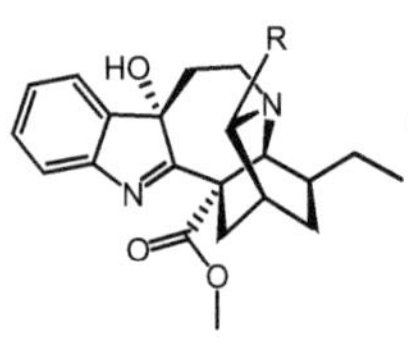

41 R = OH ((3*R*)-hydroxy-(7*S*)-coronaridine hydroxyindolenine)
42 R = CN ((3*S*)-cyano-(7*S*)-coronaridine hydroxyindolenine)

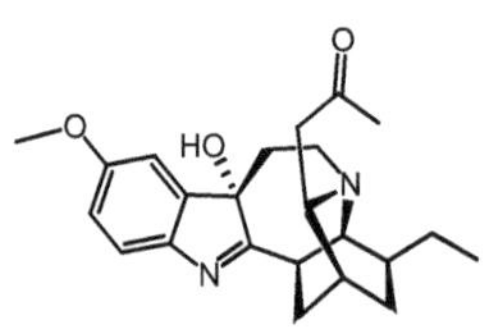

43 (tabernaricatine F)

Fig. 15 Further hydroxy indolenines (**41–43**)

further oxidation at C-3. Compounds **41** and **42** share the same origin as **39**, from which the hydroxylated derivative **41** is an obvious precursor [39]. The cyano and acetonyl derivatives **42** and **43** arise from substitution of their respective 3-hydroxy precursors and while cyano derivatives are common in Nature, the origin of the acetone fragment is more dubious. It is not unreasonable to think that it comes from the solvent used for chromatography, with silica gel acting as an acid catalyst [40]. Observation of the suitable Cotton effects for **41** and **42** allowed their configuration to be established as (7*S*). The configuration at C-3 was deduced from NOE correlations observed between H-3↔H-6β↔H-17β, placing the H-3 to the "left", towards the respective indole ring.

Besides these cyano, hydroxy, and acetonyl derivatives, there are a large number of monomeric iboga alkaloids with miscellaneous alkyl or alkoxy substituents on C-3. The 3-alkoxy derivatives, (3*R*/*S*)-ethoxyheyneanine (**44**) and (3*R*/*S*)-ethoxy-19-*epi*-heyneanine (**45**) (Fig. 16), were isolated as mixtures of diastereomers, and could be artefacts formed by addition of ethanol on an iminium intermediate during the extraction steps [32]. Similar NOE effects were observed for four 3-alkyl derivatives: ervatamine G (**46**) [16], tabernaricatine G (**47**) [40], (3*S*,24*S*)- and (3*S*,24*R*)-hydroxyethyl-coronaridine (**48** and **49**) [39], suggesting a (3*S**) configuration. It is worth noting that these last two compounds were isolated from *Ervatamia hainanensis* roots in 1982 but their configuration could not be established at that time [41]. An X-ray structure determination for the (24*S*) compound **48** secured the relative and absolute configurations for the two isomers. Their ECD spectrum exhibited similar Cotton effects as observed for coronaridine, confirming their common (3*S*,14*R*,16*S*,20*S*,21*S*)-configuration. This was confirmed by preparing Mosher esters of the C-24 alcohols.

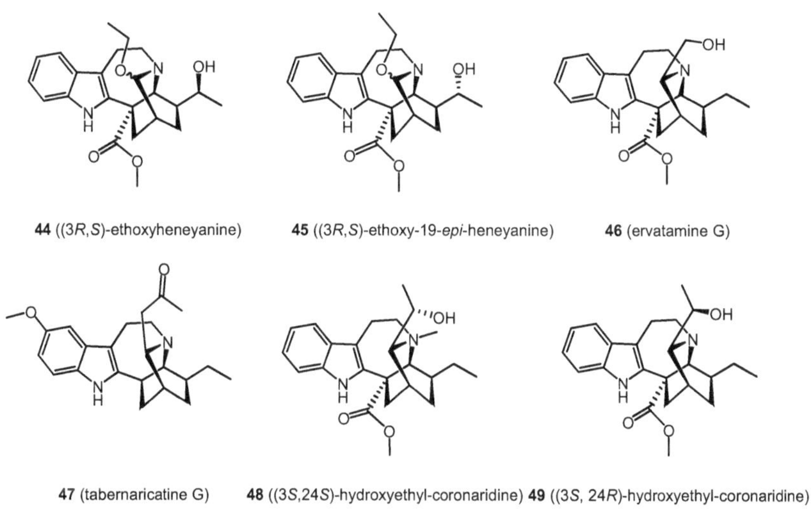

44 ((3*R*,*S*)-ethoxyheneyanine) **45** ((3*R*,*S*)-ethoxy-19-*epi*-heneyanine) **46** (ervatamine G)

47 (tabernaricatine G) **48** ((3*S*,24*S*)-hydroxyethyl-coronaridine) **49** ((3*S*, 24*R*)-hydroxyethyl-coronaridine)

Fig. 16 The 3-substituted iboga alkaloids (**44**–**49**)

4.1.3 5- and/or 6-Oxo-ibogamine/-coronaridine Derivatives

Positions α to the basic nitrogen atom are prone to oxidation and in the iboga alkaloids, this most often occurs at C-3. 19-*epi*-5-Oxovoacristine (**50**) from *Ervatamia officinalis* is an exception, and the location of the amide carbonyl is based on HMBC correlations and on the presence of an isolated AB quartet for CH_2-6. The remainder of the asymmetric centers was determined by ^{13}C NMR chemical shifts (for C-19) and NOE correlations, and the absolute configuration was established by ECD (experimental vs. calculated). The location of the carbonyls in the 5,6-diones of ibogamine and ibogaine (**51**) [42] and (**52**) [40] was based on observation of HMBC correlations between H-3 and H-21 with O = C-5 (lactam near 169 ppm). The ion at *m/z* 170 (*m/z* 156 in ibogamine) in their mass spectra also suggested the presence of an oxygen atom at C-6. The ECD spectrum of **52** was measured and compared to the calculated value, which confirmed the (14*R*,16*R*,20*S*,21*S*)-configuration. 5-Oxo-(6*S*)-hydroxy- and 5-oxo-(6*S*)-methoxycoronaridine (**53** and **54**) are two other new products obtained from *Ervatamia hainanensis* [39]. An X-ray structure was obtained for the alcohol **53**, which allowed its full structural determination, and, as a consequence, establishment of the structure of **54**. The simplest compound **55** (6-oxo-ibogaine) is unique and unexpected. There are two arguments in favor of oxidation at C-6: a downfield shift of H-9 due to carbonyl anisotropy and HMBC correlations of H-5 (two d, $J = 18.1$ Hz) with C-3, C-7, and C-21. This compound has been an intermediate in a synthesis of ibogaine, but this is the first time that it was detected as a natural product [43] (Fig. 17).

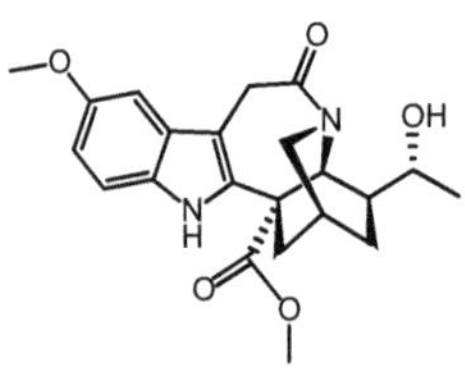

50 (19-*epi*-5-oxovoacristine)

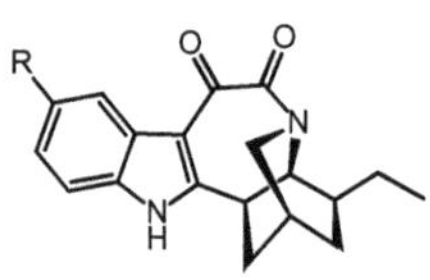

51 R = H (ibogamine-5,6-dione)
52 R = OMe (ibogaine-5,6-dione)

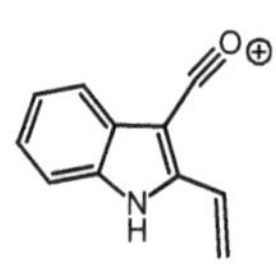

m/z 170

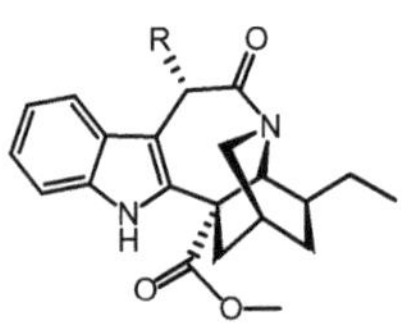

53 R = H (5-oxo-(6*S*)-hydroxycoronaridine)
54 R = OMe (5-oxo-(6*S*)-methoxycoronaridine

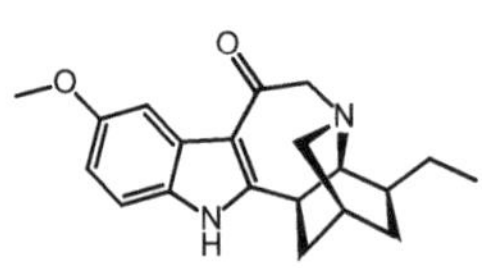

55 (6-oxo-ibogaine)

Fig. 17 Alkaloids oxidized at positions C-5 and C-6

4.1.4 Rearranged Ibogamine/Coronaridine Alkaloids

In this group of alkaloids, the present authors have chosen to place those in which one bond of the skeleton is cleaved, thus giving four rearranged types of structure: the 2,7-*seco*-, 6,7-*seco*-, 2,16-*seco*-, and the N(4),21-*seco*-iboga derivatives.

Ervaoffine D (**56**) (Fig. 18) is the only described iboga alkaloid with the 2,7-bond cleaved into a ketone and an amide [24]. Its structure was proved unequivocally by X-ray crystallography inclusive of the absolute configuration. Among the 6,7-*seco* derivatives, the structures of ervaoffine A (**57**) [24] and ervatamine F (**12**) [16] were confirmed by X-ray crystallography and the absolute configurations were found to be (2*S*,3*S*,6*S*,14*R*,16*R*,20*S*,21*S*) and (2*S*,14*R*,16*S*,19*S*,20*S*,21*S*) (ibogaine series). The priority order of C-3 and C-6 is modified by the introduction of the hydroxy group on position C-6 and these two same (2*S*)-configured compounds indeed have opposite spatial configurations at C-2. Iboluteine (**58**), also produced by the same plant, and the deoxy analog of ervaoffine A (**57**), have also been shown by X-ray crystallography to have the opposite configuration, and so **58** is not the precursor of **57**. In addition, measurements and calculations of ECD spectra for **57** were performed, allowing comparison to be made with alkaloids bearing the same chromophore. Mosher esters of **12** were prepared to confirm the C-19 stereochemistry. The third alkaloid, **59**, was isolated by two different groups in 2014 and 2015, and although published in the same journal, it was given different trivial names: ervaoffine B [24] and ervaluteine [36], when it could be simply have been named (6*R*)-hydroxy-(2*S*)-pseudo-indoxyl-ibogaine. It is interesting to note that the absolute configuration was given in the first paper, since the ECD spectrum was similar of that of ervaoffine A (**57**) [24] and a NOE was detected between OH-6 and H-3α and H-17β. In the second paper, the absolute configuration was deduced painfully from NOE correlations, comparison with (2*R*)-iboluteine (**58**), and examination of molecular models. Only one 2,16-*seco*-ibogamine, ervaoffine C (**60**), was characterized [24], and, as for the other ervaoffines, its absolute configuration (3*S*,6*R*,7*R*,14*R*,16*S*,20*S*,21*S*) was established

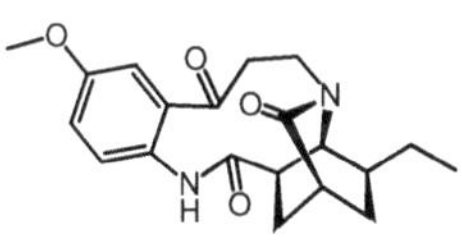

56 (ervaoffine D)

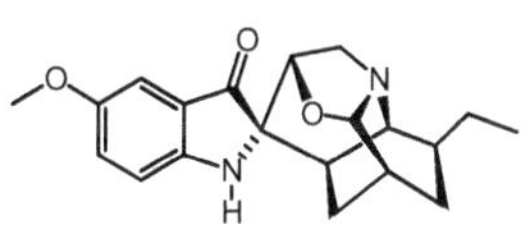

57 (ervaoffine A)

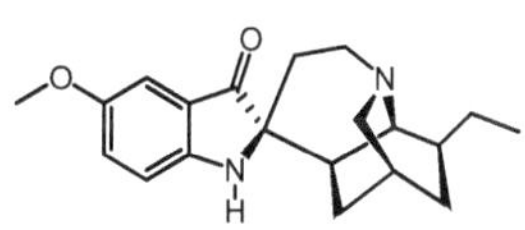

58 (iboluteine)

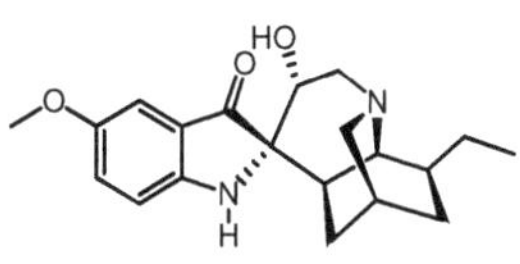

59 (ervaoffine B = ervaluteine)

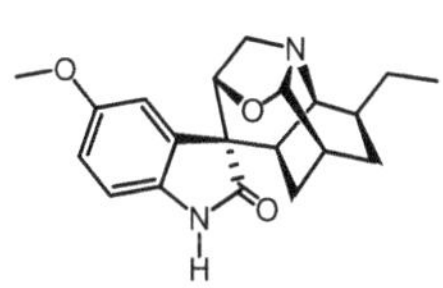

60 (ervaoffine C)

Fig. 18 *seco*-Iboga alkaloids **56–60**

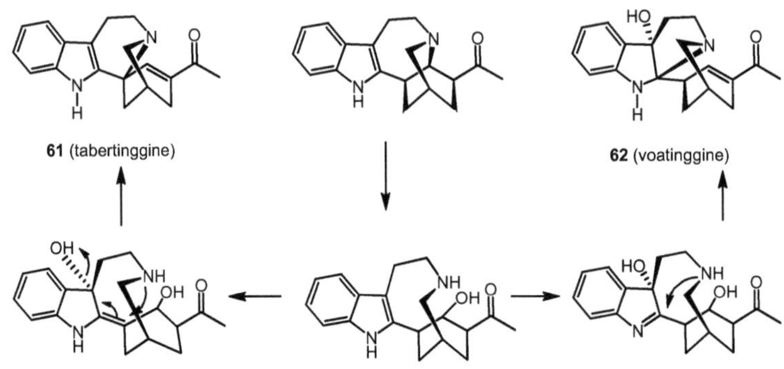

Fig. 19 Tabertinggine (**61**), voatinggine (**62**), and a proposal for their biosynthesis

from the similarity of Cotton effects with those obtained by calculated ECD for both enantiomers.

Tabertinggine (**61**) and voatinggine (**62**) (Fig. 19) are two exceptionally rearranged iboga alkaloids, for which their skeletons remain unique, and the X-ray crystallography used for both removed all doubt on their structures. A common biogenetic pathway was proposed for these compounds with initial cleavage of the C-21 to N-4 bond of conodusine A (**30**) and subsequent ring closure onto C-2 or C-3 [44].

4.1.5 Catharanthine and Pseudoeburnamonine Derivatives

All the monomers described until now in this Section are assumed to belong to the ibogaine/coronaridine series, whether demonstrated or not. There is but a single possible exception, alioline (**63**) (Fig. 20), isolated from *Catharanthus roseus* [45]. This is a strange and unique molecule possessing the features of catharanthine (**1**) to which is added a C_9 unit. However, if structure **63a** is proposed in the original article, the "Dictionary of Natural Products" changed it to structure **63b** under the same name and reference. Fragments I and II display the possibilities for the extra carbon atoms. The origin of the C_9 fragment is unclear: it could be a truncated terpene, although highly irregular, or an iridoid, but in this case, one of the methyl groups is missing.

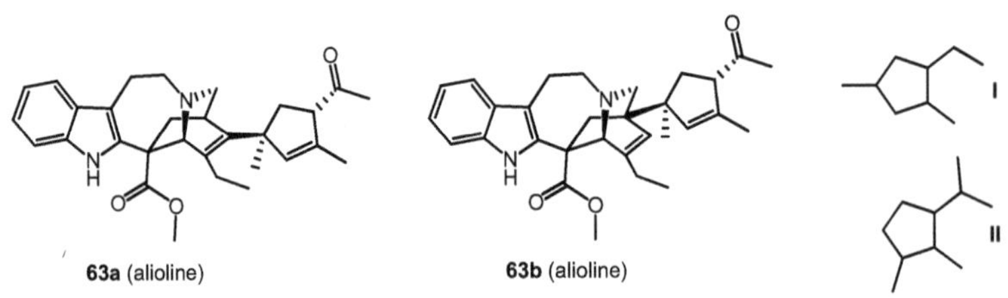

Fig. 20 Structural hypotheses for alioline (**63a** and **63b**). Comparison of the C_9 unit of alioline (**I**) and C_{10} of common iridoids (**II**)

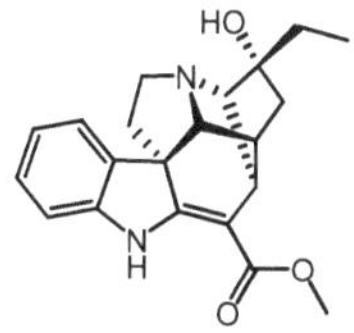

64 (pandine)

Fig. 21 Structure of pandine (**64**)

There is a single example of a new molecule with the isoeburnan skeleton, named tacamonidine (**11**), while just two reports of the isolation of pandine (**64**) (Fig. 21) are found in the isoplumeran categories [16]. Tacamonidine (**11**) belongs to the rare isoplumeran series and differs from tacamonine (**10**) by an OH group, for which the configuration was proposed by NMR spectroscopy [11].

4.1.6 Miscellaneous Representatives and Another Enigma

Four new alkaloids were isolated from *Tabernaemontana corymbosa* and named tabercarpamines G–J (**65–68**) (Fig. 22). They belong to the small family of the chippiines and are characterized by an N-1 to C-3 bond instead of the N-4 to C-3 bond [46]. Their structures were established by NMR spectroscopy without any attempt at the determination of their absolute configurations. In tabercarpamine H, there is an OH at position C-19, the configuration of which was proposed on the basis of a ROE measurement. It would have been safer to make this determination on a partially synthetic derivative of **66**, in which, for example, the free OH and the secondary amine would be engaged in a six-membered ring.

The structures of the tabercarpamines are not exceptional but their mere existence is puzzling. It is not difficult to conceive a biosynthesis scheme starting from isovoacangine or coronaridine (**3**) and giving the tabercarpamine skeleton through the intermediate of a 3-hydroxy derivative and of an elusive aldehyde, **69** (Fig. 23). Whereas one would expect some stability of this aldehyde, it has never been identified and one hypothesis would be that it exists in such a high energy state that the seemingly more strained isoquinuclidine forms are predominant.

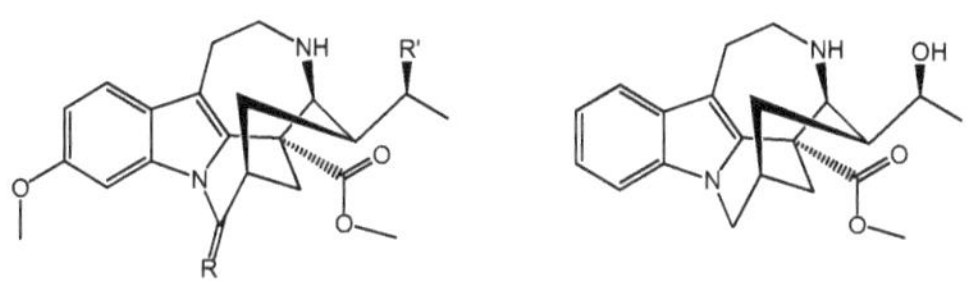

65 R = H_2, R' = H (tabercarpamine G)
66 R = H_2, R' = OH (tabercarpamine H)
67 R = O, R' = H (tabercarpamine I)

68 (tabercarpamine J)

Fig. 22 Structures of the tabercarpamines G–J (**65–68**)

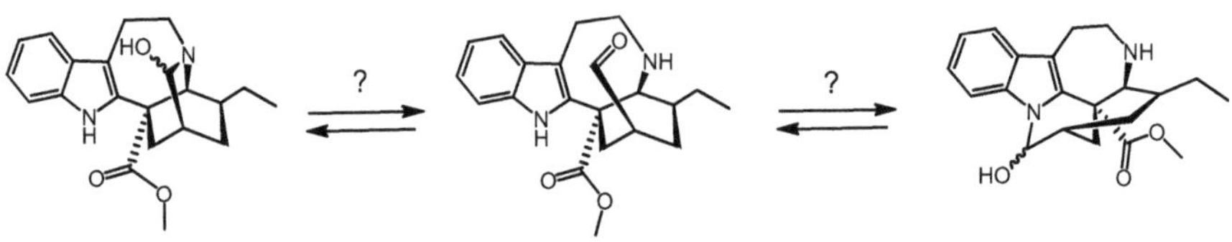

Fig. 23 A possible pathway between coronaridine-type and chippiine-type alkaloids

4.2 Dimers

The count of new dimeric alkaloids containing an iboga moiety amounts to 49 (Table 2), which adds to the previous 40 such compounds described in an earlier review [47]. This is therefore quite a well-represented class of alkaloids, and it appears that all but one contain the vobasinyl residue always substituted at position C-3′ (vobasine numbering). In all examples also, the ibogan moiety is linked on the α face, which is deduced from the multiplicity of H-3′ (dd, $J = 13$ and 3 Hz), and from its NOE with H-15′ and the indole NH. Nature utilizes a large variety of reactions to couple indole alkaloids, among which, coupling between an electron-rich indole nucleus and a carbenium ion is a privileged route. The reason why the vobasinyl cation is so common in couplings remains to be explained. It could be that, being stabilized by N-4, it is an especially long-lived species lending itself to slow kinetics reactions. Most of the newly isolated bisindoles are derived from five vobasan-type monomers (Fig. 24): vobasinol (**70**) (syn. perivine), dregamine (**71**), vincadiffine (**72**) (3-oxo tautomeric form), pagicerine (**73**), and difforlemenine (**74**). For the iboga moiety, the chemical diversity is more extensive but it is possible that chemical modifications in this moiety occur after coupling. The ibogan skeleton is equally represented by the voacangine or ibogaine series (with or without a carbomethoxy group at C-16) and substitution preferentially occurs *ortho* to the aromatic group if present. Compounds without an OH- or OMe-directing group on the aromatic part of the ibogan molecule are scarce but over the years several have been isolated.

Two alkaloids stand alone in the gallery of the iboga bisindoles (Fig. 25). One is biscarpamontine A (**75**), which is the result of a unique coupling between the iboga and aspidosperma units [48]. A further particularity is the presence of a methylene bridging the two moieties and assigned to the iboga moiety, according to a biogenetic hypothesis from the original authors. As far as we know, there are no natural iboga alkaloids possessing N-1-methyl or formyl groups and it is our belief that this supernumerary carbon atom belongs to the aspidosperma moiety (as the one found in the vobtusines is also present in the same plant). The overall structure of **75** was established by NMR spectroscopy and only relative configurations are given. Vobatensine E (**76**) is the only dimeric iboga alkaloid to have a linkage with the C-9 of ibogaine and this was deduced from analysis of the ^{1}H NMR spectroscopic features of the aromatic part of the molecule [49]. An attempt to prepare this

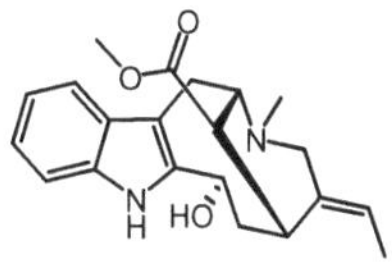

70 (vobasinol)

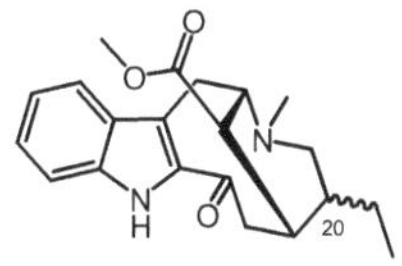

71 (20β-H dregamine)
(20α-H tabernaemontanine)

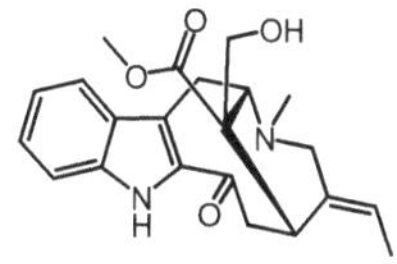

72 (vincadiffine)

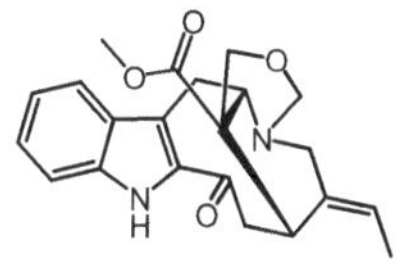

73 (pagicerine)

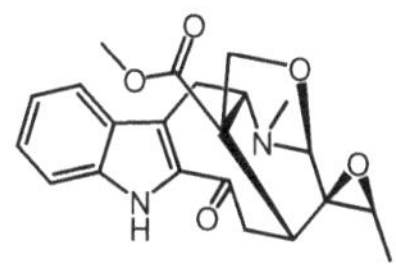

74 (difforlemenine)

Fig. 24 Most usual vobasan-type monomers found in iboga dimers

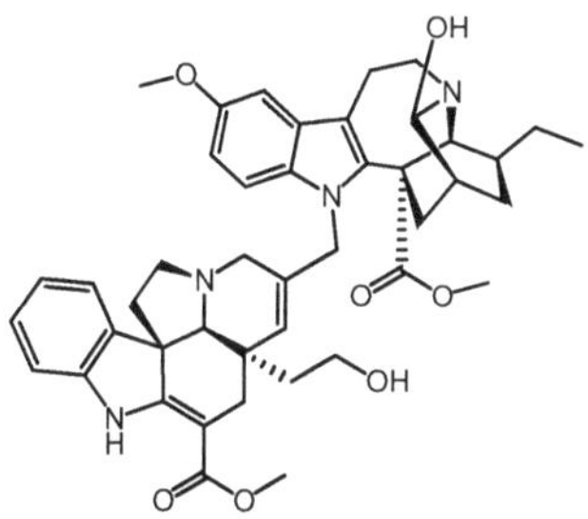

75 (biscarpamontine A)

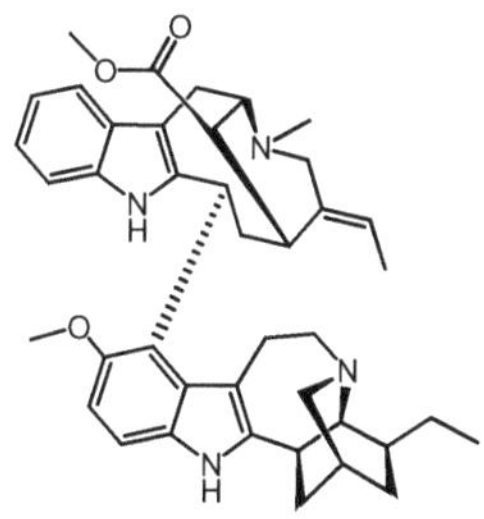

76 (vobatensine F)

Fig. 25 Bisindoles **75** and **76**

compound by condensing vobasinol and ibogaine under acidic conditions was unsuccessful. This substitution pattern is so far unique. In the C-11-OMe series, however, condensations occur on both sides of the ArOMe (C-10 and C-12).

Generally speaking, the structural elucidation strategy used for all bisindoles has been similar. The first step consists of recording high-resolution mass spectra (HR-ESI-TOF or HR-FAB-MS) and carbon NMR spectra to determine the nature of two moieties. Following this, a combination of COSY, HSQC, and HMBC NMR experiments allows the determination of the structures of the two monomers, which often are known, or by may be proposed by comparison with similar monomers and dimers. Finally, analysis of NOE and further HMBC correlations helps determine the linkage positions between the two monomeric moieties and the relative stereochemistry of some chiral centers. Very few bisindoles have had their absolute configuration determined (see below).

4.2.1 Bisindoles with an Ibogamine Moiety

There are only four bisindoles that lack an alkoxy substituent on the aromatic ring of ibogamine: (19*R*)- and (19*S*)-hydroxytabernamine (**77** and **78**), and tabernamidines A and B (**79** and **80**) (Fig. 26) [36,50]. Comparison of the NMR spectra of **77** and **78** with those of the heyneanines was key to the C-19 configuration determinations. A compound named originally oxotabernamine [50] and depicted as **79** was revised to **80** following a new isolation and comparison with the spectra of the corresponding monomers, conodusines A and B. Enolization of C-19 could explain the formation of the tabernamidines as seen in these last compounds. Vobatensine A (**81**) is the C-11-OMe analogue of (19*R*)-hydroxytabernamine and its structure was proven definitively by a partial synthesis from vobasinol and 19-*epi*-iboxygaine [49]. The same plant, *Tabernaemontana corymbosa*, yielded the dihydro derivative **82**, which was named vobatensine B, and was also prepared from tabernaemontaninol and 19-*epi*-iboxygaine. The configuration at C-20′ was deduced from the following NOE correlations: H-19′↔H-16′, H-18′↔H-21′, and H-20′↔H-14′.

Ervachinines B (**83**) and D (**84**) (Fig. 27) are two positional isomers, composed of vincadiffine (**72**) linked to 10-methoxy-ibogamine (ibogaine (**2**)) or to 11-methoxy-ibogamine (tabernanthine) [19]. Their absolute configurations were

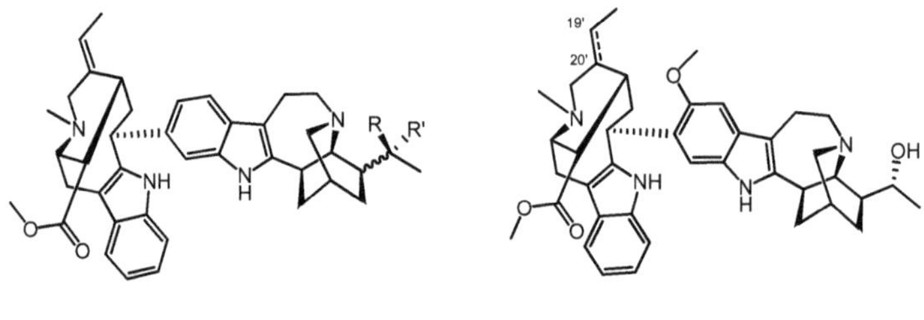

77 (R = H, R' = OH, 20α-H: (19*R*)-hydroxytabernamine)
78 (R = OH, R' = H, 20α-H: (19*S*)-hydroxytabernamine)
79 (R = R' = O, 20α-H: tabernamidine A)
80 (R = R' = O, 20β-H: tabernamidine B)

81 (Δ-19' : vobatensine A)
82 (20'α-H : vobatensine B)

Fig. 26 The hydroxytabernamines, tabernamidines A and B, and vobatensines A and B (**77–82**)

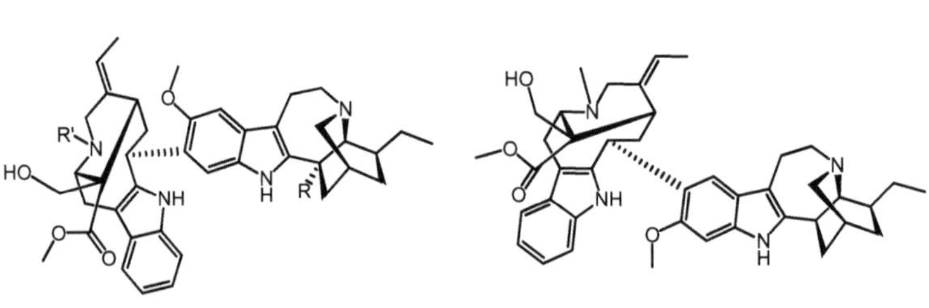

85 R = CO_2Me, R'= Me (ervachinine A)
83 R = H, R' = Me (ervachinine B = ervatensine A)

84 (ervachinine D)

Fig. 27 Ervachinines A, B, and D (**83–85**)

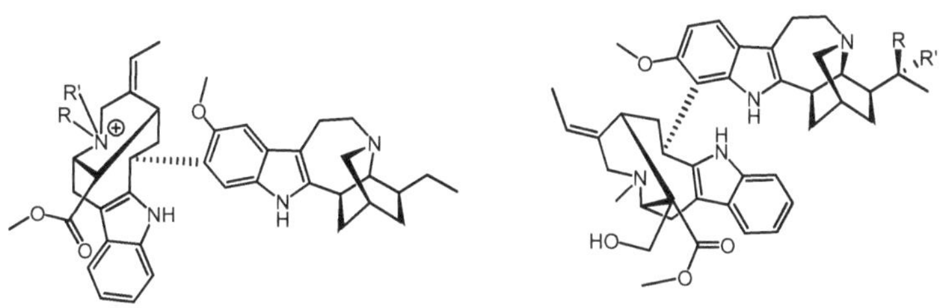

86 R = H, R' = : (ervatensine B)
87 R = CH_3, R' = O^- (vobatensine F)

88 R = OH, R' = H (conodutarine A)
89 R = R' = O (conodutarine B)

Fig. 28 Ervatensine B, vobatensine F, and condutarines A and B (**86–89**)

deduced from the similarity of their CD curves with that of ervachinine A (**85**) to which was applied, for the first time in bisindoles, the exciton chirality rule [19]. Its indole chromophores were found to be oriented in a clockwise manner with Cotton effects at 223 (−) and 243 (+) nm. Ervachinine B (**83**) had been previously isolated from the stem bark of *Tabernaemontana corymbosa* (syn. *Ervatamia chinensis*), and the name "ervatensine A" given in the 2008 Ph.D. thesis of K.-H. Lim was maintained, even though the work was not published until 2015 [13]. Discrepancies between the ^{13}C NMR spectra of ervatensine A and of ervachinine B led the Malaysian authors to perform an X-ray structure determination to ascertain the structure of ervatensine A and to establish once and for all the absolute configuration. The series was completed with ervatensine B (**86**) [13] and vobatensine F (**87**) [49], which are the *nor* analogue and the *N*-oxide of decarbomethoxyvoacamine, respectively. Two other similar compounds, conodutarines A and B (**88** and **89**) both have a less typical linkage involving C-12 instead of C-10 [51] (Fig. 28).

4.2.2 Bisindoles with a Voacangine (10-Methoxy-coronaridine) Moiety

Besides ervachinine A (**85**) (see above), there has been only one other bisindole reported with a voacangine unit, conodusarine (**90**), which is the 3-oxo derivative of voacamine (**91**) [52]. Its absolute configuration was not, strictly speaking, proven, but rather assumed, following the concomitant isolation of the parent compound (Fig. 29).

4.2.3 Bisindoles with an Isovoacangine (11-Methoxy-coronaridine) Moiety

There is a wealth of alkaloids of this type, with vincadiffine (**72**) and derivatives in the iboga category, or vobasinol (**70**) and derivatives in a non-iboga group. Variations of the iboga type are unsurprising and represent a similar diversity observed for the monomers. However, the authors of this chapter do have a concern in the

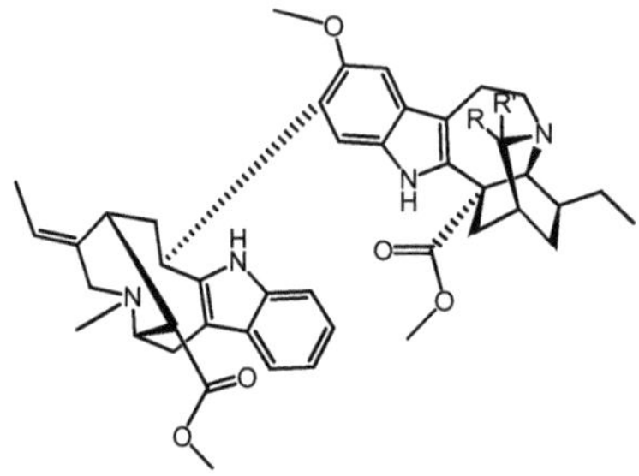

90 R = R' = O (conodusarine)
91 R = R' = H (voacamine)

Fig. 29 Bisindoles with a voacangine moiety

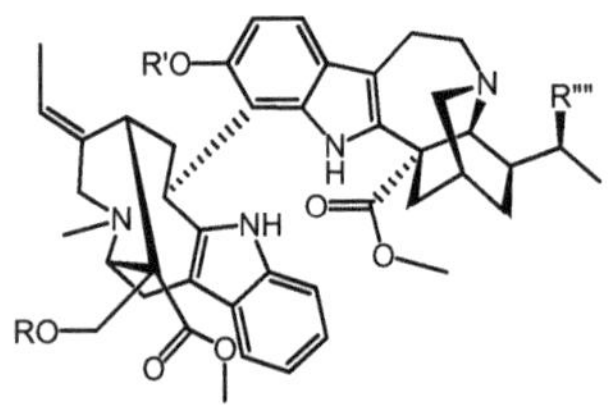

94 R = Ac R' = Me R'' = H (17-acetyl-tabernaecorymbosine A)
92 R = R' = R'' = H (cononitarine B)
97 R = R' = H R'' = OH (cononitarine A)

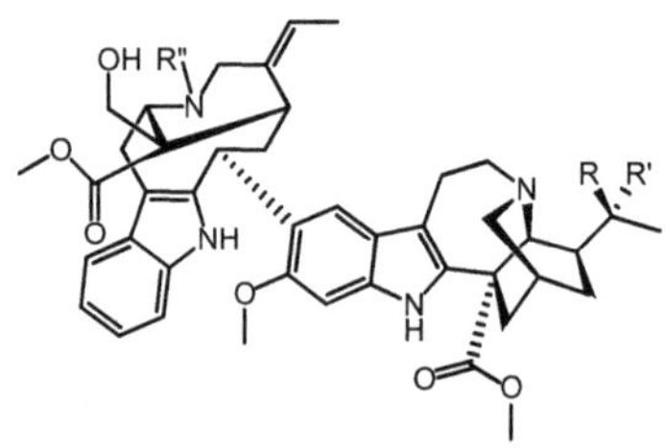

93 R = R' = H, R'' = Me (ervachinine C)
95 R = R' = O, R'' = Me (conodiparine C)
98 R = OH, R' = H, R'' = Me (conodiparine A)
99 R = OH, R' = R'' = H (conodiparine E)

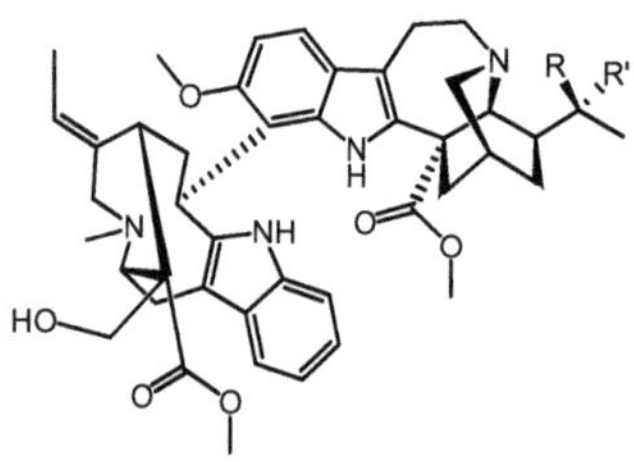

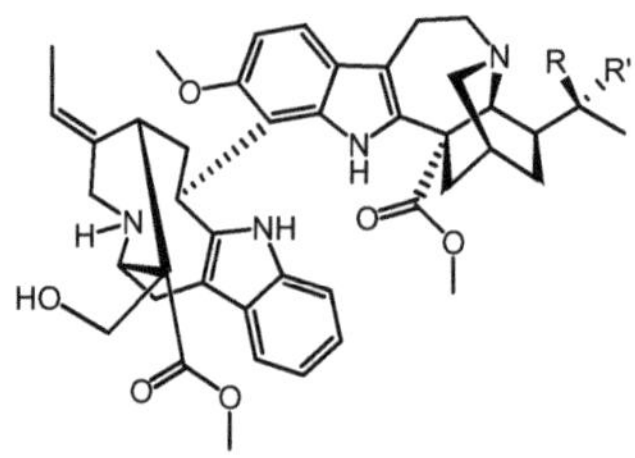

96 R = R' = O (conodiparine D)
100 R = OH, R' = H (conodiparine B)

101 R = OH, R' = H (conodiparine F)

Fig. 30 Bisindoles with a isovoacangine moiety (**92–101**)

past lack of care in the choice of trivial names for new compounds, with little or no effort apparently having been made to correlate those names to literature references.

Cononitarine B (**92**) [51] and ervachinine C (**93**) (Fig. 30) [19] are two bisindoles composed of a 11-hydroxy- or a 11-methoxy-coronaridine moiety linked to a vincadiffine unit. As for other ervachinines, the absolute configuration of **93** was deduced from the similarity of its CD spectrum to that of ervachinine A (**85**). In the same series, 17-acetyl-tabernaecorymbosine A (**94**) was isolated and its absolute configuration was confirmed by the CD exciton chirality method [53]. Conodiparine C (**95**) is the 19-oxo derivative of ervachinine C, and its isomer conodiparine D (**96**) simply differs in its location of attachment on the aromatic ring (C-12 instead of C-10) [51]. Cononitarine A (**97**) is (19*S*)-hydroxycononitarine B [51]. The other conodiparines (A (**98**), B (**100**), E (**99**), and F (**101**)) are pairs differing in the substitution pattern of N-4 in the affinisine moiety: NH vs. N-Me [51].

A peculiarity of the four alkaloids, conodirinines A and B (**102** and **103**) [54], and tabernaricatines C and D (**104** and **105**) (Fig. 31) [40], is the presence of an extra tetrahydro-1,3-oxazine ring bridging C-16 and N-4. The configuration assignment of the hydroxyethyl side chain was based on the ^{13}C NMR chemical shifts. As discussed by the authors, the signal for the N-Me group was conspicuously missing and it was the first reason to propose that it was embedded in a ring. The original paper [54] suggested that the methylene bridge is the result of a formaldehyde (or equivalent) condensation, but this also may be the result of the oxidation of the *N*-methyl into an iminium and subsequent trapping in a Polonovski fashion, as suggested by Zhang et al. [53].

Like the set of preceding compounds, tabernaricatines A and B (**106**, **107**) [40], and tabercorine C (**108**) (Fig. 32) [53], also possess a tetrahydro-1,3-oxazine ring, this time incorporating C-21 rather than a N-Me functionality. The structures of **106** and **107** were proposed mainly after comparison of their NMR data and observed NOE correlations with those of conodiparine B (**100**). No chiroptical data were reported and the structures were considered as flat although configurations were drawn for all centers including the epoxy ring. The absolute configuration of **108** was discussed after its CD data were measured and differences between the spectra of **106** and **108** led to the conclusion that they had an opposite epoxide configuration, but unfortunately taking the configuration of **106** for granted (which is not the case). The biosynthesis scheme proposed for **108** [53] is more appealing and has the merit of explaining the origin of the other compounds as well.

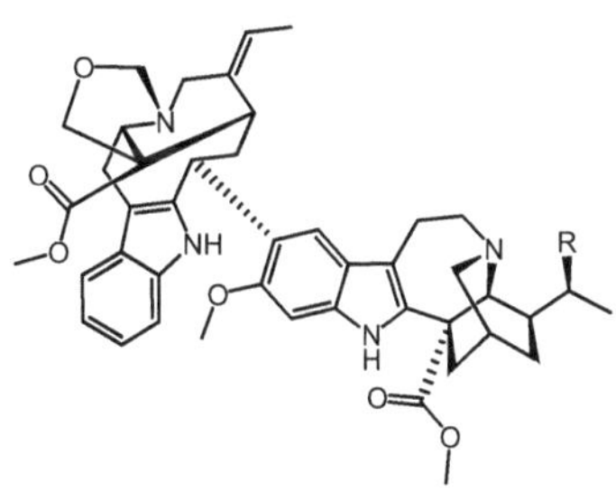

102 R = OH (conodirinine A)
104 R = H (tabernaricatine C)

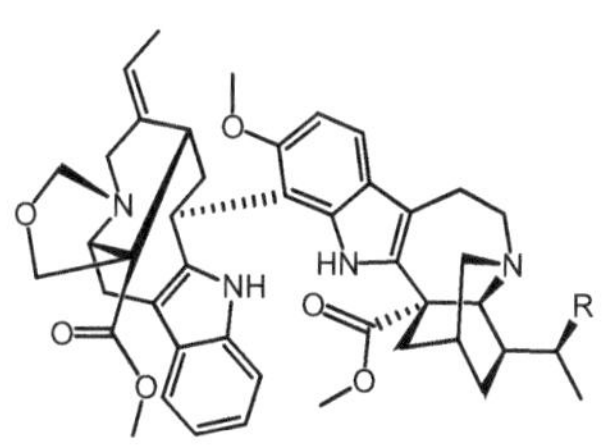

103 R = OH (conodirinine B)
105 R = H (tabernaricatine D)

Fig. 31 The bridged compounds **102–105**

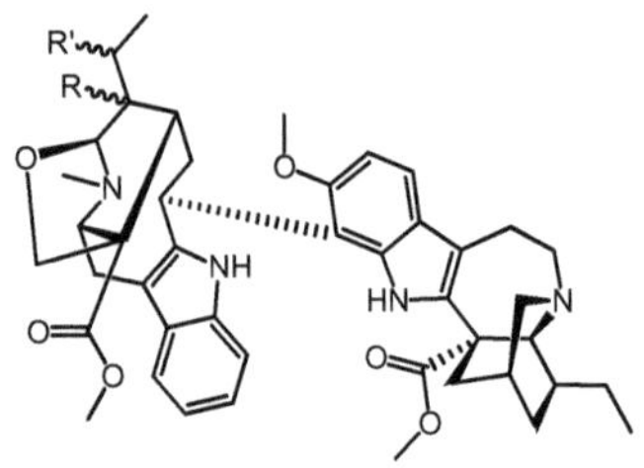

106 R, R' = oxido (tabernaricatine A)
107 R = R' = H (tabernaricatine B)

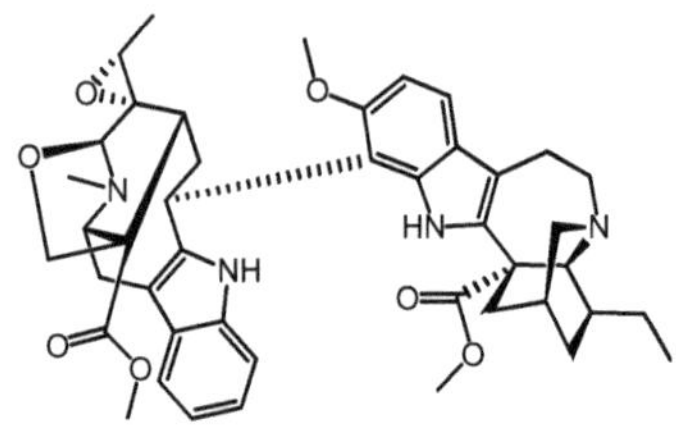

108 (tabercorine C)

Fig. 32 Tabernaricatines A and B, and tabercorine C (**106–108**)

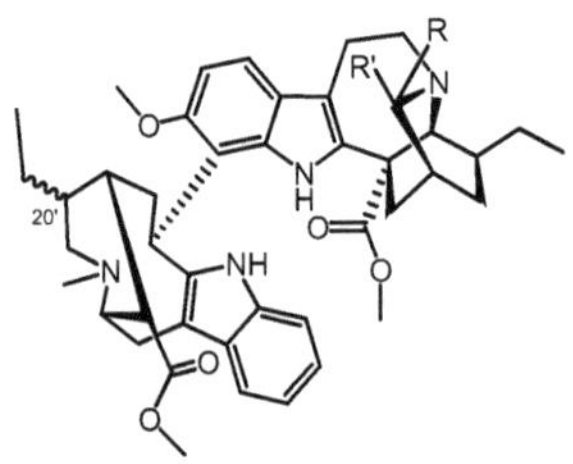

109 R = R' = O, H-20'α (3-oxo-tabernaelegantine A)
111 R = OH, R' = H, H-20'α ((3*R*)-hydroxytabernaelegantine A)
112 R = OH, R' = H, H-20'β ((3*R*)-hydroxytabernaelegantine C)
113 R = H, R' = OH, H-20'α ((3*S*)-hydroxytabernaelegantine A)
114 R = H, R' = OH, H-20'β ((3*S*)-hydroxytabernaelegantine C)
115 R = CN, R' = H, H-20'β (tabernaelegantinine C)

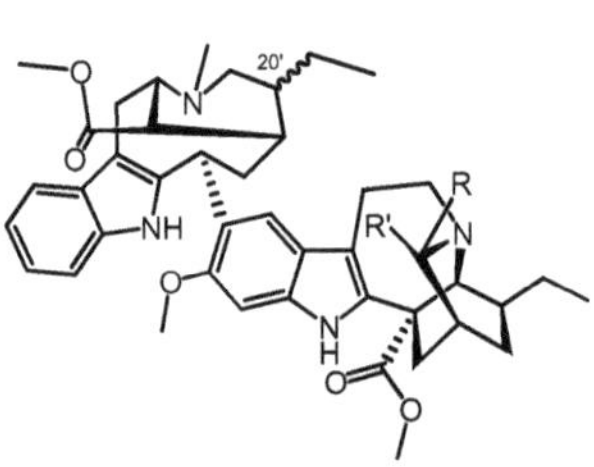

110 R = R' = O, H-20'α (3-oxo-tabernaelegantine B)
116 R = OH, R' = H, H-20'β ((3*R*)-hydroxytabernaelegantine D)

Fig. 33 Oxo and hydroxy tabernaelegantines A, B, C, and D (**109–116**)

Muntafara sessifolia (syn. *T. sessifolia*) yielded 3′-oxo-tabernaelegantines A and B (**109** and **110**) (Fig. 33), which are regioisomers differing in the substitution pattern on the aromatic ring [55]. Their structures, including the relative configurations within the two moieties, were determined mainly by NMR spectroscopy. These are the oxidation products of the well-known tabernaelegantines, although this has not been proven. A particular feature of the NMR spectra was a line-broadening phenomenon, said by the authors to disappear upon cooling at 273 K. It was even stated that "lowering the temperature favored one conformer" [55], but this is not reasonable from a thermodynamic standpoint.

Four 3-hydroxy-tabernaelegantines were characterized, with three of these having a C-12—C-3′ bridge: (3*R*)-hydroxytabernaelegantine A (**111**), (3*R*)-hydroxytabernaelegantine C (**112**) [56], (3*S*)-hydroxytabernaelegantine A (**113**) [56], and (3*S*)-hydroxytabernaelegantine C (**114**) [56]. (3*R*)-Hydroxytabernaelegantine C (**112**) was known previously as a semisynthetic bisindole obtained by acidic hydrolysis of tabernaelegantinine C (**115**) [57]. The configuration of C-3 was deduced from its ^{13}C NMR chemical shift value (i.e. near 86 ppm in the (*S*)-configuration and up to 93 ppm in the (*R*)-configuration). The interconversion of these carbinolamines under acidic conditions such as on the silica gel used for purification could explain the fact that hydroxytabernaelegantine A was isolated initially as a mixture of epimers, **111** and **113**. The configuration of the ethyl chains in the vobasinyl units was determined from the chemical shifts of C-14′, C-16′, C-18′, C-19′, and C-20′, and a NOE correlation observed between H-3′↔H-20′. (3*R*)-Hydroxytabernaelegantine D (**116**) was the sole carbinolamine bisindole containing a C-10—C-3′ bond [56].

The propensity of C-3 of the iboga alkaloids to capture nucleophiles was illustrated by the isolation of the three tabernaelegantinals A, B, and E (**117–119**) (Fig. 34) [56]. The (*R*)-configuration of C-3 was determined by the observation of a strong NOE interaction between the aldehyde proton and H-5b. It is assumed that the precursor, 3-formyl-isovoacangine, was formed by reduction of a 3-cyanoisovoacangine, itself produced from a 3-hydroxyisovoacangine via an iminium form in a Strecker-like reaction.

The same iminium intermediate could be scavenged by nucleophiles such as the acetylacetate anion to form 3-alkyl-isovoacangine. Examples among the bisindoles are tabercorine A (**120**) [54] and tabernaricatine E (**121**) [40], which is 3-oxopropyl-ervachinine C. A NOE or ROE correlation was observed between H-3↔H-17*β*, establishing that H-3 is *β*-(“left”)-oriented. The absolute

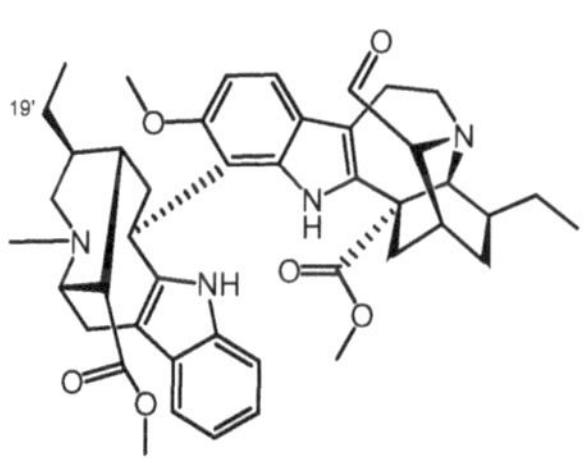

117 ((3*R*)-tabernaelegantinal A)
119 Δ-19′ ((3*R*)-tabernaelegantinal E)

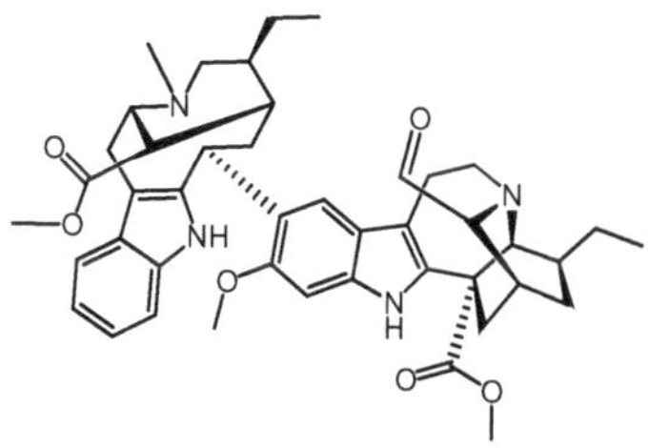

118 ((3*R*)-tabernaelegantinal B)

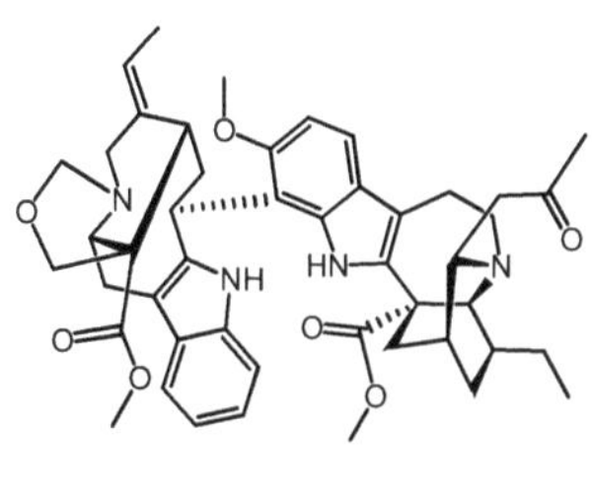

120 (tabercorine A)

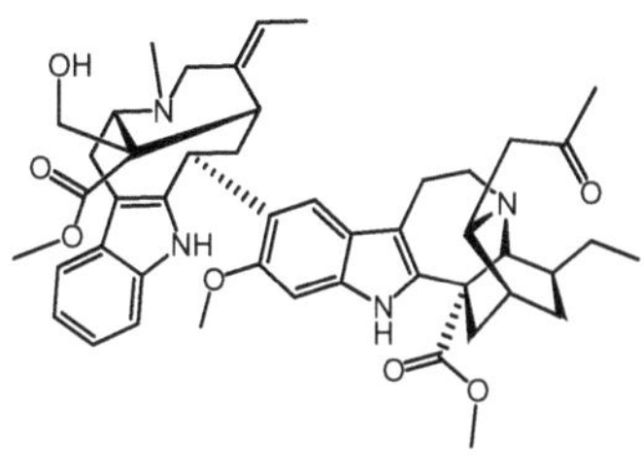

121 (tabernaricatine E)

Fig. 34 C-3 substituted bisindoles (**117–121**)

configuration of tabercorine A was established by means of the CD exciton chirality method with the two indoles oriented in a clockwise manner.

4.2.4 Bisindoles with an Iboga-Indolenine or Rearranged Moiety

Dimers in the iboga series with an oxidized indole nucleus are rather rare compared to the situation observed in the monomers. Only two 7-hydroxy-indolenines, tabercorine B (**122**) [53] and vobatensine D (**123**) (Fig. 35) [49], corresponding to tabernaricatine D (**105**) and ervatensine B (**86**), have been isolated recently. The (7*R*)-configuration of 7-OH in **122** was deduced by comparison of the NMR data with those of the hydroxyindolenine of voacangine. No ROE correlation was observed for the OH, which is not surprising since the measurements were done in acetone-d_6, but the absolute configuration was established by the exciton chirality CD rule. Vobatensine D (**123**) was claimed to have the inverted (7*S*)-configuration (α-OH) on the basis of a dubious argument: the co-occurrence of the (7*S*)-hydroxyindolenine of ibogaine of known absolute configuration in the plant [58].

A bisindole has been reported with a pseudo-indoxyl chromophore, namely, (2*R*)-vobatensine C (**124**) [49]. The configuration of its spirocyclic C-2 was deduced from the chemical shifts of C-2, C-7, C-16, and C-21 as performed for iboluteine (**58**). A NOE correlation was observed between the indole NH and H-17β.

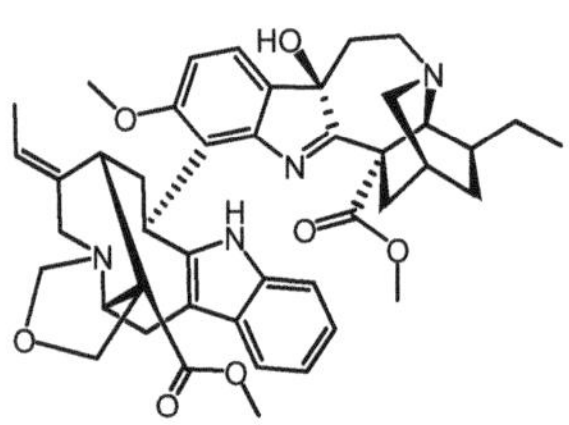

122 (tabercorine B)

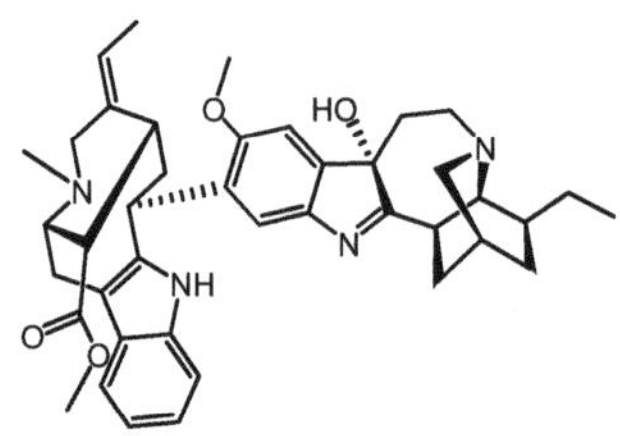

123 (vobatensine D)

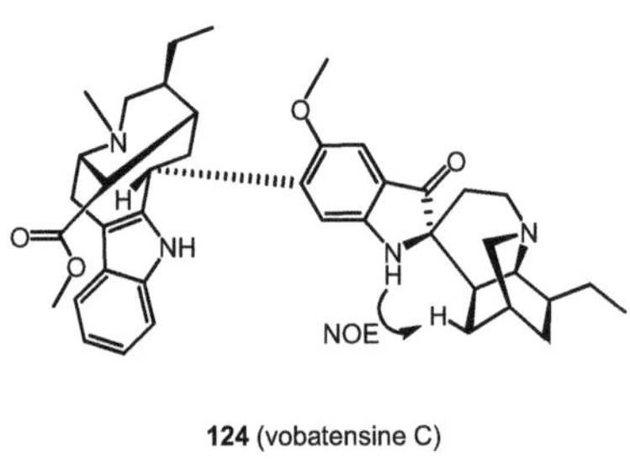

124 (vobatensine C)

Fig. 35 Hydroxy indolenines and pseudo indoxyl bisindoles (**122–124**)

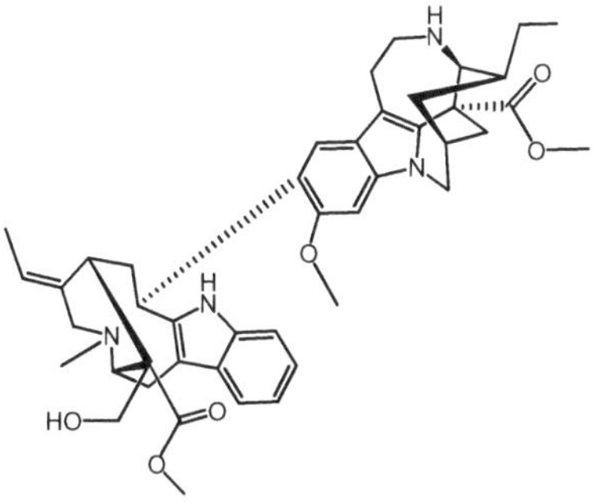

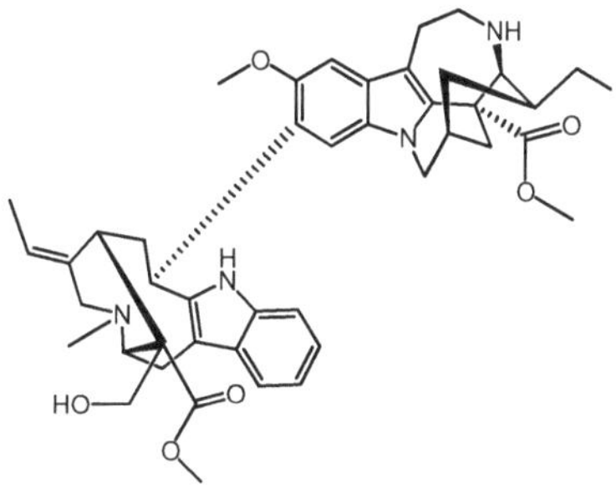

Fig. 36 Vobasinol-chippiine-type bisindoles (**125–126**)

4.2.5 Bisindoles with a Chippiine Moiety

Tabernaemontana corymbosa has yielded two dimeric alkaloids, tabercarpamines A (**125**) and B (**126**), possessing the usual vobasinol moiety and a chippiine portion [46] (Fig. 36). Their structures were established by NMR spectroscopy and mass spectrometry and a proposal was made for their absolute configurations based on the application of the exciton chirality rule. While it is perfectly acceptable to determine the helicity around the junction between moieties, this is based on an assumption that the configurations of the monomers have been determined already. In this example, as in others described previously, these are simply assumed based on biogenetic hypotheses rather than on any rigorous experiments.

5 Synthesis

As mentioned earlier in this chapter, total synthesis is almost no longer in use in the structural elucidation domain. Unless major changes occur in the marketplace, most industrially or pharmacologically important alkaloids (e.g. catharanthine (**1**) and ibogaine (**2**)) will continue to be available competitively on a large scale starting from natural materials. An exception is the 18-OMe derivative of coronaridine (**127**) (Chart 1), which is in the pre-development phase and, in the absence of a reliable source of its precursor, albiflorine (18-OH coronaridine), will remain a fully synthetic compound. Total synthesis in the area is justified by the preparation of unnatural analogues or derivatives and by the development of methods of a more general application.

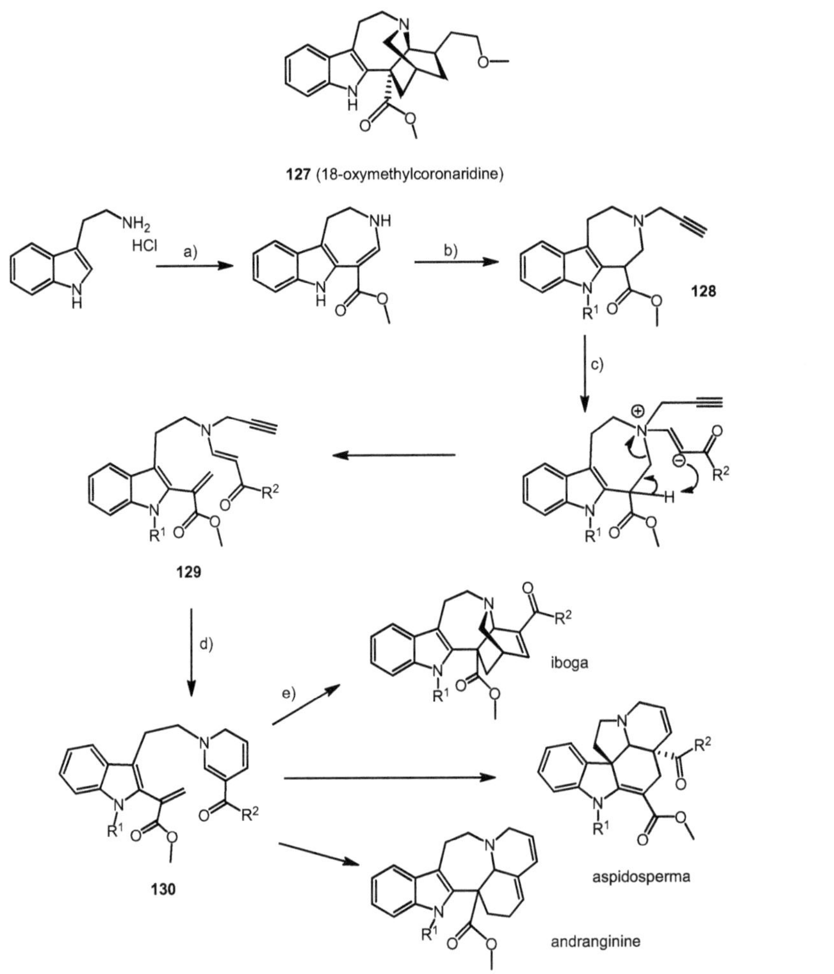

a) 1. methyl bromopyruvate, MeOH, reflux, 2. pyridine, reflux, b) 1. $NaBH_3CN$, AcOH, 2. Et_3N, propargyl bromide
c) $HCCCOR_2$, $ClCH_2CH_2Cl$, CF_3CH_2OH, d) $Cu(dppf)(MeCN)PF_6$, e) R^1= H, R^2 = OMe, $ClCH_2CH_2Cl$, 60°C, 48%

Chart 1 General principles of the unified synthesis for the preparation of the iboga, aspidosperma, and andranginine skeletons

Most of the first generation of iboga alkaloids syntheses were based on the assumed biosynthesis, and the main challenges were to build a molecule containing a reactive acrylate and an unstable dihydropyridine such as found in compound **130**. A quite innovative and general route to **130** inspired by biosynthesis was recently

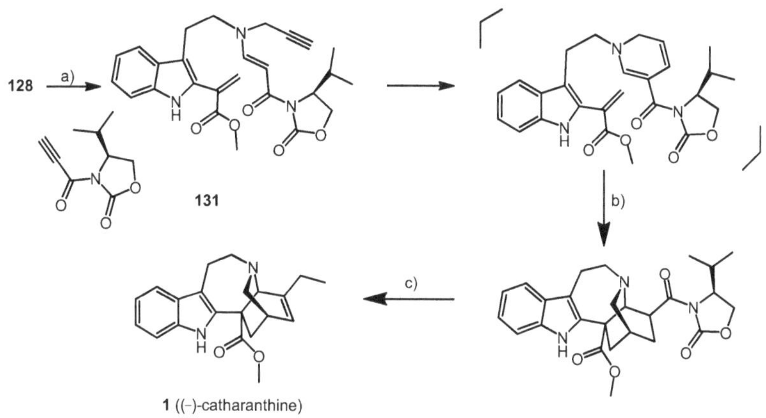

a) AcOH, CH_2Cl_2, b) Cu(dppf)(MeCN)PF_6, c) 1.$CeCl_3$, MeOH, 2. DiBAL, 3. $(PhO)_2P(O)Cl$, Et_3N, 4. $Fe(acac)_3$, MeMgBr

Chart 2 Asymmetric total synthesis of (−)-catharanthine (**1**)

proposed by a group at Hokkaido University (Charts 1 and 2) [59]. The synthesis starts from azepinoindole **128**, prepared from tryptamine according to a process similar to the one used by Kuehne. It then takes advantage of a Kuehne-type fragmentation to generate the 2-indolyl acrylate **129** while the dihydropyridine is generated by a Cu-(I) catalyzed 6-*endo* cyclization of an ene-yne (step d, Chart 1). One of the characteristics of this very clever approach is the stabilization of the dihydropyridine in the presence of the acrylate function in **130** allowing all possible Diels-Alder adducts to be obtained, thus leading to the iboga, aspidosperma, or even andranginine skeletons. As a bonus, chirality was also obtained with the introduction of a chiral oxazolidinone in **131**, the precursor of the dihydropyridine (Chart 2). The synthesis follows a route similar to that used in the racemic synthesis and was illustrated by the preparation of (−)-catharanthine.

Another approach follows the historical route developed by Trost, in which a bond is formed between C-2 of indole and the isoquinuclidine bearing a suitably placed double bond [60]. The last cyclization stage (viz. **132** → **133** in Chart 3) requires a full equivalent of palladium and another full equivalent of silver, which makes the end product more valuable than these two metals together. This is clearly a disadvantage and attempts have been made to circumvent this difficulty and make this reaction catalytic. After using the Pd/Ag sequence to prepare analogues of ibogamine such as **134** (Chart 3) [61–63], Sinha introduced the Heck reaction to cyclize the azepinoindole system (Chart 4) [64,65]. Practically speaking, the precursor was a 2-iodinated indole either

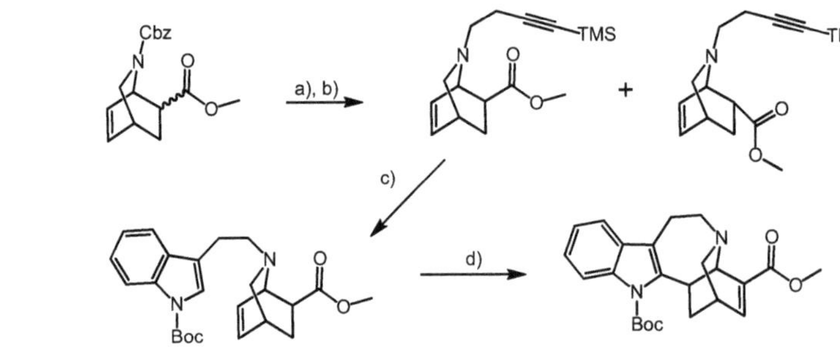

132 **133**

a) HBr/HOAc, b) $TMSCCCH_2CH_2OTs$, K_2CO_3, c) $Pd(OAc)_2$/ PPh_3, DIPEA
d) 1. $Pd(CH_3CN)_2Cl_2$, $AgBF_4$, Et_3N, 2. $NaBH_4$, MeOH,3. TFA 20% CH_2Cl_2

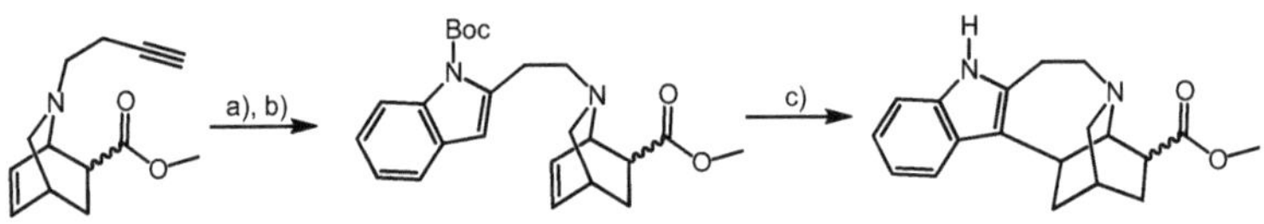

134

a) $Pd(PPh_3)_2Cl_2$, CuI, Boc-2-iodoaniline, b) TBAF,
c) 1. $Pd(CH_3CN)_2Cl_2$, $AgBF_4$, Et_3N, 2. $NaBH_4$, MeOH, 3. TFA 20% CH_2Cl_2

Chart 3 Syntheses of the iboga skeleton (top) and of iboga analogs (bottom) by the Sinha stoichiometric Pd/Ag route

prepared directly by iodination of the corresponding indole or in two steps via the triethylsilyl intermediate **135**. As an alternative, the same authors synthesized a 2-iodo tryptophyl iodide, which they condensed with the appropriate isoquinuclidine. Almost simultaneously, and as part of a program aiming at alleviating the metal stoichiometry problem, Sames at Columbia University also made use of a Heck cyclization, this time with the 2-bromo intermediate **136** (Chart 4) [66]. However, the most innovative part in this article is the search for a true catalytic reaction to perform the desired cyclization. It indeed worked with a Ni catalyst, but on a benzofuran analogue **137** (Chart 4). So far, the genuine nucleus of the iboga skeleton did not surrender to a fully catalytic CH insertion process.

The question of the crucial C-2 to C-16 ring closure is more obviously solved when C-16 is at the ketone oxidation level. It is illustrated in the synthesis of (19*R*)-ibogaminol (**138**) by Höck and Borschberg, where a single acid treatment suffices to close the seven-membered ring (step g in Chart 5) [67]. The key step of

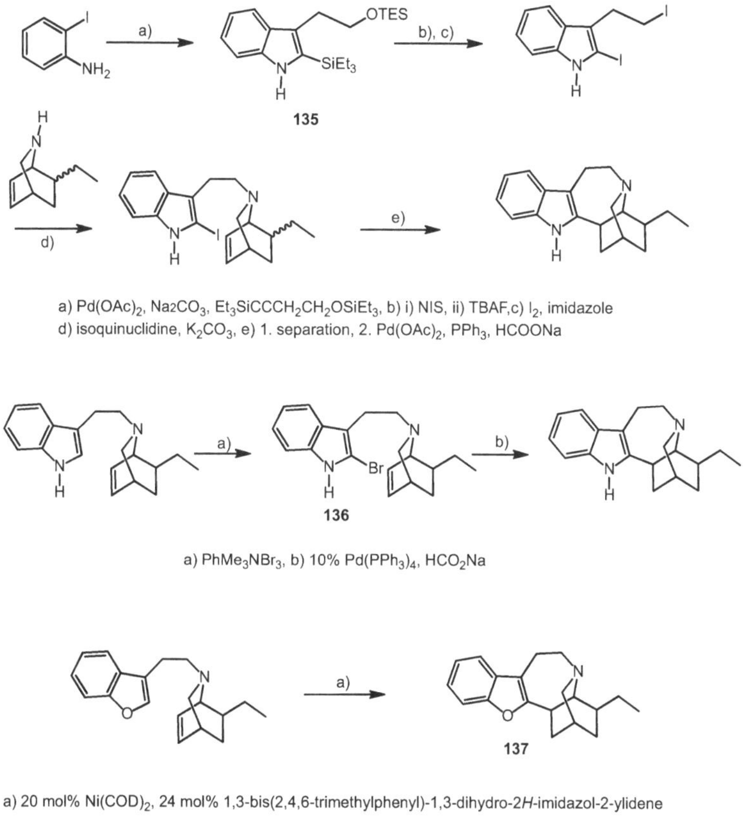

Chart 4 Sinha's (top) and Sames's (middle) routes to ibogamine via Heck cyclization. Sames's fully catalytic route to the iboga alkaloid analog **137** (bottom)

the procedure, however, was the diastereo- and enantio-selective synthesis of the isoquinuclidine **139** through a nitrone to olefin [3 + 2] cycloaddition, allowing the control of the C-19 configuration.

Asymmetric construction of the isoquinuclidine system was devised by Yang and Carter using organocatalysis based on the proline derivative **140** (Chart 6) [68]. The enantioselectivities obtained are excellent, but the presence of an aryl group and the location of the ketone on the isoquinuclidine will require several steps before reaching the natural products. There are not many articles associating medicinal chemistry and the iboga alkaloids. One of them by Sun et al. uses some of Kuehne's chemistry to build azepino-indoles like **141** adorned with functional groups, such as oxo-, thio-, and seleno-hydantoin units (Chart 6)

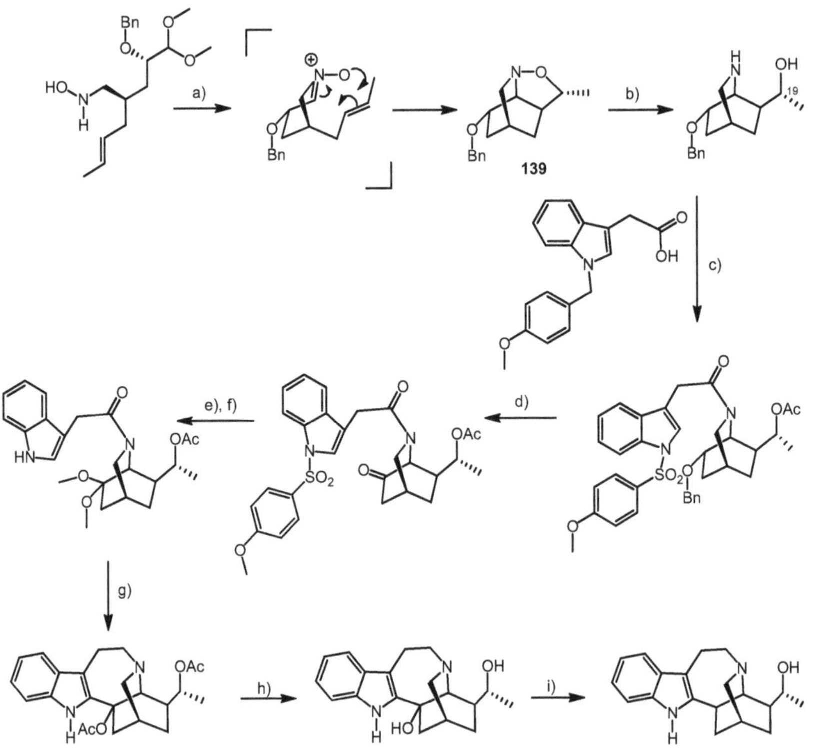

a) H_2SO_4, 47°C, b) Zn, AcOH, c) i) DCC, ii) Ac_2O, d) i) H_2, Pd/C, ii) DMSO, $(COCl)_2$, Et_3N, e) $HC(OMe)_3$, TsOH, f) Na/Hg, KH_2PO_4, THF/MeOH, g) AcCl, AcOH, cat MeOH, h) 1. $LiAlH_4$, 2. BF_3-Et_2O, i) $LiAlH_4$, $AlCl_3$, THF

Chart 5 The synthesis of (19*R*)-ibogaminol (**138**) by Höck and Borschberg

[69]. Unfortunately, no biological data were provided for the compounds synthesized.

The synthesis of voacangalactone (**16**) by Harada et al. qualifies as a high-yielding multiple-step total synthesis, for which the number of steps involved seems not to be an issue since every single step is high yielding (Chart 7) [20]. There are many innovations in this synthesis, and, for instance, the two-carbon atom chain of the tryptamine is installed in the last steps via an oxalyl chloride Friedel-Crafts reaction followed by a diborane reduction, a sequence used by Woodward to make methoxytryptamine in his approach to reserpine. To the best of our knowledge, this is the first time this reaction has been used to

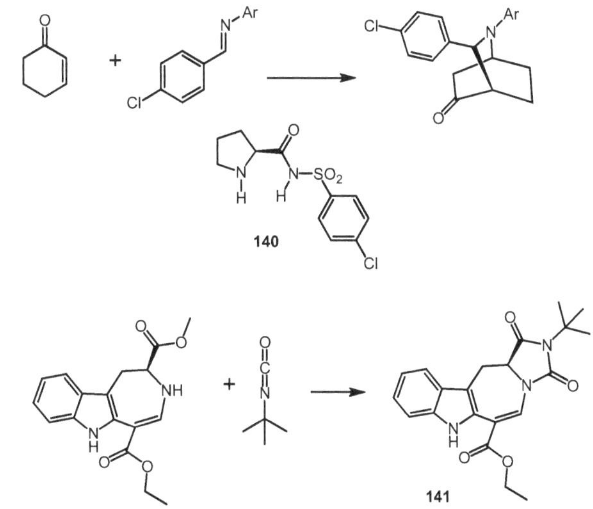

Chart 6 Asymmetric synthesis of the isoquinoline skeleton via organocatalysis (top) and synthesis of hydanthoin analogs of iboga alkaloids (bottom)

close the seven-membered ring of an azepino-indole, which is one of the difficulties in the synthesis of iboga alkaloids. Other salient features of this synthesis are the use of a Diels-Alder reaction to form the isoquinuclidine **142**, with a chiral induction and a gold-mediated indole synthesis (**143** → **144**). It took 14 steps for the synthesis of compound **142** and another 13 steps to make voacangalactone (**16**), for which 2 mg were added to the overall world resources of this compound!

Until now, 18-OMe-coronaridine (**127**) has not been obtained as a natural product, but from a pharmacological point of view it is a very promising candidate for pharmaceutical development (vide infra). Its synthesis is what may be considered as state of the art: short, convergent, enantioselective, and high yielding (see Chart 8 for an example of the synthesis of **127**). This starts with an azepino-indole, **145**, made in two steps from tryptamine, and which reacts with a wide variety of aldehydes to give intermediates of the secodine type, to then undergo Diels-Alder type cyclization. When a chiral auxiliary is placed on the nitrogen atom, chirality induction is obtained, with the efficiency of the process depending on the match between the chiral azepino-indole and the aldehyde. In the synthesis of 18-OMe-coronaridine (**127**), the aldehyde contains provisos for the methoxyethyl side chain, and, after the reaction, a masked aldehyde tetracycle (**146**) is obtained. The auxiliary group was then "exchanged" to a "handier" benzyl group and this was followed by reductive opening to a nine-membered ring to provide **147**. Upon debenzylation and removal of the aldehyde protection, a cyclization

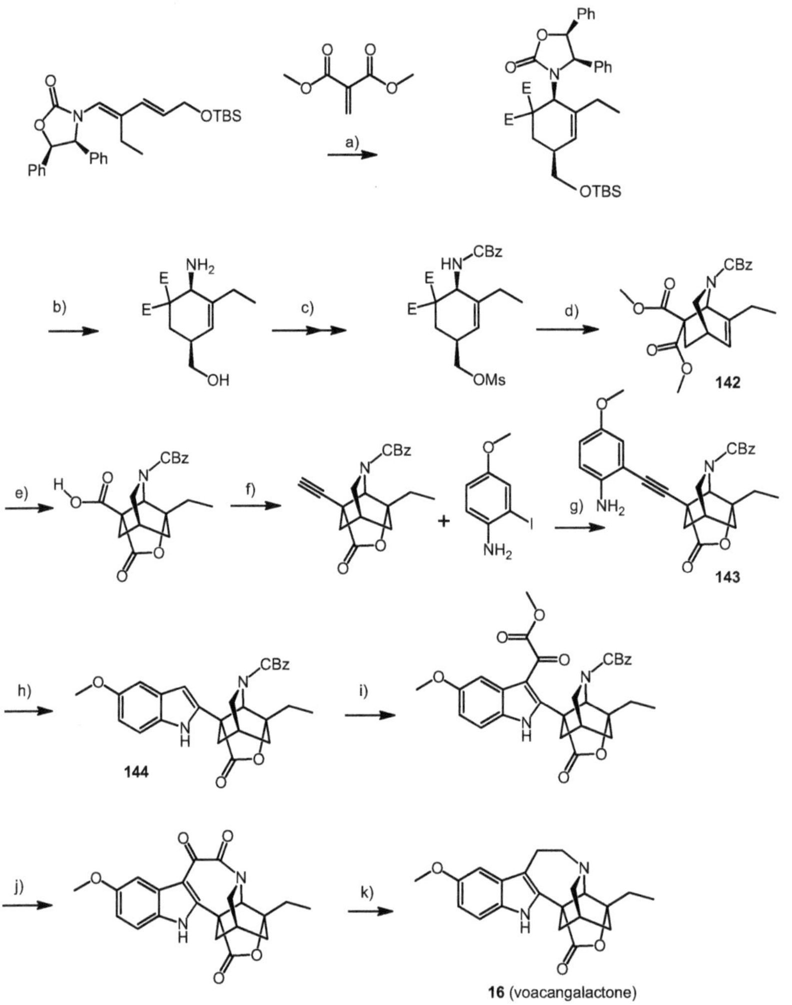

a) CH_2Cl_2, rt, 60h, b) 1. H_2, Pd(OH), 2. CSA, 3. H_2, Pd, all in one pot, c) 1. CBzCl, $NaHCO_3$, 2. MsCl, Et_3N, d) $CsCO_3$, DMF, 100°, e) 1. KOH, 2. I_2, $NaHCO_3$, 3. BH_3-Me_2S, 4. I_2, $NaHCO_3$, 5. AIBN, Bu_3SnH, f) 1. DMP, 2. $MeCOCN_2PO(OMe)_2$, g) $Pd(PPh_3)_4$, $CuSO_4$, sodium ascorbate, h) $NaAuCl_4$, 2 H_2O, i) $(COCl)_2$, then MeOH, j) TMSCl, NaI, k) BH_3-THF

Chart 7 Harada's total synthesis of voacangalactone (**16**)

occurred uneventfully to produce the cleavamine derivative **148**. Simple heating of enamine **148** under reflux in toluene produced a "miracle", with an exquisite CH shift and ring closure to give the title compound in overall good yield and less than ten steps [70].

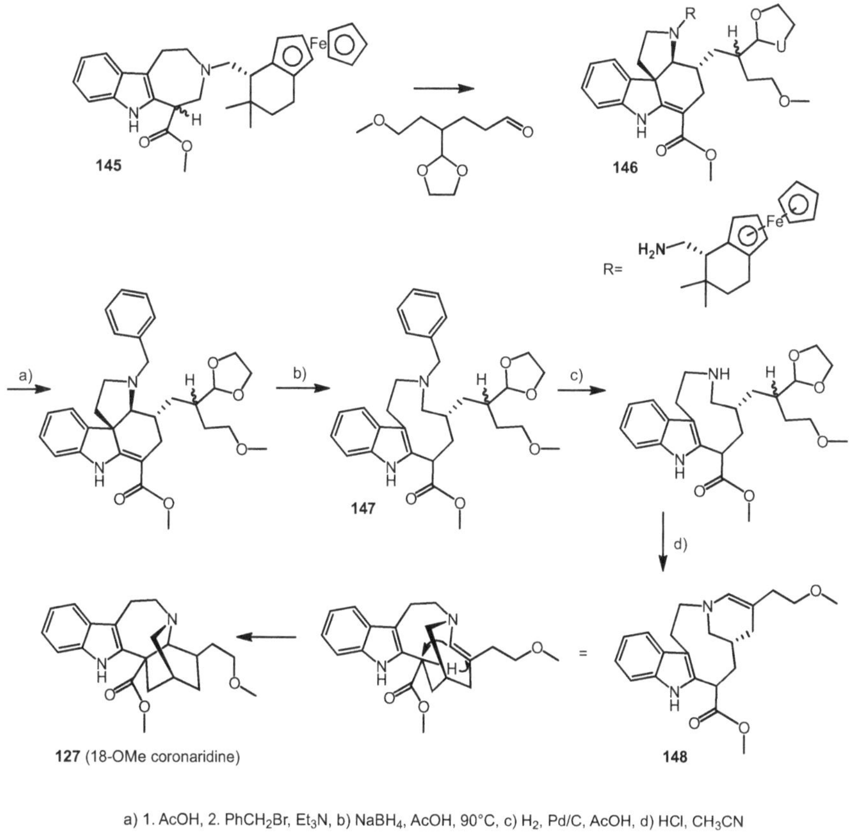

Chart 8 Kuehne's synthesis of 18-methoxy-coronaridine (**127**)

6 Biological Activities

6.1 Ibogaine and Noribogaine

6.1.1 Elements of Context

Ibogaine (**2**) is certainly included in any top list of alkaloids for its mystique and aura. It has its roots in the mythic Bwiti cult of Gabon and it has been for the past three decades the object of a fierce battle between its pros and cons when used as a drug abuse treatment. A full volume in the series "The Alkaloids" was dedicated to ibogaine, and, in the Introduction, G.A. Cordell wrote: "is ibogaine an alkaloid that can save the world from addiction? Probably not" [3]. That was in 2001, so why keep writing on this topic? Two of the reasons are that advances have been made in the understanding of the drug addiction phenomenon, and, also because, in the absence of an alternative treatment, **2** is still more or less openly in use.

The story of ibogaine (**2**) has been written several times including its most recent developments as a treatment for drug abuse [71–74]. It is less well-known, however, that iboga root extracts were at one time available on the market as an anti-fatigue and stimulant agent. Formulated tablets were sold by the Laboratoires Houdé under the name of Lambarene®, after the city in Gabon where Albert Schweitzer had his dispensary. The drug was withdrawn from the market in 1970 but **2** reappeared on the scene when Howard Lotsof was granted a patent in 1985 for "a rapid method for interrupting the narcotic addiction syndrome". This was followed by a period of conflicts and experimental work that culminated in the first ibogaine conference held in New York in 1999 [75]. The story is still ongoing since a further ibogaine conference was convened in Mexico in 2016 [76].

6.1.2 Security Problems and Fatalities

There is at least one major reason to explain the reluctance of health authorities to pursue ibogaine (**2**) as a potential therapeutic agent: security of use. Toxicity has always been a concern and it cannot be disclaimed that several "patients" have passed away some time after ingesting **2**. One can find in the forensic literature articles investigating the reasons for these fatalities [77–81]. However, there does not appear to be a general pattern linking such deaths to the intake of **2**. Neurotoxicity and cardiotoxicity are the most often suspected causes of death, and also activity on Purkinje cells has been observed in female rats under ibogaine treatment. Not all of these effects have been ascertained during patient autopsies, and the known tremorigenic effects of **2** seem to be unrelated to such fatalities.

6.1.3 Analytics

During forensic investigations, the levels of ibogaine (**2**) and related products in body fluids were determined. In the human body, the main metabolite of ibogaine (**2**) is noribogaine (12-hydroxy-ibogamine, **149**) produced by demethylation of **2** with cytochrome CYP2D6. It is also a constituent of the crude alkaloid mixture of the root bark of *Tabernanthe iboga*. The major alkaloids from the plant (also involved in poisoning cases) are ibogaine, noribogaine, voacangine (**150**), iboluteine (**58**), ibochine (**151**), ibogaline (**152**), and ibogamine (**153**) [82] (Fig. 37). In the same paper, the stability of the two main alkaloids ibogaine and noribogaine under daylight exposure was investigated and it was found that they rapidly transformed into iboluteine (**58**) and ibochine for **2**, and into their desmethoxy counterparts of **149**, with respective half-lives of 81.5 and 11 min. Liquid chromatography-MS with electrospray ionization is the method of choice to rapidly discriminate these alkaloids and many examples of the use of this technique are found in recent literature, as e.g., in Refs. [83] and [84].

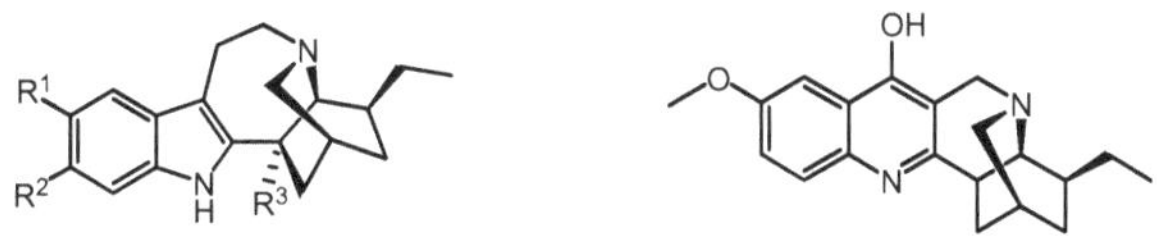

149 R^1 = OH, R^2 = R^3 = H (noribogaine)
150 R^1 = OCH_3, R^2 = H, R^3 = CO_2CH_3 (voacangine)
152 R^1 = R^2 = OCH_3, R^3 = H (ibogaline)
153 R^1 = R^2 = R^3 = H (ibogamine)

151 (ibochine)

Fig. 37 Main alkaloids from the roots of *Tabernanthe iboga* (**149–153**)

6.1.4 Mechanism of Action

The mechanism of action of ibogaine (**2**) is very complex and a glimpse at this complexity is given in an article where the affinity of psychedelic drugs towards a large series of receptors and transporters is documented [85]. To make a long story short, ibogaine activates μ and κ opioid receptors as well as $\sigma 1$ and $\sigma 2$ receptors. Unsurprisingly, because of its structural similarities with serotonin, it also inhibits to a certain extent serotonin and dopamine receptors. The action of **2** on the μ opioid receptor (MOR) was investigated in detail by Alper et al. using hypothalamus cells overexpressing this receptor [86]. It was eventually found that **2** is a weak MOR antagonist without any agonist effect in rat thalamic membranes. The inhibition of acetylcholinesterase (AChE) by **2** was invoked at one time to explain some properties of the molecule and in particular the potentiation of morphine analgesia. This was reinvestigated and an IC_{50} of 520 ± 40 μM was found, a very high value suggesting that the physiological effect of **2** on AChE is negligible [87]. These findings led to the conclusion that the mechanism of action of ibogaine is different from that of other molecules with effects on opioid tolerance and withdrawal, indicating a novel mechanism with targets still to be discovered.

In a series of intriguing papers, Paškulin et al. raised the question of the duration of the effects of **2**, which is much longer than its pharmacokinetic parameters would allow. In particular, the mood-elevating effects appear generally a day after ingestion, a lapse of time where most of the drug and its metabolites are almost no longer observable. Two proteomic studies were performed on whole rat brain and on the yeast *Saccharomyces cerevisiae* subjected or not to **2**, and the overexpressed proteins were detected by 2D gel electrophoresis [88,89]. These were involved with cell metabolism: glyceraldehyde-3-phosphate dehydrogenase, phosphoglycerate kinase, enolase, and alcohol dehydrogenase, all related to compensation of the ATP pool decrease. Ibogaine (**2**) induces a new metabolic equilibrium aiming at saving energy [90] and at the same time increases the activity of antioxidative enzymes as an adaptation to oxidative stress [91].

The complex pharmacological profile of ibogaine was highlighted recently in a zebrafish study which, inter alia, allowed observation of its effects on social behavior (suggesting a wider use of this model for hallucinogenic and drug abuse research [92]). It must also be realized that due to the rapid transformation of **2** into

149, which possesses a different pharmacological profile, it is not easy to understand the origin of the effects (beneficial or deleterious) of the iboga alkaloids on human beings. This is probably the second major reason why it seems so difficult to bring one of these molecules to the market.

6.2 18-Methoxy-coronaridine (18-MC)

Faced with the difficulties associated with ibogaine (**2**), a search for an alternative molecule in the series was undertaken. Ibogamine (**143**) and coronaridine (**3**) were eventually considered as potential candidates since they retained the action on drug self-administration without any tremorigenic side effects. These properties were shared by many other similar compounds, among which 18-OMe-coronaridine (**127**) proved to be the most interesting [93,94]. Compound **127** was fully synthetic and therefore patentable, which has been an advantage over ibogaine and other natural products, and it has its own particular mechanism of action. It was thus found to have micromolar activity on κ, μ, and $\sigma 2$ opioid receptors and on 5-HT_3 receptor without affinity for the NMDA receptor and serotonin transporter. Particularly interesting was an affinity for the $\alpha 3\beta 4$ nicotinic receptor, which was the object of pharmacological experimentation by intracerebral administration in rats in a studies related to drug self-administration [95,96]. Compound **127** was also shown to act on the sucrose reward circuit and therefore may be considered as a possible antiobesity agent [97].

Also exciting is the observation that coronaridine (**3**) and particularly 18-MC (**127**) show promise against *Leishmania amazonensis*, a causative agent of leishmaniasis [98].

A small pharmaceutical company named Savant HWP (San Carlos, CA, USA) has commenced preclinical testing comprising IND-enabling studies and GMP scale-up work on 18-MC in the hope of gaining official approval as a drug cessation treatment. The company is also conducting human clinical safety studies on compound **127** in Brazil, so that it can be investigated further for the potential therapy of leishmaniasis.

6.3 Miscellaneous Biological Properties of Iboga Alkaloids

6.3.1 Cytotoxicity and Antiproliferative Activity

As often is the case currently, the isolation of new natural products may be accompanied by investigation of their biological properties, using in-house available assays. In this respect, cytotoxicity screens against one or several cancer cell lines is a common asset of many laboratories. Due to compound quantity limitations, or the inavailability of follow-up in vivo assays, it may not always be possible to perform studies to determine antineoplastic activity using experimental tumor-bearing mouse models.

In a general screening procedure for inhibitors of the Wnt pathway, *T divaricata* gave a positive response that led subsequently lead to four bioactive iboga alkaloids: voacangine, isovoacangine, coronaridine (**3**), and coronaridine hydroxyindolenine (**41**), with IC_{50} values in the 10 μ*M* range [99]. The best compound was **3**, shown to down-regulate mRNA and therefore decrease the *β*-catenin protein level and inhibit the Wnt signaling pathway, which controls, among other factors, cell proliferation. Compound **3** has been known for a long time to be cytotoxic [100] and this may be a possible mechanism of action. 19-Oxocoronaridine, also named ervatamine I and conodusine E (**33**), was found to be ca. 40 times less potent [16]. A good level of activity was observed for tabercarpamine A (**125**) against the MCF-7, HepG2, and SMMC-7721 cancer cell lines [46]. Tabercarpamine A (**125**) was found to induce apoptosis, a biological activity shared with ervatensines A (**83**) and B (**86**) [13]. Vobatensines A–F (**81**, **82**, **124**, **123**, **76**, **87**) were assayed against a variety of cancer cell lines and showed cytotoxic effects in the 10 micromolar range [49]. No indication was given on their possible mechanism of action. Tabercorine A (**120**) and 17-acetyl-tabernaecorymbosine A (**94**) also displayed cytotoxic activities in the low micromolar range and were shown to be as potent in this regard as cisplatin, which was used as a positive control [53]. However, it is unlikely that they proceed through the same mechanism of action.

6.3.2 Central Nervous System Effects

Given the very potent and profound effects of ibogaine (**2**) on the central nervous system, it is not surprising to find reports of further new investigations of this type. One of the most promising arose from the total synthesis work of Sinha et al. and concerns a benzofuran analog of ibogaine, which possesses dual affinity for both the *μ* and *κ* opioid receptors (MOR and KOR). This compound was also nontremorigenic, and showed antinociception in mice in a standard hot-plate test of comparable potency to morphine. The identification of a new pharmacophore in this study may lead to the development of a new treatment for pain [101]. The roots of *Tabernaemontana divaricata* yielded two known alkaloids with a promising level of inhibition of acetylcholinesterase (a demonstrated target for Alzheimer's disease) [102]. These compounds were 19,20-dihydrotabernamine [103] and 19,20-dihydroervahanine A [104]. Voacamine, 3,6-oxidovoacangine, and 5-hydroxy-3,6-oxidovoacangine from *Voacanga africana* were found to be potent antagonists of the cannabinoid receptor CB1, with IC_{50} values in the submicromolar range [25]. This target is involved in memory, pain, and appetite and inhibitory compounds may show promise in the treatment of obesity, metabolic syndrome, and related disorders inter alia. Coronaridine (**3**), alone among a series of 13 iboga alkaloids, was found to have protective effects in MPP+ (1-methyl-4-phenylpyridinium)-injured primary cortical neurons [39]. This may be of relevance in a treatment of Parkinson's disease but one should not underestimate the cytotoxicity of the molecule (ca. 2 μ*M*).

6.3.3 Miscellaneous Biological Activities

The so-called ervatamines A-I, among which iboga alkaloids (**12**, **27**, **28**, **33**, **46**), as well as coronaridine (**3**), heyneanine, their 2′-oxo derivatives and pandine (**64**), were subjected to an anti-inflammatory in vitro test, and inhibition of lipopolysaccharide-induced NO production in RAW 264.7 macrophages was demonstrated for **3** and **64** [16]. The tabernaeelegantinals (**117–119**) from *Muntafara sessilifolia* were assayed against the chloroquine-resistant strain FcB1 of *Plasmodium falciparum* and showed moderate to good activity [55,56]. Selectivity indexes were determined after assessment of their cytotoxicity and it was concluded that their activity was the result of a general toxicity. However, when evaluated in vitro using the same strain of *P. falciparum*, 3′-oxo-tabernaelegantine A (**109**), also isolated from *M. sessiloflora*, was more promising, and showed antiplasmodial activity (IC_{50} 4.4 μ*M*) with non-significant cytotoxicity for the L-6 rat skeletal muscle cell line [55].

7 Conclusion and Perspectives

Iboga-type alkaloids are one of the many types of indolomonoterpene alkaloids, but they are set apart owing to their unique biological properties. Notably, one of these is their possible use for drug cessation treatment. It must be kept in mind that cocaine addiction is so far without any reliable treatment and that the substitutes to opioids are not without any drawbacks, with one of them being their illicit diversion [105]. Could Howard Lotsof's dream come true and a true drug cessation treatment be introduced to the market in the future, hopefully inspired by *Tabernanthe iboga*?

Acknowledgments It is our pleasure to thank the many individuals with whom we were able to exchange correspondence during the writing of this review: Ken Alper, Laurence Balas, Françoise Bressolle, Geoffrey Cordell, Mohammed Damak, Françoise Guéritte, Toh Seok Kam, Muriel Le Roux, Dalibor Sames, Surajit Sinha, and Hiromitsu Takayama. For pictures of plants, we are indebted to Philomène Akoua Kouassi, University of Cocody (Ivory Coast) and Bruno David, Pierre Fabre Laboratories, Toulouse (France).

References

1. Le Men J, Taylor WI (1965) A uniform numbering system for indole alkaloids. Experientia 21:508
2. Szabó LF (2008) Rigorous biogenetic network for a group of indole alkaloids derived from strictosidine. Molecules 13:1875
3. Alper KR, Glick SD (2001). In: Cordell GA (ed) The alkaloids, vol 56. Academic Press, San Diego, San Francisco, New York, Boston, London, Sydney, Tokyo, p 1
4. Sundberg RJ, Smith SQ (2002) In: Cordell GA (ed) The alkaloids, vol 59. Elsevier, Amsterdam, p 281
5. Saxton JE (1994) The ibogamine-catharanthine group. In: Saxton JE (ed) The monoterpenoid indole alkaloids, supplement to part 4. The chemistry of heterocyclic compounds, Taylor EC (ed) vol 25. Wiley, Chichester, New York, Brisbane, Toronto, Singapore, p 487
6. O'Connor SE, Maresh JJ (2006) Chemistry and biology of monoterpene indole alkaloids biosynthesis. Nat Prod Rep 23:532
7. Miettinen K, Dong L, Navrot N, Schneider T, Burlat V, Pollier J, Woittiez L, van der Krol S, Lugan R, Ilc T, Verpoorte R, Oksman-Caldentey KM, Martinoia E, Bouwmeester H, Goossens A, Memelink J, Werck-Reichhart D (2014) The *seco*-iridoid pathway from *Catharanthus roseus*. Nat Commun 5:3606
8. Kries H, O'Connor SE (2016) Biocatalysts from alkaloid producing plants. Curr Opin Chem Biol 31:22
9. El-Sayed M, Choi YH, Frédérich M, Roytrakul S, Verpoorte R (2004) Alkaloid accumulation in *Catharanthus roseus* cell suspension culture fed with stemmadenine. Biotechnol Lett 26:79310
10. Townsend CA (2011) A "Diels-Alderase" at last. ChemBioChem 12:2267
11. Fage CD, Alsiorho E, Liu Y, Wagner DT, Liu HW, Keatinge-Clay AT (2011) The structure of SpnF, a standalone enzyme that catalyzes [4 + 2] cycloaddition. Nat Chem Biol 11:256
12. Saxton JE (1994) The ibogamine-catharanthine group. In: Saxton JE (ed) The monoterpenoid indole alkaloids, supplement to part 4. The chemistry of heterocyclic compounds, Taylor EC (ed) vol 25. Wiley, Chichester, New York, Brisbane, Toronto, Singapore, p 495
13. Lim K-H, Raja VJ, Bradshaw TD, Lim SH, Low Y-Y, Kam T-S (2015) Ibogan, tacaman, and cytotoxic bisindole alkaloids from *Tabernaemontana*. Cononusine, an iboga alkaloid with unusual incorporation of a pyrrolidinone moiety. J Nat Prod 78:1129
14. Moisan L, Doris E, Rousseau B, Thuéry P (2004) Catharanthinol and dihydrocatharanthinol: two iboga-class alkaloids. Acta Cryst C 60:792
15. Jarraya RM, Bouaziz A, Hamdi B, Ben Salah A, Damak M (2008) *N*-(Hydroxymethyl) ibogaine. Acta Cryst E 64:1739
16. Zhang D-B, Yu D-G, Sun M, Zhu X-X, Yao X-J, Zhou SY, Chen J-J, Gao K (2015) Ervatamines A–I, anti-inflammatory monoterpenoid indole alkaloids with diverse skeletons from *Ervatamia hainanensis*. J Nat Prod 78:1253
17. Blaha K, Koblicova Z, Trojanek J (1974) Chiroptical investigations on the iboga and voacanga alkaloids. Collect Czech Chem Commun 39:2258
18. Harada N, Nakanishi K (1983) Circular dichroism spectroscopy-exciton coupling in organic stereochemistry. University Sciences Books and Oxford University Press, Mill Valley, CA and Oxford, UK
19. Guo L-L, He H-P, Di Y-T, Li S-F, Cheng Y-Y, Yang W, Li Y, Yu J-P, Zhang Y, Hao J (2012) Indole alkaloids from *Ervatamia chinensis*. Phytochemistry 74:140
20. Harada M, Asaba KN, Iwai M, Kogure N, Kitajima M, Takayama H (2012) Asymmetric total synthesis of an iboga-type indole alkaloid, voacangalactone, newly isolated from *Voacanga africana*. Org Lett 14:5800
21. Bartlett MF, Dickel DF, Taylor WI (1958) The alkaloids of *Tabernanthe iboga*. Part IV.1 The structures of ibogamine, ibogaine, tabernanthine and voacangine. J Am Chem Soc 80:126

22. Le Men J, Potier P, Le Men-Olivier L, Panas, JM, Richard B, Potron C (1974) Alkaloids of *Gabonia eglandulosa*–eglandine and eglandulosine. Bull Soc Chim Fr:1369
23. Le Men-Olivier L, Le Men J, Massiot G, Richard B, Mulamba T, Potier P, Husson HP, Van Beek T, Verpoorte R (1984) Révision des structures de l'églandine et de l'églandulosine. Bull Soc Chim Fr:94
24. Tang B-Q, Wang W-J, Huang X-J, Li G-Q, Wang L, Jiang R-W, Yang T-T, Shi L, Zhang X-Q, Ye W-C (2014) Iboga-type alkaloids from *Ervatamia officinalis*. J Nat Prod 77:1839
25. Kitajima M, Iwai M, Kikura-Hanajiri R, Goda Y, Iida M, Yabushita H, Takayama H (2011) Discovery of indole alkaloids with cannabinoid CB1 receptor antagonistic activity. Bioorg Med Chem Lett 21:1962
26. Jin YS, Du Jing-Ling JL, Chen HS, Jin L, Liang S (2010) A new indole alkaloid from *Ervatamia yunnanensis*. Fitoterapia 81:63
27. Mak JYW, Pouwer RH, Williams CM (2014) Natural products with anti-Bredt and bridgehead double bonds. Angew Chem Int Ed 53:13664
28. Langlois N, Langlois Y (2015) La découverte de la navelbine: histoire vécue. Imprimerie de Bretagne, Morlaix, France
29. Guéritte F, Le Roux M (2016) La Navelbine et le Taxotere: histoires de sciences. Iste Wiley Elsevier, London
30. Beatty JW, Stephenson CRJ (2014) Synthesis of (−)-pseudotabersonine, (−)-pseudovincadifformine, and (+)-coronaridine enabled by photoredox catalysis in flow. J Am Chem Soc 136:10270
31. The Plant List (2015) A working list of all plant species. http://www.theplantlist.org/
32. Kam T-S, Sim K-M (2002) Five new indole alkaloids from *Tabernaemontana corymbosa*. J Nat Prod 65:669
33. Wenkert E, Cochran DW, Gottlieb HE, Hagaman EW, Braz Filho R, Matos FJA, Madruga MILM (1976) ^{13}C NMR spectroscopy of naturally occurring substances. 45: Iboga alkaloids. Helv Chim Acta 59:2437
34. Takayama H, Suda S, Chen IS, Kitajima M, Aimi N, Sakai S (1994) Two new dimeric indole alkaloids from *Tabernaemontana subglogosa* Merr. from Taiwan. Chem Pharm Bull 42:280
35. Chen HS, Huang M, Liang S, Zhang XD, Jin L, Chen MY, Zheng JM, Liu JG (2008) Indole alkaloids from *Ervatamia* and their use as detoxifying agents. Chin Pat 200610031038.0
36. Nge CE, Kam WC, Thomas NF, Lim S-H, Low Y-Y, Kam T-S (2016) Ibogan, aspidosperman, vincamine, and bisindole alkaloids from a Malayan *Tabernaemontana corymbosa*: iboga alkaloids with C-20α substitution. J Nat Prod 79:1388
37. Lim K-H, Goh T-S (2009) Conoliferine and isoconoliferine, structurally novel alkaloid-lignan conjugates from *Tabernaemontana corymbosa*. Tetrahedron Lett 50:3756
38. Lim K-H, Goh T-S (2009) Conomicidines A and B, unusual alkaloid—hydroxycinnamyl alcohol conjugates from *Tabernaemontana corymbosa*. Helv Chim Acta 92:1895
39. Liu Z-W, Huang X-J, Xiao H-L, Liu G, Zhang J, Shi L, Jiang R-W, Zhang X-Q, Ye W-C (2016) New iboga-type alkaloids from *Ervatamia hainanensis*. RSC Adv 6:30277
40. Bao M-F, Yan J-M, Cheng G-G, Li X-Y, Liu Y-P, Li Y, Cai X-H, Luo X-D (2013) Cytotoxic indole alkaloids from *Tabernaemontana divaricata*. J Nat Prod 76:1406
41. Feng XZ, Kan C, Potier P, Kan SK, Lounasmaa M (1982) Monomeric indole alkaloids from *Ervatamia hainanensis*. Planta Med 44:212
42. de Souza JJ, Mathias L, Braz-Filho R, Carcino Vieira IJ (2010) Two new indole alkaloids from *Tabernaemontana hustrix* Steud. (Apocynaceae). Helv Chim Acta 93:422
43. Rosenmund P, Haase WH, Bauer J, Frische R (1975) Contribution to chemistry of indole. 7. Syntheses in iboga series 3. Ibogamine, ibogaine and epiibogamine. Chem Ber 108:1871
44. Nge C-E, Gan C-Y, Low Y-Y, Thomas NF, Kam T-S (2013) Voatinggine and tabertinggine, pentacyclic indole alkaloids derived from an iboga precursor via a common cleavamine-type intermediate. Org Lett 15:4774
45. Habib-ur-Rehman, Atta-ur-Rahman (2005) NMR studies on alioline—a novel vinca alkaloid. Z Naturforsch 60b:870

46. Ma K, Wang J-S, Luo J, Yang M-H, Kong L (2014) Tabercarpamines A–J, apoptosis-inducing indole alkaloids from the leaves of *Tabernaemontana corymbosa*. J Nat Prod 77:1156
47. Sapi J, Massiot G (1994) Bisindole alkaloids. In: Saxton JE (ed) The monoterpenoid indole alkaloids, vol 4. Wiley, Chichester, New York, Brisbane, Toronto, Singapore, p 523
48. Zaima K, Hirata T, Hosoya T, Hirasawa Y, Koyama K, Rahman A, Kusumawati I, Zaini NC, Shiro M, Morita H (2009) Biscarpamontamines A and B, an aspidosperma-iboga bisindole alkaloid and an aspidosperma-aspidosperma bisindole alkaloid, from *Tabernaemontana sphaerocarpa*. J Nat Prod 72:1686
49. Sim DS-Y, Teoh W-Y, Sim K-S, Lim S-H, Thomas NF, Low Y-Y, Kam T-S (2016) Vobatensines A – F, cytotoxic iboga-vobasine bisindoles from *Tabernaemontana corymbosa*. J Nat Prod 79:1048
50. Kam T-S, Sim K-M (2002) New tabernamine derivatives from *Tabernaemontana*. Heterocycles 57:2137
51. Kam T-S, Sim K-M, Pang H-S (2003) New bisindole alkaloids from *Tabernaemontana corymbosa*. J Nat Prod 66:11
52. Kam T-S, Pang H-S (2004) Conodusarine, a new biologically active bisindole alkaloid from *Tabernaemontana divaricata*. Heterocycles 63:845
53. Zhang Y, Guo L, Yang G, Guo F, Di Y, Li S, Chen D, Hao X (2015) New vobasinyl-ibogan type bisindole alkaloids from *Tabernaemontana corymbosa*. Fitoterapia 100:150
54. Kam T-S, Sim K-M (2003) Conodirinines A and B, novel vobasine-iboga bisindoles incorporating an additional tetrahydro-1,3-oxazine unit on the vobasinyl moiety. Helv Chim Acta 86:122
55. Girardot M, Deregnaucourt C, Deville A, Dubost L, Joyeau R, Allorge L, Rasoanaivo P, Mambu L (2012) Indole alkaloids from *Muntafara sessilifolia* with antiplasmodial and cytotoxic activities. Phytochemistry 73:65
56. Girardot M, Gadea A, Deregnaucourt C, Deville A, Dubost L, Nay B, Maciuk A, Rasoanaivo P, Mambu L (2012) Tabernaelegantinals: unprecedented cytotoxic bisindole alkaloids from *Muntafara sessilifolia*. Eur J Org Chem:2816
57. Danieli B, Palmisano G, Gabetta B, Martinelli EM (1980) Tabernaelegantinines C and D, two new bisindole alkaloids containing a cyano group from *Tabernaemontana elegans* Stapf. J Chem Soc Perkin Trans 1:601
58. Madinaveitia A, de la Fuente G, González A (1998) The absolute configuration at C(7) of voacangine hydroxyindolenine. Helv Chim Acta 81:1645
59. Mizoguchi H, Oikawa H, Oguri H (2014) Biogenetically inspired synthesis and skeletal diversification of indole alkaloids. Nat Chem 6:57
60. Trost BM, Godleski SA, Genet JP (1978) Total synthesis of racemic and optically active ibogamine: utilization and mechanism of a new silver ion assisted palladium cyclization. J Am Chem Soc 100:3930
61. Jana GK, Sinha S (2010) Synthesis of iboga alkaloids by Pd-catalyzed heteroannulation of 2-iodoaniline with an internal alkyne as the key step. Tetrahedron Lett 51:1441
62. Jana GK, Sinha S (2010) A concise route to iboga-analogues via the formation of suitably substituted-2-indoles. Tetrahedron Lett 51:1994
63. Paul S, Pattanayak S, Sinha S (2011) Synthesis of a new series of iboga-analogues. Tetrahedron Lett 52:6166
64. Jana GK, Sinha S (2012) Reductive Heck coupling: an efficient approach toward the iboga alkaloids. Synthesis of ibogamine, epiibogamine and iboga analogs. Tetrahedron Lett 53:1671
65. Jana GK, Sinha S (2012) Total synthesis of ibogaine, epiibogaine and their analogues. Tetrahedron 68:7155
66. Kruegel AC, Rakshit S, Li X, Sames D (2015) Constructing iboga alkaloids via C–H bond functionalization: examination of the direct and catalytic union of heteroarenes and isoquinuclidine alkenes. J Org Chem 80:2062

67. Höck S, Borschberg H-J (2006) Enantioselective synthesis of (−)-(19*R*)-ibogamin-19-ol. Helv Chim Acta 89:542
68. Yang H, Carter RG (2009) Asymmetric construction of nitrogen-containing [2.2.2] bicyclic scaffolds using *N*-(*p*-dodecylphenylsulfonyl)-2-pyrrolidinecarboxamide. J Org Chem 75:5151
69. Barve IJ, Dalvi PB, Thikekar TU, Chanda K, Liu Y-L, Fang C-P, Liu C-C, Sun C-M (2015) Design, synthesis and diversification of natural product-inspired hydantoin-fused tetrahydroazepino indoles. RSC Adv 5:73169
70. Kuehne ME, Wilson TE, Bandarage UK, Dai W, Yu Q (2001) Enantioselective synthesis of coronaridine and 18-methoxycoronaridine. Tetrahedron 57:2085
71. Alper KR, Lotsof HS, Frenken GMN, Luciano DJ, Bastiaans J (1999) Treatment of acute opioid withdrawal with ibogaine. Am J Addict 8:234
72. Alper KR, Lotsof HS, Kaplan CD (2008) The ibogaine medical subculture. J Ethnopharmacol 115:9
73. Mačiulaitis R, Kontrimavičiūtė V, Bressolle FMM, Briedis V (2008) Ibogaine, an anti-addictive drug: pharmacology and time to go further in development. A narrative review. Hum Exp Toxicol 27:181
74. Brown TK (2013) Ibogaine in the treatment of substance dependence. Curr Drug Abuse Rev 6:3
75. For a historical timeline see Alper KR, Glick SD (2001) In: Cordell GA (ed) The Alkaloids, vol 56. Academic Press, San Diego, San Francisco, New York, Boston, London, Sydney, Tokyo, p 4
76. http://www.ibogaineconference.org/
77. Alper KR, Stajić M, Gill JR (2012) Fatalities temporally associated with the ingestion of ibogaine. J Forensic Sci 57:398
78. Maas U, Strubelt S (2006) Fatalities after taking ibogaine in addiction treatment could be related to sudden cardiac death caused by autonomic dysfunction. Med Hypotheses 67:960
79. Chèze M, Lenoan A, Deveaux M, Pépin G (2008) Determination of ibogaine and noribogaine in biological fluids and hair by LC–MS/MS after *Tabernanthe iboga* abuse. Iboga alkaloids distribution in a drowning death case. Forensic Sci Int 176:58
80. Papadodima SA, Dona A, Evaggelakos CI, Goutas N, Athanaselis SA (2013) Ibogaine related sudden death: a case report. J Forensic Leg Med 20:809
81. Mazoyer C, Carlier J, Boucher A, Peoc'h M, Lemeur C, Gaillard Y (2013) Fatal case of a 27-year-old male after taking iboga in withdrawal treatment: GC-MS/MS determination of ibogaine and ibogamine in iboga roots and postmortem biological material. J Forensic Sci 58:1666
82. Kontrimavičiūtė V, Mathieu O, Balas L, Escale R, Blayac JP, Bressolle FMM (2007) Ibogaine and noribogaine: structural analysis and stability studies. Use of LC-MS to determine alkaloid contents of the root bark of *Tabernanthe iboga*. J Liq Chromatogr Relat Technol 30:1077
83. Kontrimavičiūtė V, Breton H, Mathieu O, Mathieu-Daudé J-C, Bressolle FMM (2006) Liquid chromatography–electrospray mass spectrometry determination of ibogaine and noribogaine in human plasma and whole blood. Application to a poisoning involving *Tabernanthe iboga* root. J Chromatogr B 843:131
84. Kontrimavičiūtė V, Breton H, Barnay F, Mathieu-Daudé J-C, Bressolle FMM (2006) Liquid chromatography-electrospray mass spectrometry determination of ibogaine and 12-hydroxy-ibogamine in human urine. Chromatographia 63:533
85. Ray TS (2010) Psychedelics and the human receptorome. PLoS One 5, e9019
86. Antonio T, Childers SR, Rothman RB, Dersch CM, King C, Kuehne M, Bornmann WG, Eshleman AJ, Janowsky A, Simon ER, Reith MAE, Alper K (2013) Effect of iboga alkaloids on µ-opioid receptor-coupled G protein activation. PLoS One 8, e77262
87. Alper K, Reith MEA, Sershenc H (2012) Ibogaine and the inhibition of acetylcholinesterase. J Ethnopharmacol 139:879

88. Paškulin R, Jamnik P, Živin M, Raspor P, Štrukelj B (2006) Ibogaine affects brain energy metabolism. Eur J Pharmacol 552:11
89. Paškulin R, Jamnik P, Obermajer N, Slavić M, Štrukelj B (2010) Induction of energy metabolism related enzymes in yeast *Saccharomyces cerevisiae* exposed to ibogaine is adaptation to acute decrease in ATP energy pool. Eur J Pharmacol 627:131
90. Paškulin R, Jamnik P, Danevčič T, Koželj G, Krašovec R, Krstić-Milošević D, Blagojević D, Štrukelj B (2012) Metabolic plasticity and the energy economizing effect of ibogaine, the principal alkaloid of *Tabernanthe iboga*. J Ethnopharmacol 143:319
91. Nikolić-Kokić A, Oreščanin-Dušić Z, Spasojević I, Slavić M, Mijušković A, Paškulin R, Miljević Č, Spasić MB, Blagojević DP (2015) Ex vivo effects of ibogaine on the activity of antioxidative enzymes in human erythrocytes. J Ethnopharmacol 164:64
92. Cachat J, Kyzara EJ, Collins C, Gaikwada S, Green J, Roth A, El-Ounsia M, Davis A, Pham M, Landsman S, Stewart AM, Kalueff AV (2013) Unique and potent effects of acute ibogaine on zebrafish: the developing utility of novel aquatic models for hallucinogenic drug research. Behav Brain Res 236:258
93. Maisonneuve IM, Glick SD (2003) Anti-addictive actions of an iboga alkaloid congener: a novel mechanism for a novel treatment. Pharmacol Biochem Behav 75:607
94. Pace CJ, Glick SD, Maisonneuve IM, He L-W, Jokiel PA, Kuehne ME, Fleck MW (2004) Novel iboga alkaloid congeners block nicotinic receptors and reduce drug self-administration. Eur J Pharmacol 492:159
95. Panchal V, Taraschenko OD, Maisonneuve IM, Glick SD (2005) Attenuation of morphine withdrawal signs by intracerebral administration of 18-methoxycoronaridine. Eur J Pharmacol 525:98
96. Glick SD, Ramirez RL, Livi JM, Maisonneuve IM (2006) 18-Methoxycoronaridine acts in the medial habenula and/or interpeduncular nucleus to decrease morphine self-administration in rats. Eur J Pharmacol 537:94
97. Taraschenko OD, Maisonneuve IM, Glick SD (2010) 18-Methoxycoronaridine, a potential anti-obesity agent, does not produce a conditioned taste aversion in rats. Pharmacol Biochem Behav 96:247
98. Delorenzi JC, Freire-de-Lima L, Gattass CR, de Andrade CD, He L, Kuehne ME, Saraiva EMB (2002) In vitro activities of iboga alkaloid congeners coronaridine and 18-methoxycoronaridine against *Leishmania amazonensis*. Antimicrob Agents Chemother 46:2011
99. Ohishi K, Toume K, Arai MA, Sadhu SK, Ahmed F, Ishibashi M (2015) Coronaridine, an iboga type alkaloid from *Tabernaemontana divaricata*, inhibits the Wnt signaling pathway by decreasing β-catenin mRNA expression. Bioorg Med Chem Lett 25:3937
100. Gunasekera SP, Cordell GA, Farnsworth NR (1980) Anticancer indole alkaloids of *Ervatamia heyneana*. Phytochemistry 19:1213
101. Banerjee TS, Paul S, Sinha S, Das S (2014) Synthesis of iboga-like isoquinuclidines: dual opioid receptors agonists having antinociceptive properties. Bioorg Med Chem 22:6062
102. Ingkaninan K, Changwijit K, Suwanborirux K (2006) Vobasinyl-iboga bisindole alkaloids, potent acetylcholinesterase inhibitors from *Tabernaemontana divaricata* root. J Pharm Pharmacol 58:847
103. Henriques AT, Melo AA, Moreno PR, Ene LL, Henriques JA, Schapoval EE (1996) *Ervatamia coronaria*: chemical constituents and some pharmacological activities. J Ethnopharmacol 50:19
104. van Beek TA, Verpoorte R, Baerheim Svendsen A, Leeuwenberg AJM, Bisset NG (1984) *Tabernaemontana* L. (Apocynaceae): a review of its taxonomy, phytochemistry, ethnobotany and pharmacology. J Ethnophamacol 10:1
105. Bretteville-Jensen AL, Lillehagen M, Gjersing L, Burdzovic Andreas J (2015) Illicit use of opioid substitution drugs: prevalence, user characteristics, and the association with non-fatal overdoses. Drug Alcohol Depend 147:89

Catherine Lavaud is a pharmacist from the University of Champagne Ardenne (France). She obtained a Ph.D. degree in 1986 under the supervision of Dr. Georges Massiot at the University of Rheims. She was recruited as an Assistant Professor in 1984 and further promoted to Full Professor in 1998 at the Faculty of Pharmacy in the same University. She teaches pharmacognosy and miscellaneous topics related to natural products. She is presently Vice-Dean of the Faculty of Pharmacy. Her research interests are saponins related to food and herbal medicines, and other natural products as alkaloids, terpenes and heteroside metabolites.

Georges Massiot obtained his Ph.D. degree with Professor Henri-Philippe Husson at the Institut de Chimie des Substances Naturelles in Gif sur Yvette (France) in 1975. He pursued his research as a postdoctoral fellow at the University of Wisconsin under Professor Barry M. Trost. He then joined the CNRS at the Faculty of Pharmacy in Reims. From 1998 to 2008, he was in charge of a laboratory on natural products within "Les Laboratoires Pierre Fabre", a French pharmaceutical company. From 1992 to 1998, he held administrative responsibilities as a deputy director of the chemical sciences department of the CNRS. Presently, he has retired and benefits from an emeritus status at the University of Champagne Ardenne, Reims. He participates in the scientific evaluation of research laboratories in France with the HCERES agency. His research interests are natural products with particular emphasis on alkaloids, terpenes and saponins, including structural elucidation, partial and total synthesis, and biological activities.

A Critical Evaluation of the Quality of Published ^{13}C NMR Data in Natural Product Chemistry

Wolfgang Robien

Contents

W. Robien (✉)
Institute of Organic Chemistry, University of Vienna, Währinger Straße 38, 1090 Vienna, Austria
e-mail: Wolfgang.Robien@univie.ac.at

A.D. Kinghorn, H. Falk, S. Gibbons, J. Kobayashi (eds.), *Progress in the Chemistry of Organic Natural Products 105*, DOI 10.1007/978-3-319-49712-9_3

1 Introduction

The structure elucidation of organic compounds is a complex multicomponent task that today is based mainly on spectroscopic data interpretation. Among the spectroscopic techniques applied to solve structural problems, NMR spectroscopy plays a central role because of the availability of a tremendous plethora of different techniques that can result in a deep insight into the constitution, configuration, conformation, and the dynamic behavior of a molecule of interest. Natural product chemistry is a challenging discipline located at the interface between chemical synthesis, biology, pharmacy, and medicine and has had a great impact on the development of these sciences [1]. In order to understand biosynthesis pathways or the function of enzyme-inhibitors, for example, having a detailed knowledge of the chemical structure of the molecule under investigation is crucial. Despite this common understanding of the importance of the structure elucidation process itself and the necessity of the reproducibility of the conclusions drawn from the experimental data, the results obtained thereof have been in error in many cases, leading to a great number of revisions of structure. These structural revisions have been summarized in several review articles and classified either accordingly to the change between the original structural proposal and the revised structure or by the methods used for revision or sorted by compound classes. In 2005, Nicolaou and Snyder [2] published a review of natural product structural revisions made by total synthesis over the period of 1990–2004. Suyama, Gerwick, and McPhail [3] focused on the revision of the structures of marine natural products. In their article, the revisions published within the period of 2005–2010 were classified by changes in the molecular skeleton. The structures were grouped into "revision of cyclization", "revision of regioisomers", "revision of substituents", "revision of hydrocarbon chains", "revision of dimerization", and different types of "revision of stereocenters". A similar classification scheme was used also by Maier [4], focusing on structural revisions made by chemical synthesis in the period 2005–2009. More recent reviews dealing with natural product chemistry show additional examples of structure revisions and discuss the problem of the proliferation of scientific errors [5, 6].

Elyashberg, Williams, and Blinov, in a review published in 2010, summarized the structural revisions made possible by the application of their isomer generator program when combined with spectrum prediction used for ranking the list of computed structure proposals, based on the similarity between experimental and predicted spectroscopic data [7]. Another compilation of structural revisions together with a "computer-assisted structure elucidation" system (CASE) was given by Elyashberg in 2015 [8]. It should be mentioned that these reviews summarized already acknowledged structural revisions. However, a high probability exists that there is another "treasure trove" of structures hidden in the chemical literature waiting for revision, and still used as reference information for further conclusions. A sarcastic comment was included in the introduction of [7], namely, "Figuratively speaking, it means that 40–45 issues of the imaginary 'Journal of Erroneous Chemistry' were published where all articles contained only incorrectly elucidated structures and, consequently, at least the same number of issues was necessary to describe the revision of these structures". This may well be an underestimate, even when taking into account that this review appeared in 2010. When using SciFinder® (Chemical Abstracts Service, Columbus, OH, USA) [9] for searching terms like "structure revision", "reassignment", "revised structure", or similar combinations of key terms, a steadily growing list of publications appears.

The above-mentioned reviews summarize either known revisions and analyze the conclusions leading to the wrong structural proposal or show methods to find alternative structures from the already published spectroscopic data, which fulfil the constraints much better than the given structure solution. The most effective strategy is to avoid at least the most obvious errors, an approach mentioned in the paper of Elyashberg [8] and discussed in a very short paragraph. On the one hand, there are excellent NMR techniques allowing deep insights into molecular structures, but on the other hand there have been a large number of wrong conclusions drawn from the data obtained. The central challenge in this area is to develop standardized workflow protocols in order to avoid the most trivial errors for the correct interpretation of NMR spectra and hence the proper compound structure elucidation. The importance of independent chemical synthesis for the structure revision process has been proven by many examples and will be mentioned in this chapter. However, it would be possible to reduce the number of incorrect natural product structures dramatically using a more efficient approach to process experimentally obtained spectroscopic data.

Carbon-13 NMR spectroscopy allows direct insight into the skeleton of a given natural product and is therefore a decisive method used during the structure elucidation process. Over the past four decades, many useful 2D NMR techniques have been developed in order to correlate two nuclei [10]. This distance information, in terms of the number of bonds between coupled nuclei, is an extremely efficient constraint during the isomer generation process and may reduce considerably the list of possible candidate structures to a given set of spectroscopic data. A problem with these techniques (e.g., HMBC) is the possible ambiguity of the information obtained, which might lead to incorrect structural proposals when assuming the wrong distance between coupled nuclei during the interpretation and the subsequent isomer generation process. Another advantage of ^{13}C NMR spectroscopy is the availability of large data collections allowing fast and reliable spectrum prediction using different technologies. Furthermore, such a digital repository of spectroscopic data can be evaluated by statistical methods for uncommon chemical shift values, giving hints to erroneous structures and/or wrong signal assignments. In order to predict chemical shift values, a correlation is needed between the structural property (= the structural environment of the atom of interest including stereochemical effects [11]) and the spectroscopic property (= chemical shift value, δ_C). For this reason, unassigned spectroscopic data as published in many journals are completely useless for predicting chemical shift values for a proposed structure, since they can only be utilized as fingerprint information for spectroscopic similarity searches.

In the chemical literature, there is a large amount of data available that show a very large heterogeneity with respect to their quality and their presentation. In this chapter, the quality of published reference material for ^{13}C NMR spectroscopy will be analyzed in detail by a standardized protocol based on the "CSEARCH" NMR-database [12]. This procedure has been named "CSEARCH-Robot-Referee" and is available to the community via a Web interface [13]. The starting point for all subsequent evaluations is the proposed structure and the peak list with optional assignments and experimental multiplicity information, as given in the literature.

The intention of this contribution is to provide both an overview of the possibilities of automatic "Peer-reviewing" of ^{13}C NMR data and to provide a notion of just how many incorrect structures and/or erroneous signal assignments have been published to date in the chemical literature. This automatic "Peer-reviewing" protocol proposed is in principle a very sophisticated compatibility check between a given structure and the experimentally determined peak list, applying a ranking system into four categories, namely, "Accept", "Minor Revision", "Major Revision", and "Reject". Based on this classification, additional procedures can be started automatically in order to provide a more detailed picture of the problem, including proposals for alternative structures. It is estimated that at least 80% of the errors present in the literature with respect to ^{13}C NMR data can be avoided easily by systematic integration of such a scheme into the workflow of the structure elucidation of organic compounds.

2 The "CSEARCH-Robot-Referee"

The "CSEARCH-Robot-Referee" consists of four components:

- A knowledge database that holds 321,000 assigned and curated ^{13}C NMR spectra taken from chemical literature in the public domain;
- An evaluation engine accessing this knowledge base and performing all necessary decisions;

- A "Structure Generator" program that starts from the given query, by modifying the positions where the evaluation engine has found large deviations between the experimentally obtained and predicted chemical shift values. This modification involves also exchanging atom types leading to alternative proposals that are not isomeric with the given structural query;
- A "Spectroscopic Similarity Search" engine using predicted spectra for the compounds available via PubChem (National Center for Biotechnology Information, National Library of Medicine, Bethesda, MD, USA), allowing fast dereplication. An additional analysis for common skeletons and substitution patterns in the list of structures found can identify at least general structural features of an unknown substance.

2.1 The Knowledge Base

The "CSEARCH" database contains some 694,000 assigned ^{13}C NMR spectra (as of October 2015) that were extracted from approximately 500 different journals; for legal reasons, only 321,000 spectra have been made available for use with the "CSEARCH-Robot-Referee". The selection of journals is focused on natural product chemistry, but not only the well-known journals in this field like the *Journal of Natural Products*, *Phytochemistry*, *Planta Medica*, *Fitoterapia*, and *Chemical and Pharmaceutical Bulletin* have been extracted with respect to NMR data, but also journals addressing a smaller group of more specialized readers have been included, such as the *Journal of Ginseng Research* and the *Eurasian Journal of Forest Research*. The increase of data over the past decade was representative of approximately 15,000–20,000 spectra per year [14]. Unassigned ^{13}C NMR data must be omitted, because they are regarded as having no value for spectrum prediction. Also, incorrect values are omitted, so together this has led to a strongly focused database, where the only data stored are those that are perceived initially as being representative of high quality data. Ongoing data correction further improves the quality of the stored reference material. However, despite this investment of time and technology, the "CSEARCH" database is far from being error-free. It has been shown that high-quality NMR spectroscopic data do not necessarily emanate from so-called "higher-impact" journals [15]. This finding is based on statistics and therefore has no implication for a single paper or a single author, but the average Impact Factor (Thomson Reuters, Philadelphia, PA, USA) for journals publishing spectroscopic data evaluated as being better than average is only 1.95 compared to 2.45 for journals having spectroscopic data rated as worse than average. This has served as a strong motivation to extend NMR spectroscopic data extraction from the more prominent journals, which are still covered, to more specialized journals.

2.2 The Importance of Ongoing Data Correction

For the present analysis, 74,915 entries included in the "CSEARCH" database between January 1, 2010 and October 6, 2015 were selected, representing 1,517,876 assigned chemical shift values. These 74,915 entries have been used for prediction of their ^{13}C NMR spectra and the deviation between the experimental values and the predicted values for each spectrum was taken as a quality measure. Between the initial cycle of data entry versus the actual cycle of the same entry, an average improvement of 0.218 ppm per carbon was obtained solely on the basis of continuous correction, which consists of the implementation of structure revisions and/or reassignments brought to the attention of the author.

The outcome of a more detailed investigation using the same workflow procedure on a journal-by-journal basis for their relevant publications within the period from 2000 to 2010 is summarized in Table 1. It should be mentioned that not all entries from this timeframe have been extracted and stored in the "CSEARCH" database; data that are obviously incorrect were omitted. Also transmission errors contribute to deviation of the resultant data. However, in the experience of the author, transmission errors are considered to have a minor effect on the detected size of the deviation. Furthermore, these are good indicators of the quality and style of the overall presentation of the spectroscopic data in the literature.

From Table 1, it can be seen that the improvement to be obtained by ongoing data curation is in the range of 0.1 to 0.3 ppm per carbon, which appears to be quite minor. When taking into account that some 90% of published NMR data may be correct, and, among the remaining 10% of incorrect values, the deviations are usually concentrated on three to six carbon atoms per molecular structure, an average effect is obtained of approximately 5–10 ppm per carbon induced by the correction. This value is far beyond the quality of the prediction of carbon chemical shift values, which is usually below 2 ppm, and therefore should be detectable readily during NMR spectroscopic data-processing work and/or the preparation of a manuscript for publication.

Table 1 Comparison of journals in the field of natural product chemistry with respect to the obtained improvement of the published ^{13}C-NMR data by ongoing data curation

Journal	Number of entries	Number of carbons	Improvement per carbon $\Delta\delta_C$/ppm[a]
Phytochemistry	8312	207,166	0.132
Journal of Natural Products	18,078	459,955	0.087
Planta Medica (2003–2005)	756	18,516	0.160
Chemical and Pharmaceutical Bulletin	9657	243,589	0.094
Data published 2001–2005	79,891	1,505,084	0.295
Data published 2006–2010	55,134	968,939	0.122

[a] The improvement shown is the difference between the first and the actual cycle of the same entry

Table 2 Prediction of the chemical shift values of the four phenyl carbons in the 4-hydroxycinnamoyl moiety

Carbon	1	2,6	3,5	4
Predicted chemical shift value (δ_C /ppm)	126.9	131.0	116.5	160.5
Expectation range (δ_C /ppm)	121.8–133.8	116.7–131.4	113.2–131.0	141.5–163.9
Number of examples	362	766	759	766
Minimum value from	Compound 1 from [16]	Compound 1 from [16]	Compound 1 from [17]	Compound 4 from [18]
Maximum value from	Compound 1 from [19]	Compound 1 from [20]	Compound 1 from [16]	Compound 2 from [21]
Total range ($\Delta\delta_C$ /ppm)	12.0	14.7	17.8	22.4

Query used: 5-*O*-*p*-*trans*-coumaroylquinic acid [22, 23]; reference structures found are identical within a five-bond sphere

In order to make a case for well-assigned high-quality reference NMR data to be more visible, the chemical shift values of the phenyl carbons of the 4-hydroxycinnamoyl moiety may be addressed, as found in the database. This is a frequently occurring structural unit among many natural products. The 4-hydroxycinnamoyl unit is a well separated spin-system, which is easily predictable, with narrow expectation ranges for the four different carbon positions within the phenyl ring.

From Table 2 it can be seen that the mean values are in good agreement with the known experimental data of the query [22], but the expectation ranges (between 12.0 and 22.4 ppm) are far beyond the usual prediction quality. In order to get a reliable result for this prediction, only (*E*)-cinnamoyl units are included and measurements were only selected if they were performed in common NMR solvents in order to exclude any major solvent effects. A total of 727 entries out of nearly 700,000 in the "CSEARCH" database contributed to this analysis. It was found that there was a substantial inconsistency of the signal assignments even within this well-separated spin system. Several examples of assignment errors for the ^{13}C NMR data of cinnamic- and benzoic acid derivatives, as corrected within the "CSEARCH" database, are summarized in Table 3.

The errors in Table 3 are trivial in nature, and thus, because of the frequent occurrence of hydroxylated cinnamic acid moieties in natural product molecules, only a minor influence on the mean values of the prediction can be observed by these incorrect datasets. However, these wrong and therefore inconsistent assignments increase the expectation range for each signal, leading to major problems when using these data in ASV (Automatic Structure Verification) programs [32], where the expectation ranges of the chemical shift values are of major importance.

Table 3 Assignment errors found in the given citation for hydroxylated cinnamic and benzoic acid derivatives together with the result from the "CSEARCH-Robot-Referee" system before and after data correction

Ref., journal	Compound number in Ref.	Misassigned chemical shift values (δ_C /ppm)	Result of evaluation before and after correction	
[24] Chem Pharm Bull	21	Hexahydroxydiphenyl moiety misassigned	Major	Minor
[24] Chem Pharm Bull	24	Hexahydroxydiphenyl moiety misassigned	Major	Minor
[25] Fitoterapia	2	161.0, 130.1, 116.0	Minor	Minor
[25] Fitoterapia	3	148.8, 153.4, 123.7, 112.1, 110.2	Minor	Accept
[26] J Nat Prod	3	131.7, 117.8	Minor	Accept
[27] J Nat Prod	2	130.2, 116.0	Minor	Accept
[28] Phytochemistry	1	158.1, 130.1	Minor	Accept
[28] Phytochemistry	2	160.6, 130.5	Minor	Accept
[28] Phytochemistry	3	161.2, 131.2, 115.0	Major	Minor
[29] Phytochemistry	13	130.01, 115.14	Minor	Accept
[29] Phytochemistry	14	132.35, 115.96	Minor	Accept
[30] Phytochemistry	3	162.7, 149.0	Major	Accept
[30] Phytochemistry	6	161.2, 146.9	Major	Accept
[31] Phytochemistry	2	159.8, 115.5, 129.9, 125.8, 113.6, 145.8	Minor	Accept

Such errors are always detected, and data curation usually improves the result of the evaluation

2.3 *The Scope of the ^{13}C NMR Spectroscopic Data Reassignment/Structure Revision Problem*

Within the "CSEARCH" database, and also within the external data internally available within the "CSEARCH" environment, each NMR spectroscopic data re-assignment and/or compound structural change has been documented in the "remark" field of the entry. When scanning the 694,000 datasets for the term "reassign", this gives some 7100 entries with reassigned spectroscopic data, and more than 300 structures that have been revised. These nearly 7500 corrected reference assignments have originated from approximately 3500 (out of some 80,000) literature citations. It should be emphasized that the database is strongly biased, because of the selection of the articles involved. Also, any obviously erroneous data are not extracted from the literature, and a considerable amount of information has been deleted, because this was not possible to correct on the basis of published spectroscopic data. Every entry within the "CSEARCH" collection has been processed using the "CSEARCH-Robot-Referee" function, leading to the observation that 50% of the data may be classified as "Accept", 30% as "Minor Revision", 15% as "Major Revision", and 5% as "Reject". Altogether, about 140,000 spectra (approximately 20% of the total) are dubious to some extent.

This might involve incomplete spectroscopic data being used for characterizing a new compound, despite a particular entry being a valuable reference because of the uncommon functional groups present. Also, incorrect NMR spectroscopic assignments and erroneous structural proposals could be responsible for dubious data entries. From these numbers, it can be deduced that a large problem with the existing literature on the NMR spectroscopic assignments of natural products has been identified.

2.4 Unassigned ^{13}C NMR Data

Unassigned ^{13}C NMR chemical shift values, usually represented as a sorted peak list, occur extremely frequently in the chemical literature. In many cases, only the chemical shift values without any multiplicity information are given, making these data of no value for later use using a prediction software tool. The basic principle of a prediction tool is to deduce a chemical shift value from the connectivity table of the molecule being queried. Thus, a prediction tool aids in the understanding of the relationship between the structural properties of a specific carbon and its chemical shift value. This relationship is the exact signal assignment of a specific chemical shift value to a certain carbon atom, and, from a large collection of such assignments, rules can be proposed to predict the spectroscopic attributes of a given structure being queried. An unassigned peak list is representative of only a "fingerprint" of a chemical structure allowing only a similarity search, and missing multiplicity information increases the ambiguity of a similarity search making this dataset even less useful. In order to use a structure and its ^{13}C NMR spectroscopic data as reference information for assignment of the spectroscopic lines to the carbon positions, a basic rule is that a chemical shift value should be assigned to every carbon in the structure. Incompletely assigned spectra might be useful for prediction of the position of a particular carbon atom chemical shift in a structure being queried, but incomplete spectra are much less useful for performing similarity searches. An optimum situation is to have the same number of chemical shift values assigned as there are carbon atoms in a given natural product structure.

2.5 Strategies Used by the "CSEARCH-Robot-Referee"

The "CSEARCH-Robot-Referee" consists of three independent pieces of software, with the "Robot" component performing the evaluation of the structure and the given spectroscopic data, according to the usual approach used during peer-reviewing, resulting in a classification of "Accept", "Minor Revision", "Major Revision", or "Reject". Depending on the result and the user-specific setting, two further engines are incorporated, which are able to derive alternative structural proposals to the given peak list using different strategies. The "Dereplication Engine" performs a "Spectroscopic Similarity Search" using over 74,000,000 (as of May 2016) predicted ^{13}C NMR spectra calculated for the compounds listed in PubChem [33]. Where too many or too few entries are found, additional constraints like elemental composition, standard deviation, and the number of lines for the reference substance can be varied and the request "re-queued" automatically. The resulting "hit list" of possible structural candidates will be sorted by coincidence between the experimentally determined and the predicted chemical shift values. Furthermore, the best fitting structures can be analyzed for common ring skeletons and functional group substitution patterns, giving at least some information as to the compound class to be expected. This engine has already proven its excellent performance especially when dereplicating natural products and can be directly accessed from the TopSpin™ (Bruker BioSpin GmbH, Rheinstetten, Germany) family of programs [34] and from upcoming versions of the MNova (Mestrelab Research, Santiago de Compostela, Spain) software [35].

Another approach to derive alternative structure proposals is the application of the "Structure Generator" program using the compound structure proposal of the user as a starting point, together with the peak list and the positions where the experimental chemical shift values do not fit those predicted as derived from the evaluation. This approach is not an isomer generator program according to the mathematical definition of this task, since a further set of similar structures, having a different molecular formula, may be generated. Thus, a highly restricted subset of all possible isomers is created, and heteroatoms may be added, omitted, or exchanged in position giving also non-isomeric but similar structural proposals. The restrictions and the resulting gain of processing speed occur because only positions in the structure proposed undergo variation having large differences between an assigned chemical shift value and the predicted one. The superior quality of this integrated approach can be shown since nearly all obvious errors in the line assignments and/or wrong structural proposals are detected, and in many cases reasonable alternative structures can be derived, either using the "Spectroscopic Similarity Search" or the "Structure Generation" engines. The starting point is only the 1D ^{13}C NMR peak list as provided by the peak-picking routine, together with a reasonable structure proposal. Cross-peaks from relevant 2D NMR experiments are not necessary. Accordingly, data input from 2D NMR measurements is avoided, prohibiting the widespread use of isomer generation under routine applications, because the preparation of the data is too time-consuming. With this workflow protocol, the necessary input is a simple copy and pasting of the results from the peak-picking routine together with a structural proposal.

The workflow within the decision engine, which starts the "Spectroscopic Similarity Search" and the "Structure Generator" components on an optional basis, consists of the following steps:

- A check of the chemical structure, valences, and stereocenters;
- A summary of the given chemical shift values, separated into assigned lines, exchangeable assigned lines, and unassigned lines;
- An analysis of groups of exchangeable assigned lines, and checks for the total ranges for each group, the multiplicity within groups, and the overlap between groups of exchangeable assigned lines;
- An analysis of the experimentally determined multiplicities versus multiplicities calculated from the structure;
- The performance of a symmetry analysis starting from both the query structure and the spectroscopic data;
- The conduct of spectrum estimation using "Neural Network Technology" and the "HOSE (Hierarchical Organization of Spherical Environments)-code Technique" [36];
- The automatic assignment of lines that have been entered without any assignment;
- The performance of statistical analysis of the underlying reference data for each carbon prediction, with an alert when there is only reference information available with low similarity;
- A comparison of results from prediction using HOSE-code and NN-technology and the generation of a warning when inconsistencies are found;
- The evaluation of all statistical parameters and the derivation of a quality parameter for the prediction;
- An attempt to find a better possibility for given signal assignment, having a smaller difference between the experimental and prediction values;
- The use of spectroscopic data for a similarity search over all available "CSEARCH" databases and the provision of only known alternative structures that fit the same spectroscopic data;
- The utilization of a given query structure for a search for identical structures;
- The provision of all reference data together with their literature citation and the associated Digital Object Identifier (DOI) [37], if available;
- Calculation of the International Chemical Identifier (InChiKey) [38] from the query structure and comparison against the PubChem collection in order to give information about already known structures; this feature is also used during structure generation in order to find already existing alternative structures;
- The insertion of links into resulting tables to PubChem data [33] and to structures contained in eMolecules® (La Jolla, CA, USA) [39];
- Depending on the result of the evaluation and on the user-specific settings, launch of the "Spectroscopic Similarity Search" over 74 million predicted ^{13}C NMR spectra and/or the "Structure Generator" to obtain alternative structure proposals, and
- Optionally to donate evaluated spectroscopic data to the scientific community as additional reference information.

In addition, each user is able to create her/his user-specific database from earlier requests, which is a very useful feature when structural queries are processed, since these are not well-represented in the underlying "CSEARCH" databases. It should be mentioned that only highly verified and well-assigned spectra should be used as user-specific reference material. For security reasons, any request based on user-specific data will be processed twice. Thus, the first evaluation uses all available "CSEARCH" databases together with the user-specific database, while the second evaluation is based solely on the "CSEARCH" databases. A significant difference of the results of both evaluations has to be checked in detail by the user.

2.6 Limitations, Enhancements, and Synergies

The evaluation engine itself can handle organic compounds up to the inclusion of 99 carbon atoms presented. The result of the classification is expressed as "Accept as it is", "Minor Revision", "Major Revision", and "Reject", depending on the compatibility between the proposed structure and the ^{13}C NMR data provided. In order to characterize a compound, it is necessary to supply "n" assigned chemical shift values for a query molecule having "n" carbons. Any violation of this basic rule does not prohibit the evaluation, but the result will become less reliable from any missing or unassigned ^{13}C NMR spectroscopic line. In order to launch a "Spectroscopic Similarity Search", no further preliminary steps have to be fulfilled, whereas for starting the "Structure Generator" engine, it is necessary to follow the above-stated 1:1-rule between the number of carbons and the number of chemical shift values.

As a starting point for structure generation, only a given structural proposal and the measured 1D ^{13}C NMR data are necessary, thereby not requiring the input of additional information such as the correlation signals from a HMBC spectrum. In order to make the structure generation process more efficient, only very similar structures are created and these are ranked subsequently by the deviation between the experimental peak list and the predicted chemical shift values. From prior examples in the scientific literature, it has been shown for structure revisions that the revised structure may be a different molecular formula than that of the previously proposed structure, such as by elimination of water or from an oxidation reaction. Therefore, restricting the process of structure generation to isomer generation is far below the requirements of practical problem solving.

Another valuable tool available is a comparison of each processed structure, either being provided as a query or created during the structure generation, against the PubChem collection, which provides information of already known substances. This feature is helpful in natural product chemistry, where the same set of spectroscopic data may be misassigned to a new structure instead of to an already known substance.

Despite these sophisticated algorithms and the use of a high-quality database, the responsibility for a valid compound structure elucidation remains with the chemist performing the interpretation of the spectroscopic data on hand. These "CSEARCH" database tools can be very helpful in directing the natural product scientist to inconsistencies between a given structural proposal and the ^{13}C NMR data measured, and offers the possibility of mapping the data on the structural formula to display molecular fragments, which may have to be revised. By integrating this scheme into a daily laboratory routine and as part of the writing process for manuscripts [40], a large majority of either trivial assignment errors or wrong structures could be avoided. A combination of this automated protocol and other methods of compound structure elucidation is still necessary, because very sophisticated problems such as the configurational reassignments conducted for molecules like brevenal [41] or a series of solandelactones [42], are somewhat beyond both the scope and the possibilities of this comparison between structure and ^{13}C NMR chemical shift data.

2.7 *Performance of the "CSEARCH-Robot-Referee"*

The following examples are taken from previously mentioned reviews [2, 3, 7] and are intended to show the performance of the "CSEARCH-Robot-Referee". It is worth mentioning that in many cases an incorrect structure as given in a literature report is contained within the underlying knowledge base used for structural prediction. Correction of data is only possible when a given revision of structure is known, whereas still uncorrected structures will remain among the relevant background information to be considered. Despite this, only a very minor influence on the ultimate structural proposal may occur.

The examples compiled in Fig. 1 are known structure revisions showing the past performance of the "CSEARCH-Robot-Referee" in detecting the inconsistencies between a given structure and the available experimental ^{13}C NMR chemical shift values. In many cases, a correct alternative structure proposal can be made and this is usually ranked as more likely than the original one.

3 Examples Showing Typical Types of Errors

In this section, typical types of errors are described that may occur in the natural products literature. These errors consist of trivial typographical errors, adding or omitting chemical shift values, performing trivial assignment errors based on misunderstood electron densities, using identical experimental spectroscopic data for elucidating different structures without mentioning a structural revision, and using misinterpreted experimental data leading to incorrect structural proposals. The most common sources of errors in the natural products literature available in the public domain have been based on:

- Typographical errors in tables containing ^{13}C NMR chemical shift values;
- Exchanging columns in tables presenting chemical shift values;

Original Structure	Revised Structure
Position: 209, $\Delta\delta_C = 4.65$ ppm CID: 25016148, "Minor Revision" 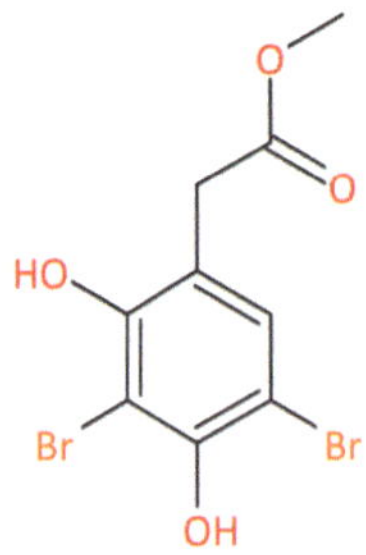 	Position: 7 of 1257, $\Delta\delta_C = 2.00$ ppm CID: 53770006 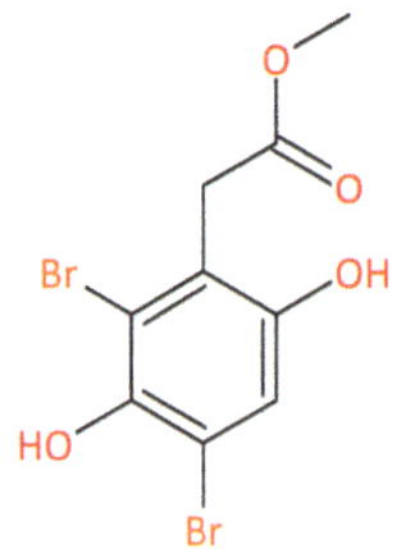
Position: 240, $\Delta\delta_C = 4.48$ ppm CID: 25016149, "Minor Revision" 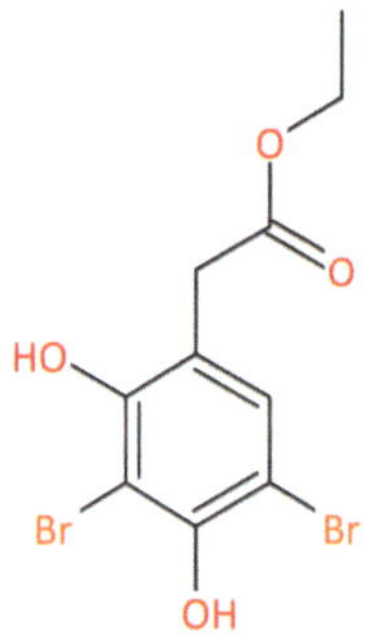 	Position: 2 of 1321, $\Delta\delta_C = 1.62$ ppm CID: unknown 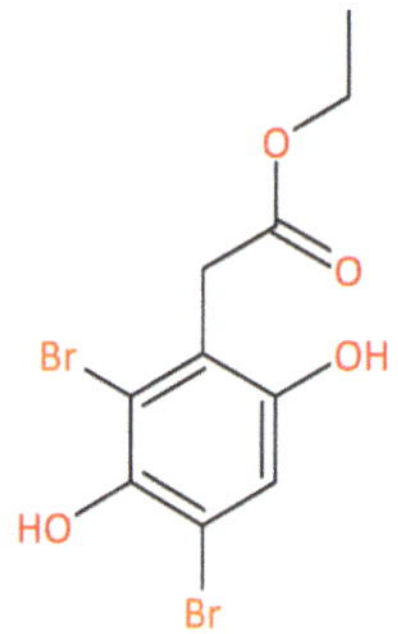
Position: 88, $\Delta\delta_C = 3.87$ ppm CID: 21589730, "Major Revision" 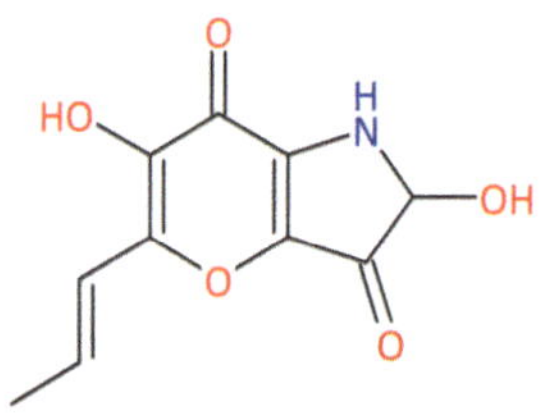 	Position: 93 of 1993, $\Delta\delta_C = 3.90$ ppm CID: 76764638 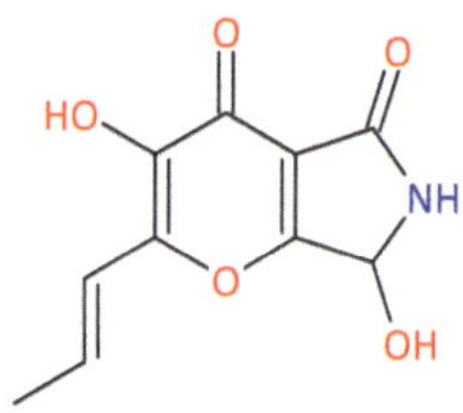

Fig. 1 Known structure revisions [2, 3, 7] used for verifying the performance of the "CSEARCH-Robot-Referee" system. The structures are given together with their position within the ranked list of structure proposals, their similarity measure, the result of the evaluation and the "Compound-Identifier" (CID), when the structure exists in the PubChem collection

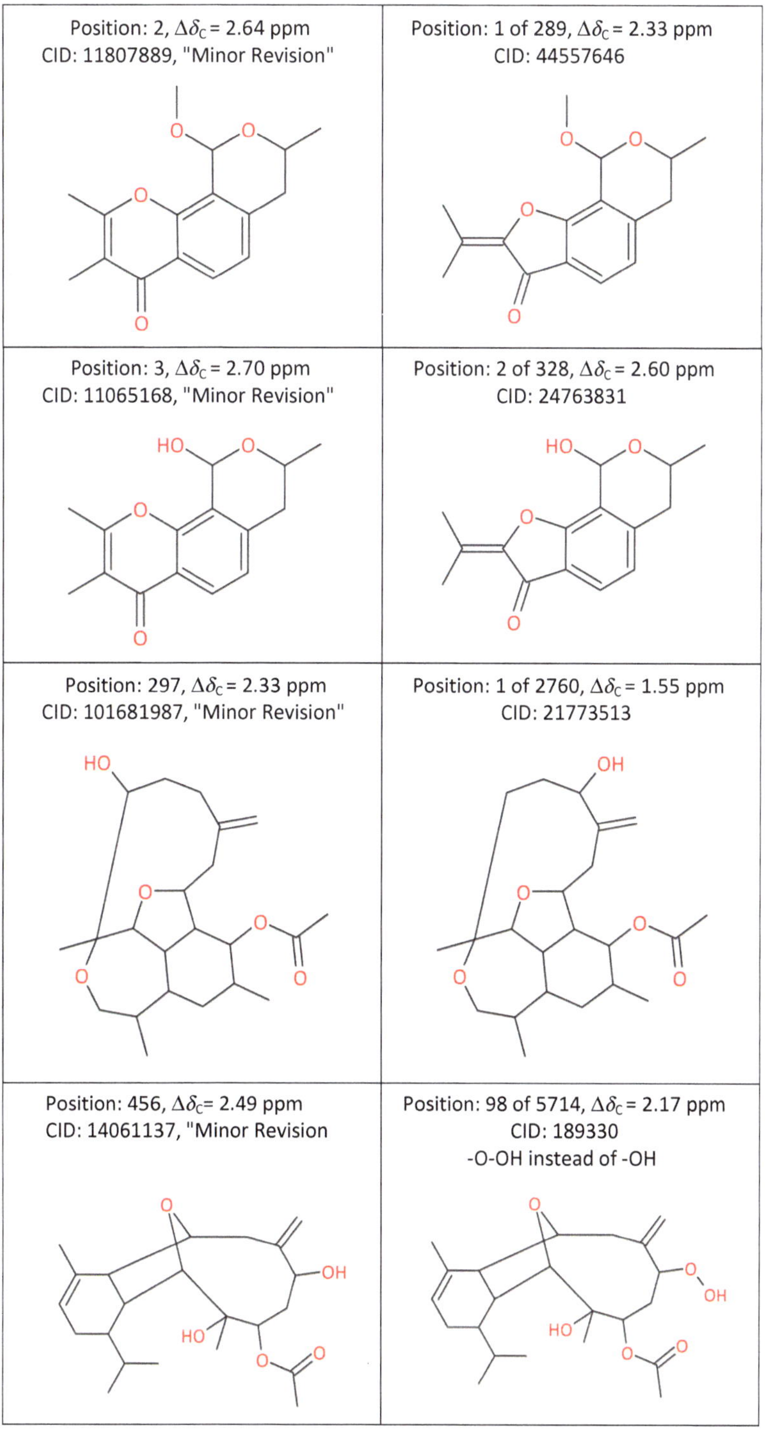

Fig. 1 (continued)

Position: 1200, $\Delta\delta_C$ = 3.11 ppm CID: 10457858, "Minor Revision" 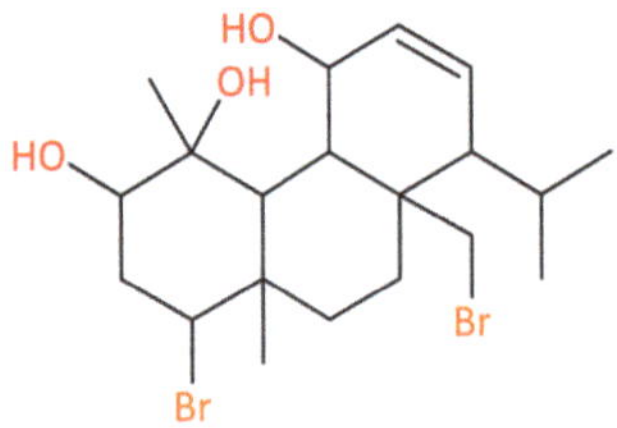	Position: 525 of 3244, $\Delta\delta_C$ = 2.72 ppm CID: 24850617 -O-OH instead of -OH 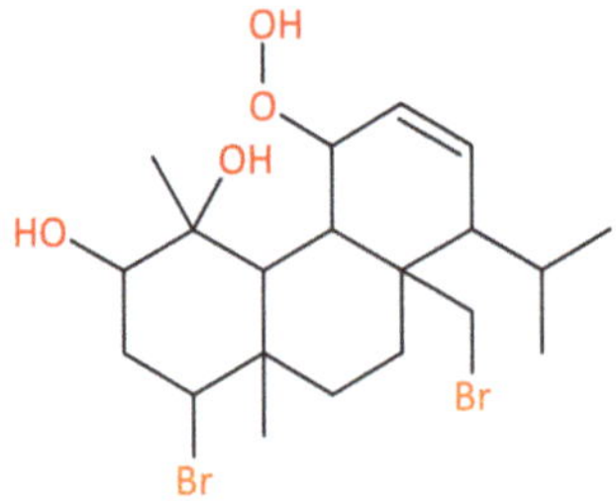
Position: 913, $\Delta\delta_C$ = 3.09 ppm CID: 21635112, "Minor Revision" 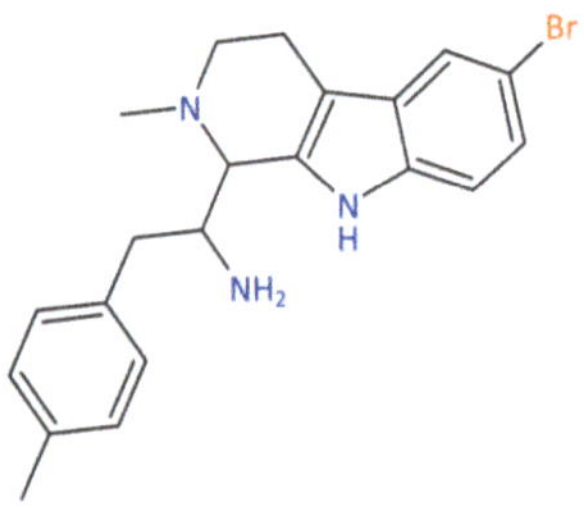	Position: 508 of 5349, $\Delta\delta_C$ = 2.71 ppm CID: unknown 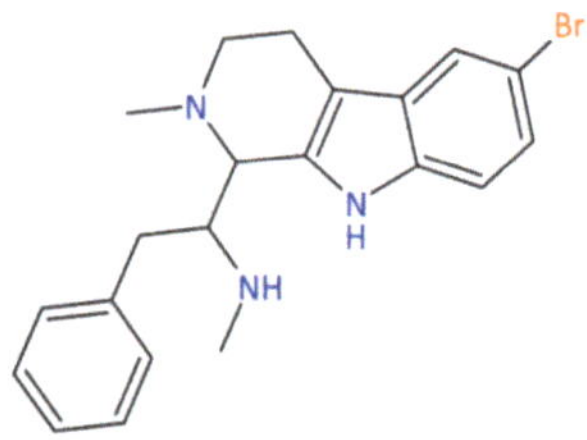
Position: 4853, $\Delta\delta_C$ = 6.23 ppm CID: 76764482, "Major Revision" 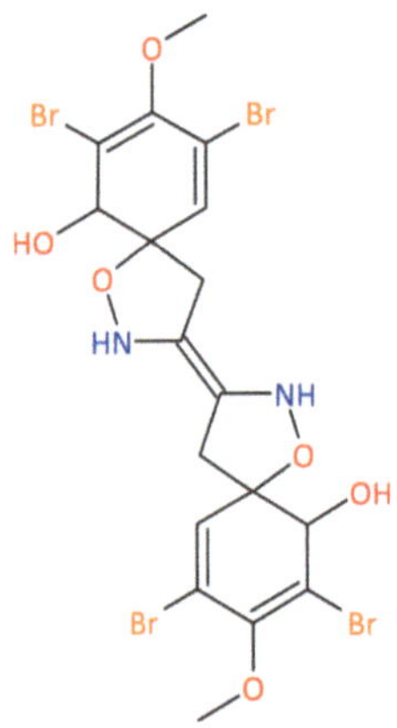	Position: 1053 of 7955, $\Delta\delta_C$= 4.67 ppm CID: 21778408 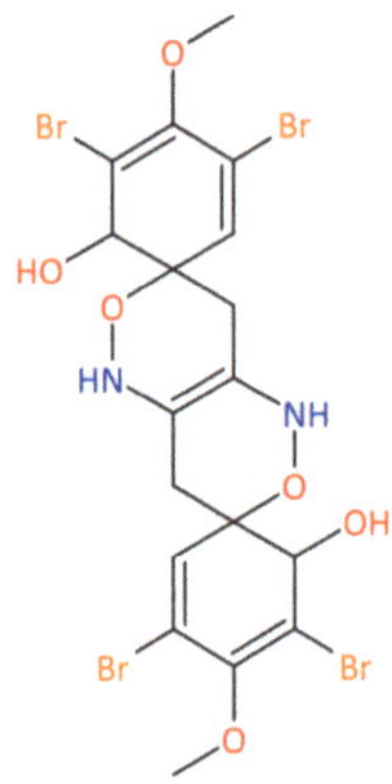

Fig. 1 (continued)

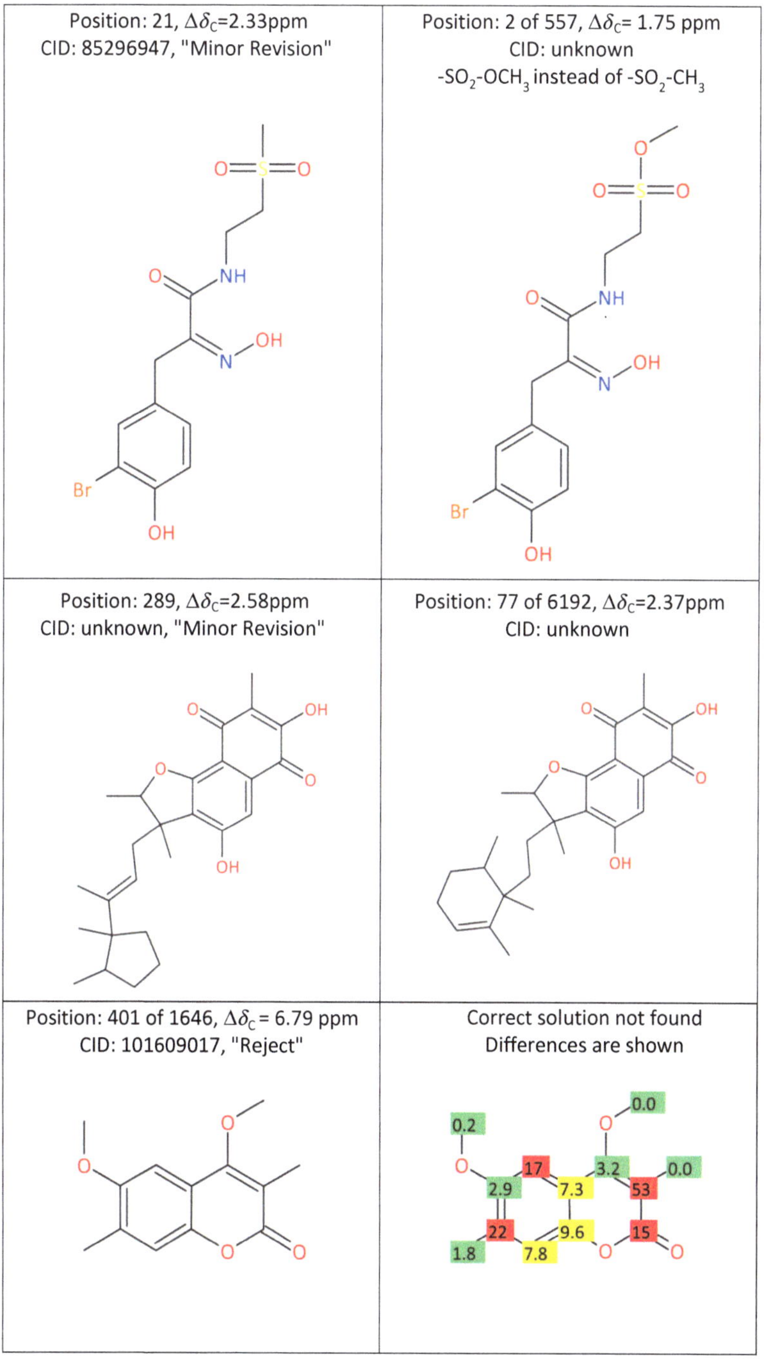

Fig. 1 (continued)

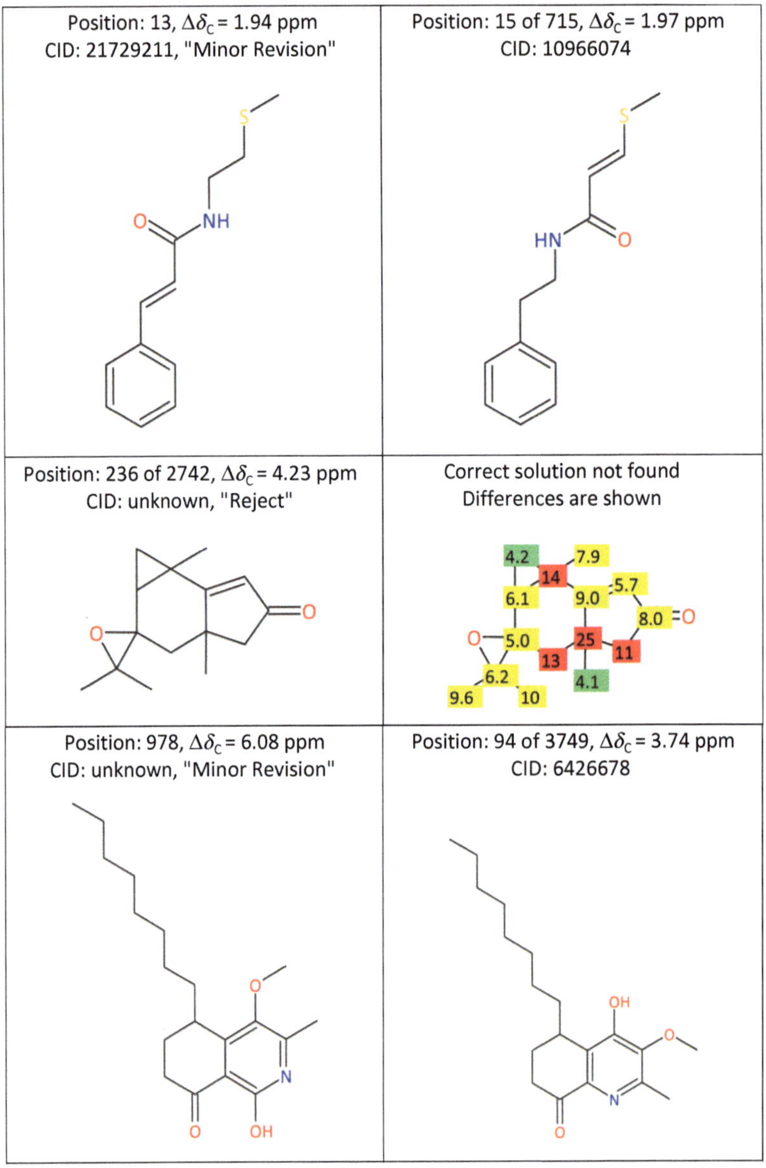

Fig. 1 (continued)

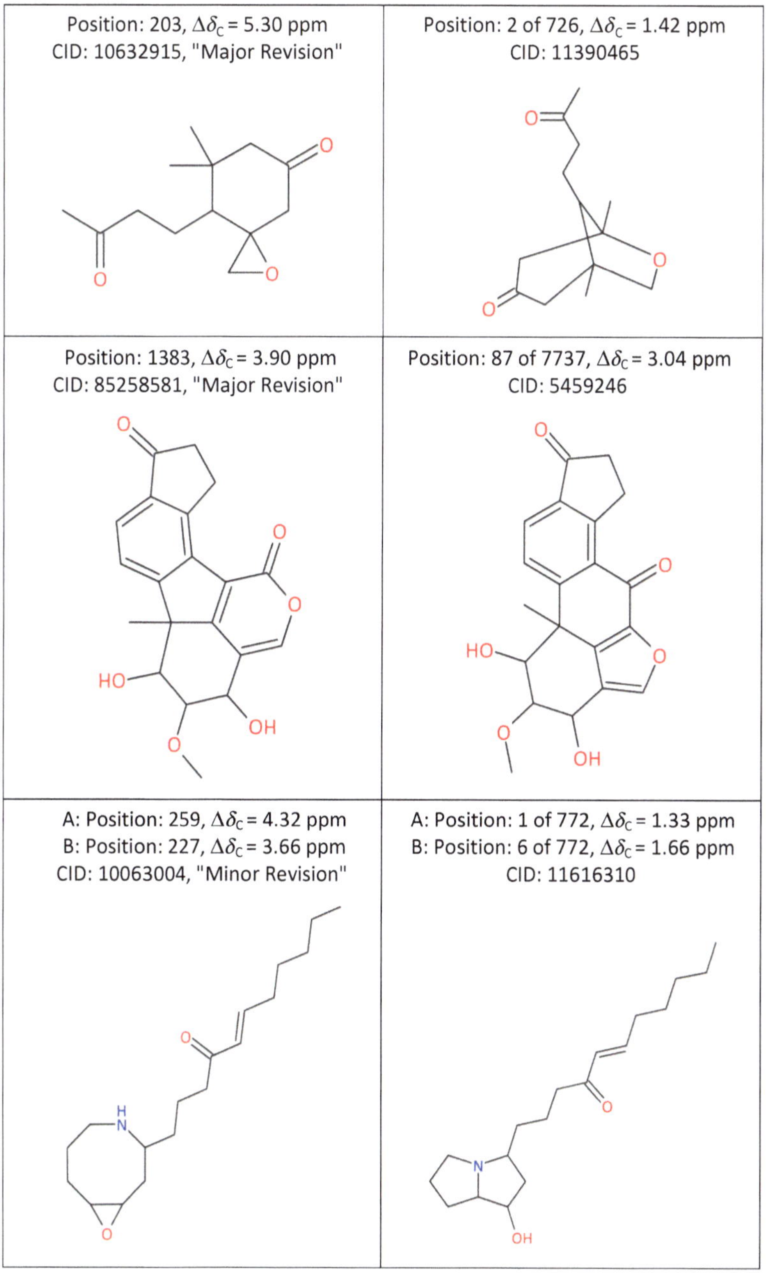

Fig. 1 (continued)

- Errors in drawing a chemical structure;
- Inconsistent numbering used in the structure proposed and in the table summarizing NMR chemical shift values;
- Assuming that lines observed in a ^{13}C NMR spectrum belong to impurities;
- Assigning lines in a ^{13}C NMR spectrum to a compound under investigation that actually belong to impurities and/or the solvent;
- Using the same spectroscopic data twice in order to "hide" a structure revision;
- Ignoring basic knowledge and hence leading to misassignments, such as misassigned enone fragments;
- Misinterpretation of 2D NMR spectroscopic data, with a typical example referring to the interpretation of cross-peaks obtained from a HMBC-type spectrum (e.g., a ^{3}J correlation versus a ^{4}J correlation).

3.1 NMR Solvent Errors

Taking an example from the chemical synthesis literature, a tamoxifen derivative was generated by Vallaint et al. [43]. The intermediate compound 8 (Fig. 2) in this paper was characterized by its ^{1}H and ^{13}C NMR, and mass spectrometric data.

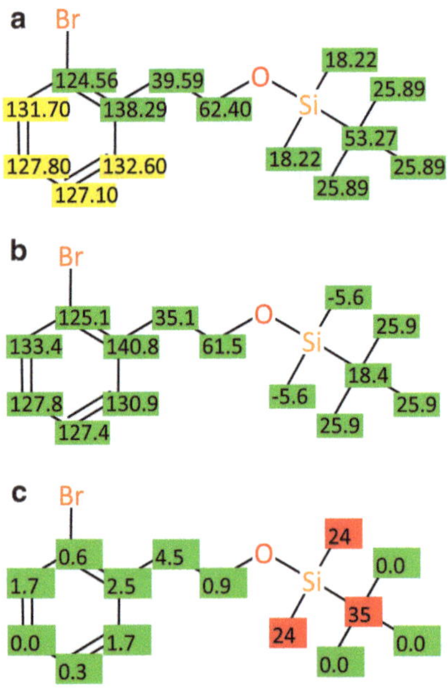

Fig. 2 (**a**) ^{13}C NMR data as given in [43]. The chemical shift values for the phenyl ring have only been partly assigned, and (**b**) predicted ^{13}C NMR chemical shift values. (**c**) Differences between experimentally obtained and predicted chemical shift values showing extremely large deviations within the protecting group

When starting the "CSEARCH-Robot-Referee" from the proposed structure together with the partially assigned carbon peak list, the resulting classification would be "Major Revision".

In Fig. 2a, the original data are visualized together with in Fig. 2b the predicted chemical shift values. The large differences located at three positions within the protecting group are displayed in Fig. 2c. From these data, conclusions can be drawn that the authors have ignored a signal in the range between –4 and –6 ppm and the line at 18.22 ppm should be assigned to the quaternary carbon within the *tert*-butyl moiety. For this quaternary carbon attached to a silicone atom, a value of 53.27 ppm is given, which is understandable when reading the experimental procedure in detail, since the reaction and the subsequent chromatography were both performed either in pure dichloromethane or in a petroleum ether/ dichloromethane mixture with subsequent evaporation. Evidently, a minor amount of dichloromethane remaining in the sample was responsible for the signal at 53.27 ppm, which was misinterpreted as a signal belonging to the quaternary carbon of the protecting group of compound 8 in [43].

3.2 *Ignoring Existing Chemical Shift Values*

Gustafsson, Saxin, and Kihlberg reported in 2003 the synthesis of a *C*-glycoside analogue of *β*-D-galactosylthreonine [44]. Intermediate 7 described in this publication contains a –CH_2-I group with an expected ^{13}C NMR chemical shift value of around 15 ppm. In the supplementary information for this article, the ^{13}C NMR spectrum of compound 7 is included, showing clearly on page S15 a signal around 18 ppm. In the range of 35 to 40 ppm, three more signals were visible. On page S16, the spectroscopic data were summarized, assigning these three signals to the side-chain, but with a value of 75.3 ppm to the –CH_2-I moiety. The signal at 18 ppm, which is clearly visible in the spectrum and definitely belonged to the compound in question, is not mentioned in the data sheet of this compound, leading to a final classification by the "CSEARCH-Robot-Referee" for this dataset of "Major Revision". Even when substituting the chemical shift value of 75.3 ppm by the experimental value of approximately 18 ppm, this dataset would be still classified as "Major Revision", because a compound having 40 carbon atoms in its structure obviously would need 40 ^{13}C NMR chemical shift values (although some of them may be identical because of symmetry), in order to follow the 1:1-relationship between the number of carbons and the number of chemical shift values. From the summary on page S16 it can be seen that the chemical shift values of the protecting groups are not tabulated. This example shows that the result of the classification is also determined by the large number of missing signals. In any case, the extreme difference between the predicted values and the experimental data as published is evident, showing an incompatibility between the structure proposed and the spectroscopic data published; in this case an existing peak clearly visible in the spectrum was simply ignored!

3.3 A Case in Which Two Different Structures Were Proposed from the Same ^{13}C NMR Spectroscopic Data

1,4-Dihydropyridine functionalities occur in natural products. Their derivatives are of pharmaceutical interest and therefore represent a well investigated class of compounds. In a paper by Morales et al. [45] the presentation of compound 6c is quite inconsistent; in the structure scheme on page 103 the aryl group was defined as "Ar = 2-Cl-C_6H_4", whereas in the figure on page 104 compound 6c had a nitro group at position C-3′. In the experimental part, the compound name is given as 4-(3-nitrophenyl)-5-oxo-1,2,3,4,5,7-hexahydrofuro[3,4-*b*]-2(1*H*)-pyridone, and also the ^{1}H and ^{13}C NMR data pointed to a C-3′ nitro derivative. Furthermore, characteristic IR data of a nitro group supported this structural proposal. From these observations, a typographical error in the structure scheme can be assumed with high probability. This error in the assignment of compound identifiers to the different substituents in the structure scheme was verified by an examination of the synthesis procedures applied. The synthesis of compound 6c started either from precursor 4c or 4f. Both compounds 4c and 4f were defined as 2-chlorophenyl derivatives according to the structure scheme and were named as 3-nitrophenyl derivatives in the experimental portion. The experimental data given for 4f supported presence of a 3-nitrophenyl moiety.

Using the ^{13}C NMR peak list given for synthetic compound 6c [45], which consisted of 13 different chemical shift values, seven out of these 13 peaks were assigned to a hexahydrofuropyridone unit, but the remaining six peaks belonging to a 3-nitrophenyl unit were unassigned. Evaluation with the "CSEARCH-Robot-Referee" gave a result of "Minor Revision", with the six unassigned lines assigned automatically during the evaluation based on the predicted chemical shift values.

During an evaluation, the structure assigned to a compound may be used as a query for a search for identical structures within the available "CSEARCH" databases. The results of such a search have shown that the same structure has been published in [46] by the same collaborative groups. This is a quite usual finding when the data from a previously performed and already published measurement are used as reference information in a subsequent publication. The great advantage of a digital repository is that one can cross-check between different datasets to search for alternative structures having the same spectroscopic pattern (see also Sect. 4.5 for further description of this approach). Surprisingly, two alternative compounds, IIIc and IIId from [46] have almost identical spectroscopic data, with one being an exact spectroscopic match and the other one differing by less than 0.4 ppm in the aryl ring despite the structural modification. In order to get greater resolution of this problem the chemical shift values of these compounds are visualized in Fig. 3.

The spectroscopic data of the 3-nitro derivative 6c from [45] was used as reference information for compound IIIb in [46]. Interestingly, the 4-nitro derivative (compound IIIc from [46]) was found to have the identical spectroscopic data, while variations of 0.1, 0.2, and 0.4 ppm occurred in the 4-methyl derivative IIId

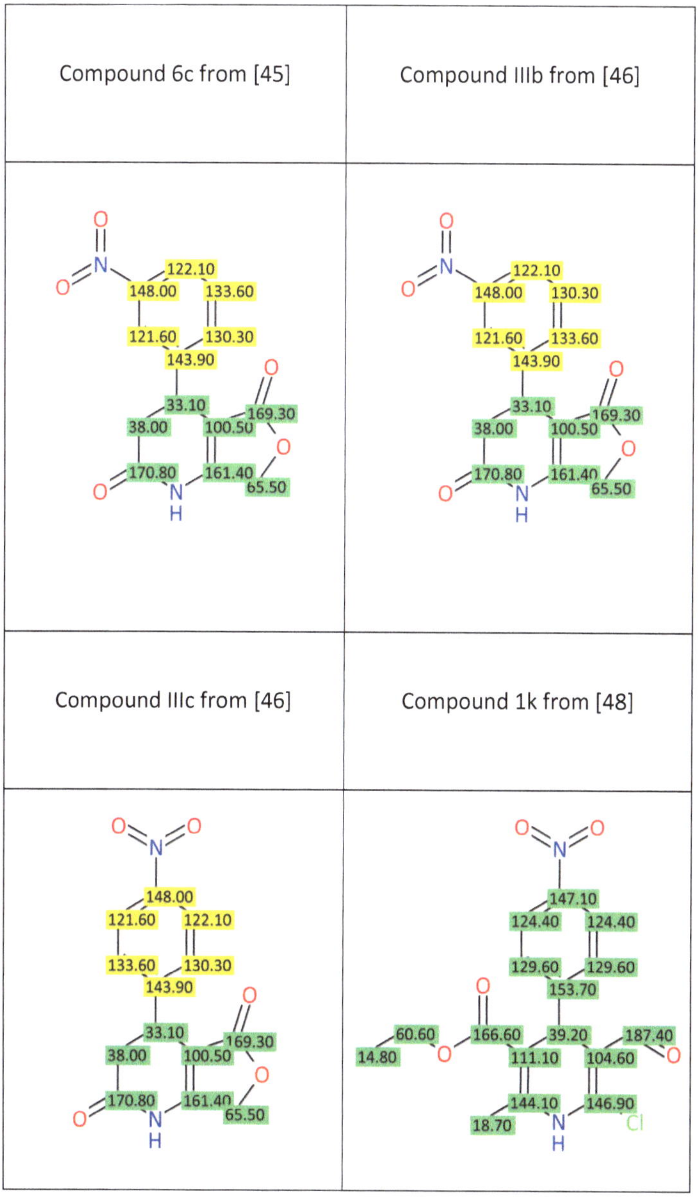

Fig. 3 Compounds 6c, IIIb, and IIIc have identical carbon spectroscopic data, but represent two different structures. Compound IIId is identical with IIIc within 0.4 ppm despite the different influence of a nitro and a methyl group on the phenyl system. The data of the 4-nitrophenyl portion in IIIc differ markedly from the data of 1k. Both compounds IIIh and 1i have a 3-methoxyphenyl moiety showing large shift differences at equivalent positions. The unsubstituted compound IIIg is given for comparison

Compound IIId from [46]

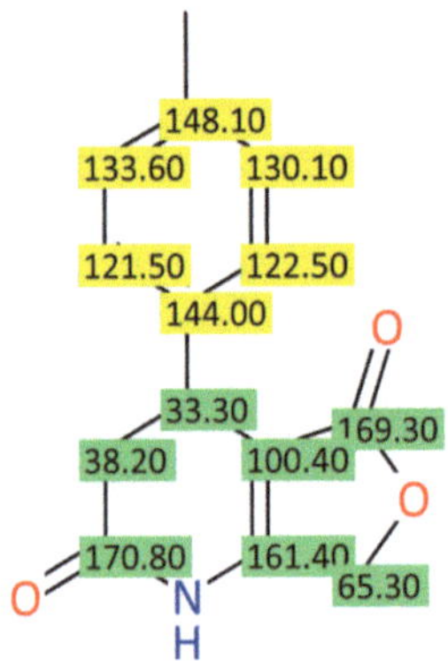

Compound IIIg from [46]

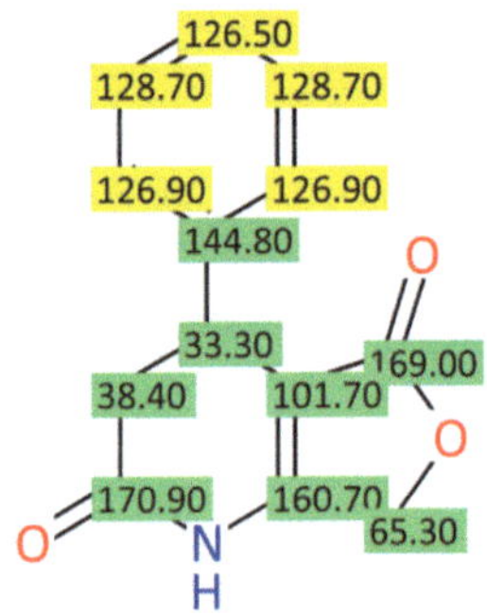

Compound IIIh from [46]

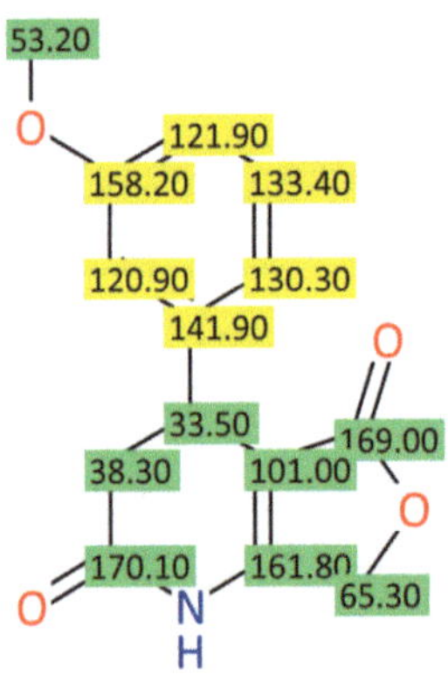

Compound 1i from [48]

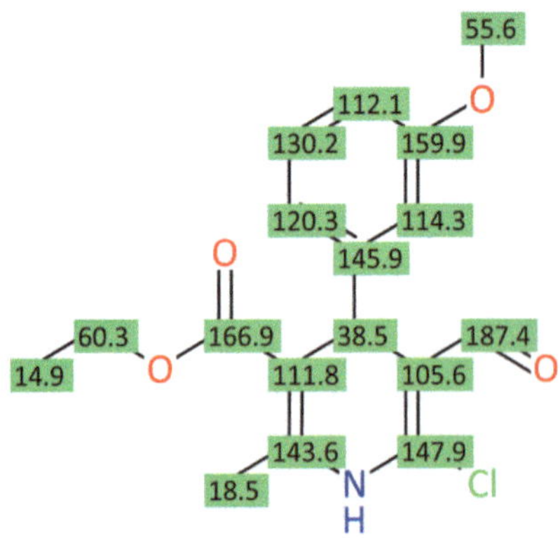

Fig. 3 (continued)

[46]). In turn, the aryl protons appeared in IIIc in the range from 7.42 to 8.15 ppm, whereas the methyl substituted analogue IIId [46] showed ^{1}H NMR signals at about the same range between 7.32 and 8.12 ppm, despite the influence of a nitro substituent and a methyl group being quite different on neighboring aryl protons. The ^{1}H and the ^{13}C NMR chemical shift values for the methyl group were omitted in the experimental portion of [46]. However, the protons of the unsubstituted phenyl ring in compound IIIg appeared in the usual range between 7.18 and 7.35 ppm, but instead of five protons only four protons were present. It was shown that there was no undue influence from any other moiety on the ^{1}H NMR chemical shift values for this type of compound. There were four signals in the carbon NMR spectrum as to be expected for a freely rotatable phenyl substituent. A comparison of these data with the 3-methoxy derivative showed large deviations between the experimental and predicted ^{13}C NMR chemical shift values. Furthermore, the unsubstituted phenyl system ^{1}H NMR proton resonances were in the usual range 7.18–7.35 ppm whereas the aryl protons of the 3-methoxy derivative, IIIh, were given within a range of 7.43–8.12 ppm [46]. However, this is opposite to that expected, as a result of the shielding effect in the *ortho-*, *meta-*, and *para-*positions induced by the methoxy group, making a chemical shift value of 8.12 ppm not credible. The predicted chemical shift values for the ^{13}C NMR spectrum could be found searching CAS REGISTRY^{SM} using SciFinder® [9, 47]. Experimentally, the ^{1}H NMR chemical shift values of the 3-methoxyphenyl moiety in IIIh ranged from 7.43 to 8.12 ppm [46], whereas the same unit for compound 1i of [48] occurred between 6.66 and 7.15 ppm. The protons of 3-methoxyphenyl derivatives would experience shift influences of up to nearly 1 ppm, whereas the protons in the unsubstituted analogues would remain nearly unchanged (7.18–7.35 ppm [46], 7.13–7.23 ppm [48]. Both papers were published by the same collaborative groups from the University of Havana (Havana, Cuba) and the Complutense University of Madrid (Madrid, Spain) [46, 48].

The application of the "CSEARCH-Robot-Referee" to verify the dataset of the synthesized compound 6c in [45] published in 1996 revealed that this peak list has been used in a later publication as reference information and assigned to the same compound. Furthermore, the identical spectrum was used as a "proof" for a different structure IIIc [46] and a slightly modified (maximum 0.4 ppm deviation) peak list was used to identify another substance, IIId [46]. A detailed comparison of this paper with later references showed also major problems with the ^{1}H NMR data used for structural verification after manual inspection. The starting point for this analysis was an automated "peer-reviewing" of one dataset of interest, and a major inconsistency was uncovered within a series of publications published by the same collaborative groups that did not mention the errors in their earlier papers.

Another example of identical spectroscopic data being assigned to two different synthesized compounds was found when analyzing the ^{13}C NMR data published in [49]. In this article, a chromene derivative (6d), was described and characterized from its spectroscopic data. An assignment of the ^{13}C NMR chemical shift values was not given in the experimental part, with only a multiplicity determination being reported. The evaluation of this entry with the "CSEARCH-Robot-Referee" gave a classification of "Minor Revision" because of the missing assignment. Moreover, an automatically performed search for an identical structure with respect to the two-dimensional topology revealed three other references [50–52] dealing with the same compound, whereas a search for an alternative structure starting from the peak list gave no result. When inspecting the spectra shown for the same compound in these four literature citations, it was seen that the chemical shift values of compound 6d in [49] were different from the other reference spectra found. The three reference spectra, for compound 1a of [50], compound 6a of [51], and compound 4a of [52], were identical, so therefore only one comparison is shown in Fig. 4. When taking the dataset of compound 1a of [50] having the identical structure, but having a different peak list as starting point for another evaluation, the same structure is found again, as to be expected, but also one alternative structure proposal, compound 9a from [53], was retrieved, as shown in Fig. 5.

A detailed analysis of the five publications [49–53] used during this analysis showed that ten synthesized derivatives containing nitrogen were published as compounds 9a–9j in [53] together with their carbon spectroscopic data. Several of these compounds also appeared in a subsequent publication [50] among a series of analogs (2a to 2g), but having different spectroscopic data. The ten compounds named 9a to 9j in [53] are heterocycles containing nitrogen, which were modified as oxygenated analogs in series 1a to 1j in [50], but having identical reported

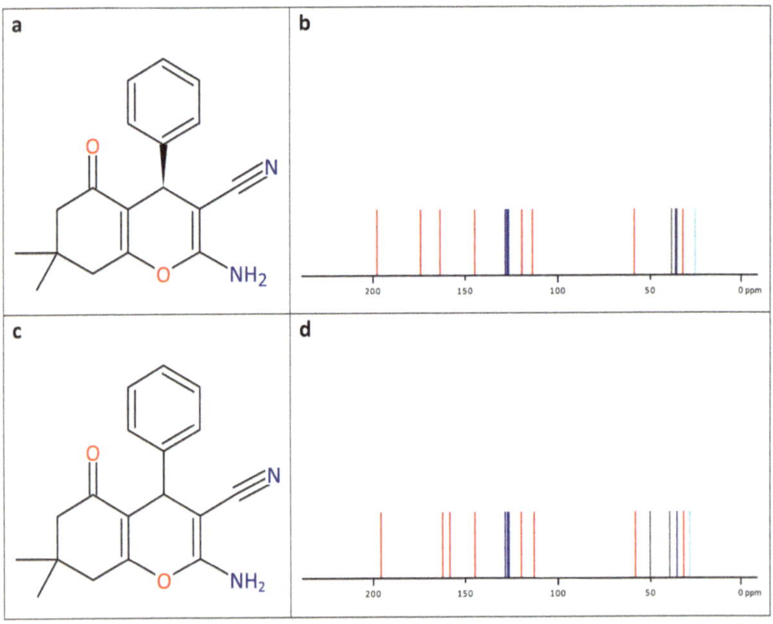

Fig. 4 Identical compounds (with respect to their two-dimensional topology) have been characterized by different spectroscopic patterns. Compound 6d (**a**) from [49]. (**b**) The published ^{13}C NMR data. (**c**) Compound 1a from [50]. (**d**) The published ^{13}C NMR data, which are contained identically in Refs. [50–52]

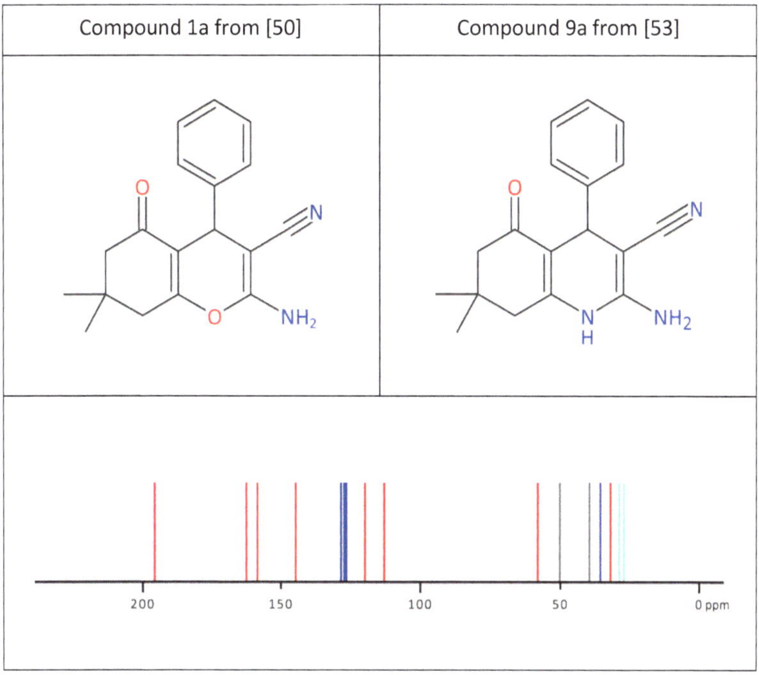

Fig. 5 Two different structures given with identical ^{13}C NMR spectroscopic data

Table 4 Timeline of the two publications where identical spectra were assigned to two different structures

Ref.	Received	Accepted	Appeared	Structure proposal
[53]	18th November 1999	Unknown date	Issue 4, July/August 2000	Hexahydroquinoline derivative
[50]	28th August 2000	3rd November 2000	Issue 2, February 2001	Pyran derivative

spectroscopic data. The timeframe when [50] and [53] were published (Table 4) is worthy of comment. In the later publication [50] the previous publication [53] was cited as "J. Heterocycl. Chem., in press" despite there being plenty of time (at least from August to November 2000) to insert the final literature citation. The citation points only to the previous paper with respect to the preparation of the compounds. In fact, a major structural revision on a complete series of compounds was performed, but this is not mentioned [50].

3.4 *Trivial Typographical Errors*

Kaiser, Basso, and Ritter [54] published a comprehensive set of reference data for 22 different C-3 substituted camphor derivatives. Twenty out of these 22 reference

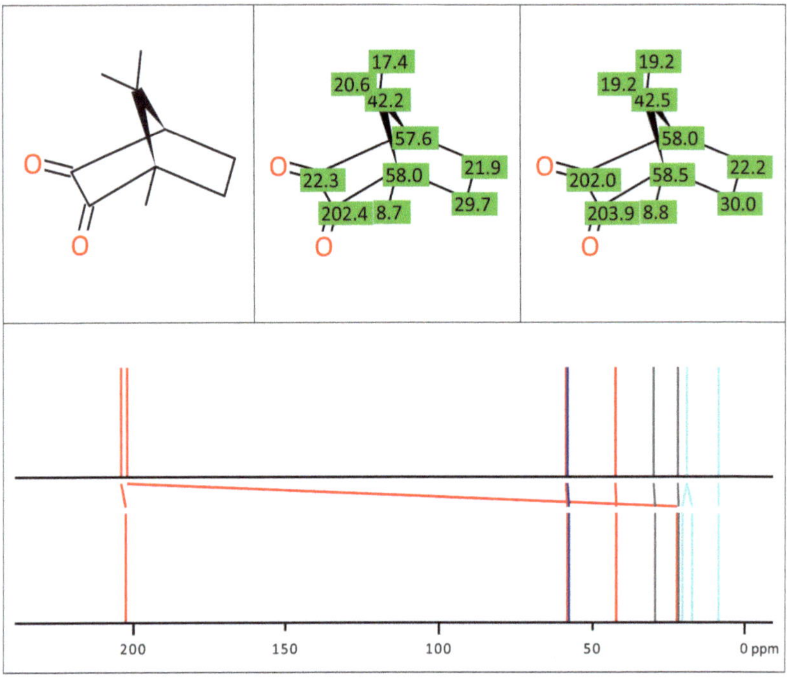

Fig. 6 Chemical shift values (middle) calculated from the given increments [54] versus predicted chemical shift values (right) inserted into the structure display; the visualization of the differences shows an extremely large increment caused by a typographical error

datasets were classified as "Accept" by the "CSEARCH-Robot-Referee". However, the data of the 3-dimethylamino derivative of camphor were classified as "Minor Revision", because of a small deviation between the predicted and experimental values at two positions. Among the data tabulated for bornane-2,3-dione, an incorrect chemical shift increment for C-3 of –20.3 ppm with respect to a basic value of 42.6 ppm is given, leading to a chemical shift value of 22.3 ppm for this carbonyl group [54]. The predicted value is somewhere around 202 ppm showing a difference of 179 ppm between the predicted and reported reference data. This is undoubtedly a typographical error in a large table holding some 220 increments. The example displayed in Fig. 6 shows the advisability of modifying the approach used in the process of publishing scientific results with two different strategies: the first one is to make use of the existing technologies like spectrum prediction and the second one is to generate verified electronic files holding data, which are then used during the whole workflow starting at the NMR instrument, where the data are produced, and ending at the stage of printing of the publication. For this purpose, a simple annotated SD-file would be sufficient to contain both the chemical structure along with the assignments of the ^{13}C NMR resonance peaks.

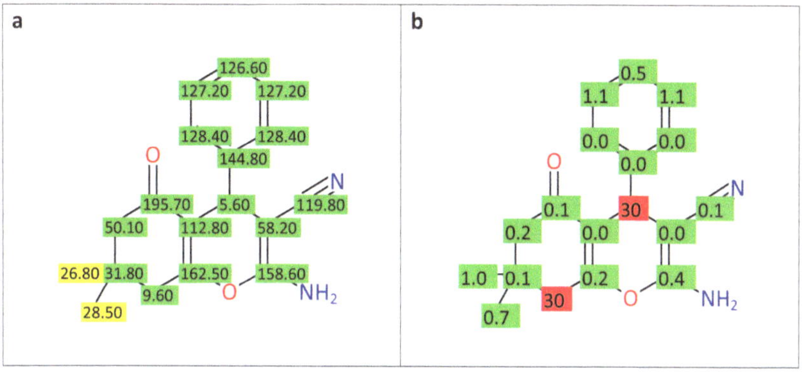

Fig. 7 (**a**) Carbon-13 NMR chemical shift values/ppm for compound 4a from [52]. (**b**) Differences between experimentally obtained and predicted chemical shift values clearly showing two typographical errors

Another typographical error can be found in [52] when analyzing the spectroscopic data of the synthesized compound 4. The "CSEARCH-Robot-Referee" detected this problem and recommended "Major Revision". In Fig. 7, a graphical representation of this typing error is given.

3.5 Assignment Errors

A quite different category of error is based on missing information on the basic effects on chemical shift values induced by different substituents. A few examples were given in Table 3; here a more detailed analysis of the nature and size of the problem will be given.

An example from [55] that is featured in Fig. 8 represents a typical assignment error occurring frequently in the chemical literature. From a standard compilation of spectroscopic data [56], it is known that a hydroxy group attached to a phenyl ring has a shielding effect at the *ortho-* and *para-* positions, whereas the *meta-* position is slightly deshielded. This effect can be easily seen when comparing, for example, the chemical shift data from benzene versus phenol, with a comprehensive summary of ^{13}C NMR substituent effects in monosubstituted benzene derivatives having been published by Ewing [57].

In order to have a more general impression of the overall problem of incorrectly assigned ^{13}C NMR data, all datasets within the "CSEARCH" databases published in the years 2010–2015 have been analyzed. A total of 31,297 datasets published within this period of time were extracted from the public domain literature, and a search for reassigned data revealed that the assignments of

Original assignment as given in [55]	Difference between the original data and the predicted chemical shift values	Corrected assignment
"Minor Revision"		"Accept as it is"

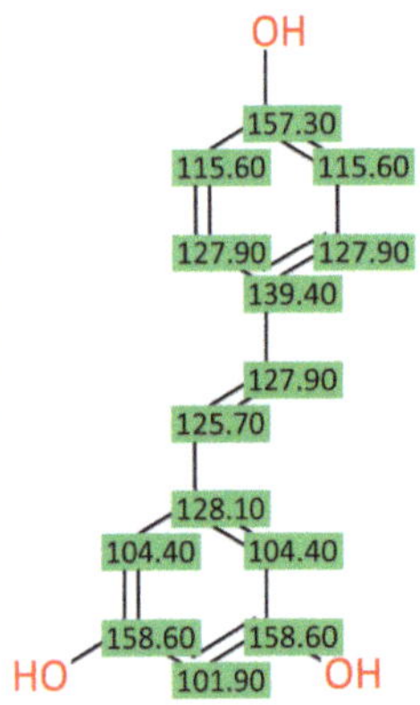

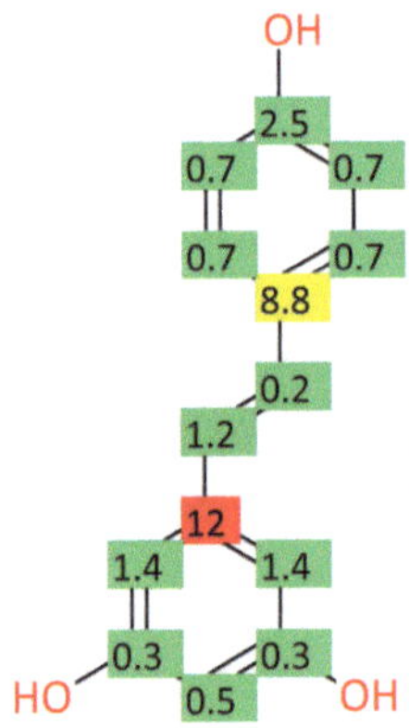

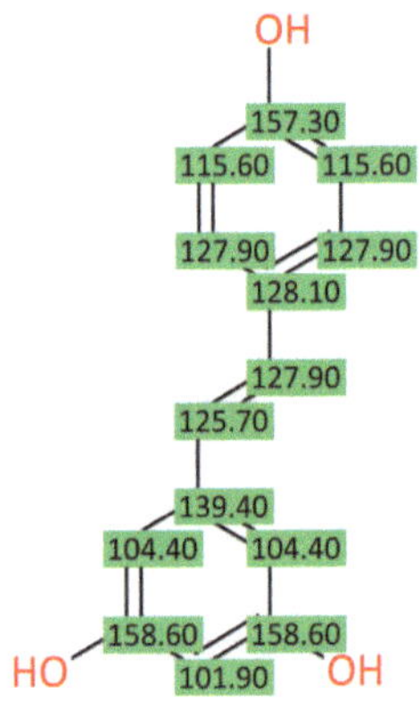

Fig. 8 (*E*)-Resveratrol as given in [55] with wrongly assigned *ipso*-carbons in both phenyl rings. This misassignment was detected readily by the "CSEARCH-Robot-Referee" system and the result of the evaluation changed from "Minor Revision" to "Accept", when correcting this error

540 entries in 286 different literature citations were revised subsequently, with each revision was documented in the "remark field" within the database. In Table 5, the literature citation, the number of the compound in the publication and a short description of the assignment error are given. The last two columns contain the results of the evaluation of the original version of the dataset with the exact assignment as given in the reference, and the result of the evaluation after the correction of the assignment error described. This table gives only a rather incomplete impression of the problem, because the data selection is restricted to a six-year period of chemical publication. Furthermore, it should be mentioned that several important journals in the field of natural product chemistry such as *Phytochemistry* and *Planta Medica* were not processed by the "CSEARCH" databases between 2010 and 2015.

Table 5 Summary of some assignment errors found by application of the "CSEARCH-Robot-Referee"

Ref., journal	Compound number in Ref.	Misassigned chemical shift values (δ_C /ppm)	Result of evaluation before and after correction	
[58] Chem Pharm Bull	1	158.4, 131.6, 129.3, 116.1	Major	Accept
[58] Chem Pharm Bull	2	158.5, 131.5, 129.4, 116.2	Major	Accept
[59] Chem Pharm Bull	1	118.0, 146.0	Major	Accept
[59] Chem Pharm Bull	2	117.8, 146.2	Major	Accept
[59] Chem Pharm Bull	3	117.4, 146.0	Major	Accept
[60] Chem Pharm Bull	3	130.6, 115.8	Major	Major
[61] Chem Pharm Bull	11e	116.0, 129.4 Structures 11d and 11e exchanged	Minor	Minor
[62] Chem Pharm Bull	6a	130.6, 115.4	Minor	Accept
[63] Chem Pharm Bull	17	163.6, 94.1	Major	Accept
[64] Fitoterapia	3	35.1, 70.3	Major	Accept
[65] Fitoterapia	1	114.9, 130.6	Accept	Accept
[65] Fitoterapia	2	114.9, 130.6	Minor	Minor
[65] Fitoterapia	3	114.9, 130.5	Minor	Accept
[66] Fitoterapia	2	Phenyl rings exchanged	Accept	Accept
[67] J Nat Prod	2, 3, 4	Me groups inconsistent	Accept	Accept
[67] J Nat Prod	5, 6	Me groups inconsistent	Accept	Accept
[68] J Nat Prod	11	Data incompatible with substitution pattern of phenyl ring	Major	Major
[69] J Nat Prod	2	114.8, 132.4	Minor	Accept
[70] J Nat Prod	5	120.1, 138.6	Major	Minor
[71] J Nat Prod	1	121.4, 146.4	Major	Major
[72] J Nat Prod	13	57.4 (CH_3)	Major	Major
[73] J Nat Prod	MeOct-S	Alkyl chain	Accept	Accept
[73] J Nat Prod	MeOct-SO	Alkyl chain	Accept	Accept
[73] J Nat Prod	MeOct-SO_2	Alkyl chain	Accept	Accept
[74] J Nat Prod	1	159.1, 109.8	Minor	Minor
[74] J Nat Prod	1a	159.1, 109.8	Minor	Minor
[74] J Nat Prod	1b	159.1, 109.8	Minor	Minor
[74] J Nat Prod	1c	158.9, 109.8	Minor	Accept
[74] J Nat Prod	1d	158.9, 109.8	Minor	Accept
[75] J Nat Prod	9	31.8, 22.7	Major	Minor
[75] J Nat Prod	7	31.9, 22.7	Major	Minor
[75] J Nat Prod	6	31.9, 22.7	Major	Minor
[75] J Nat Prod	10	31.8, 22.7	Major	Minor
[75] J Nat Prod	8	31.9, 22.7	Major	Minor
[76] J Nat Prod	2a	142.3, 127.4	Minor	Accept
[76] J Nat Prod	2b	142.5, 127.8	Minor	Accept
[76] J Nat Prod	2c	143.3, 126.8	Minor	Minor
[77] J Nat Prod	15	Data exchanged with 16	Major	Major
[77] J Nat Prod	16	Data exchanged with 15	Minor	Minor

(continued)

Table 5 (continued)

Ref., journal	Compound number in Ref.	Misassigned chemical shift values (δ_C /ppm)	Result of evaluation before and after correction	
[78] J Nat Prod	6	129.9, 115.7	Accept	Accept
[79] J Nat Prod	7	161.7, 123.6	Major	Accept
[79] J Nat Prod	10	166.3, 105.8	Major	Accept
[79] J Nat Prod	1	166.5, 107.0	Major	Accept
[79] J Nat Prod	3	168.0, 107.0	Major	Accept
[79] J Nat Prod	5	132.8, 116.6, 166.6, 106.2 Inconsistent numbering	Major	Accept
[80] Molecules	4862F	131.44, 116.91	Major	Minor
[81] Molecules	L2	128.99, 113.79	Minor	Accept
[82] Molecules	1	130.6, 109.5	Minor	Accept
[83] Molecules	1	125.12, 142.54	Minor	Minor
[84] Molecules	31	98.4, 120.0	Major	Minor
[84] Molecules	32	103.3, 128.2	Major	Minor
[85] Molecules	4	157.7, 104.10	Major	Accept
[86] Molecules	IG-1	119.2, 134.7	Minor	Minor
[86] Molecules	IM-1	172.3, 156.2	Minor	Accept
[87] Molecules	1c	56.7, 64.5	Major	Major
[88] Molecules	3	158.22, 138.08, 132.16, 115.52	Major	Accept
[89] Molecules	4, 6	Data for compounds exchanged	Major	Accept
[90] Molecules	13	127.7, 164.2	Major	Minor
[91] Molecules	1, 2, 3	Numbering error C_{18} and C_{19}	Reject	Minor
[92] Molecules	III	116.2, 130.9, 130.8	Major	Accept
[93] Molecules	11	134.7, 155.9	Minor	Accept
[94] Rec Nat Prod	1	116.1, 130.5	Minor	Minor
[95] Rec Nat Prod	1	132.6, 107.8	Major	Minor
[96] Rec Nat Prod	7	153.87, 111.41, 107.91	Major	Accept
[97] Magn Reson Chem	2	129.7, 119.4	Minor	Minor
[97] Magn Reson Chem	5	130.0, 116.5	Minor	Minor
[97] Magn Reson Chem	9	130.2, 116.6	Minor	Minor
[98] Magn Reson Chem	1e	55.7, 76.2	Minor	Accept
[99] Mar Drugs	1	121.89, 102.42	Major	Accept
[99] Mar Drugs	3	115.43, 129.48	Reject	Major
[100] Steroids	6b	132.5, 150.7, 188.1	Major	Minor
[101] Steroids	2a	113.7, 132.9	Major	Minor
[101] Steroids	2b	113.9, 132.9	Major	Minor
[101] Steroids	3a	113.7, 132.4	Major	Accept
[101] Steroids	3b	113.9, 134.9	Major	Accept

The first column gives the literature citation, the second column shows the number of the compound within the paper. The third column summarizes the misassigned chemical shift values and gives some remarks, the fourth column gives the result of the evaluation of the original dataset as published, and the last column gives the result of the evaluation after correction of the assignment

3.6 Chemical Structure Drawing Errors

Diandraflavone, compound 2 from [102], is an example where the capabilities of the "CSEARCH-Robot-Referee" can be demonstrated in some detail. The evaluation started from the structure proposed and the assigned ^{13}C NMR chemical shift values, as shown in Fig. 9a, and resulted in a rating of "Major Revision". This was due to the large difference between the experimentally determined chemical shift values and the predicted values at four positions located in the phenyl ring and the *C*-glycosidic bound sugar moiety, as presented in Fig. 9b. This comparison detected units within the molecule wherein an excellent compatibility between their structural features and the spectroscopic properties was apparent. However, questionable portions of the structure could be identified. The highlighted carbons in Fig. 9c as well as some neighboring non-carbon atoms were subjected to a structure generation process. A red color is used in Fig. 9c when the difference between the predicted and experimentally obtained ^{13}C NMR chemical shift values were larger than predefined limits, whereas highlighting with a blue color represent carbons with smaller differences adjacent to the carbon atoms questioned. Furthermore, non-carbon atoms adjacent to suspicious carbons with respect to their chemical shift value are either inserted or deleted in order to extend the structural search to non-isomeric, but still similar molecules. In this example, an oxygen atom was inserted between C-4′ of the phenyl ring and C-1″ of the attached sugar moiety leading to the structure as shown in Fig. 9d, which has a molecular formula of $C_{28}H_{32}O_{13}$ instead of $C_{28}H_{32}O_{12}$. This verified that the process used of scanning the structural space with respect to the variation of only questionable carbons, but allowing insertion/deletion/modification of non-carbon atoms, is definitely not isomer generation. This process of scanning generated 7411 alternative structure proposals. Subsequent ranking of this list of chemical structures using the similarity between the experimentally determined peak list and the predicted chemical shift values puts the (incorrect) structure query at position 1488, whereas a reasonable alternative can be found at position 8. The other seven alternative structures were either highly strained or could be excluded by a deeper analysis of the additional data given in the paper [102].

The starting point of this analysis was the incorrect structural proposal for diandraflavone [102] and the measured peak list, leading to a set of 7411 alternative structures. Another possibility to retrieve alternative structures is the use of the peak list in a "Spectroscopic Similarity Search" using either a line-by-line search and/or a more pattern-oriented technology like the SAHO (Spectral Appearance in Hierarchical Order) method [103] over a large database of reference spectra. The PubChem collection [33] was downloaded and all organic compound structures up to 99 carbons were used to build a searchable reference collection of predicted ^{13}C NMR spectra, with subsequent ranking of the retrieved hits by coincidence between the experimentally given peak list and the predicted chemical shift values. When using the experimental ^{13}C NMR data of compound 2 as given in [102] and this reference collection consisting of 74 million predicted spectra [104]

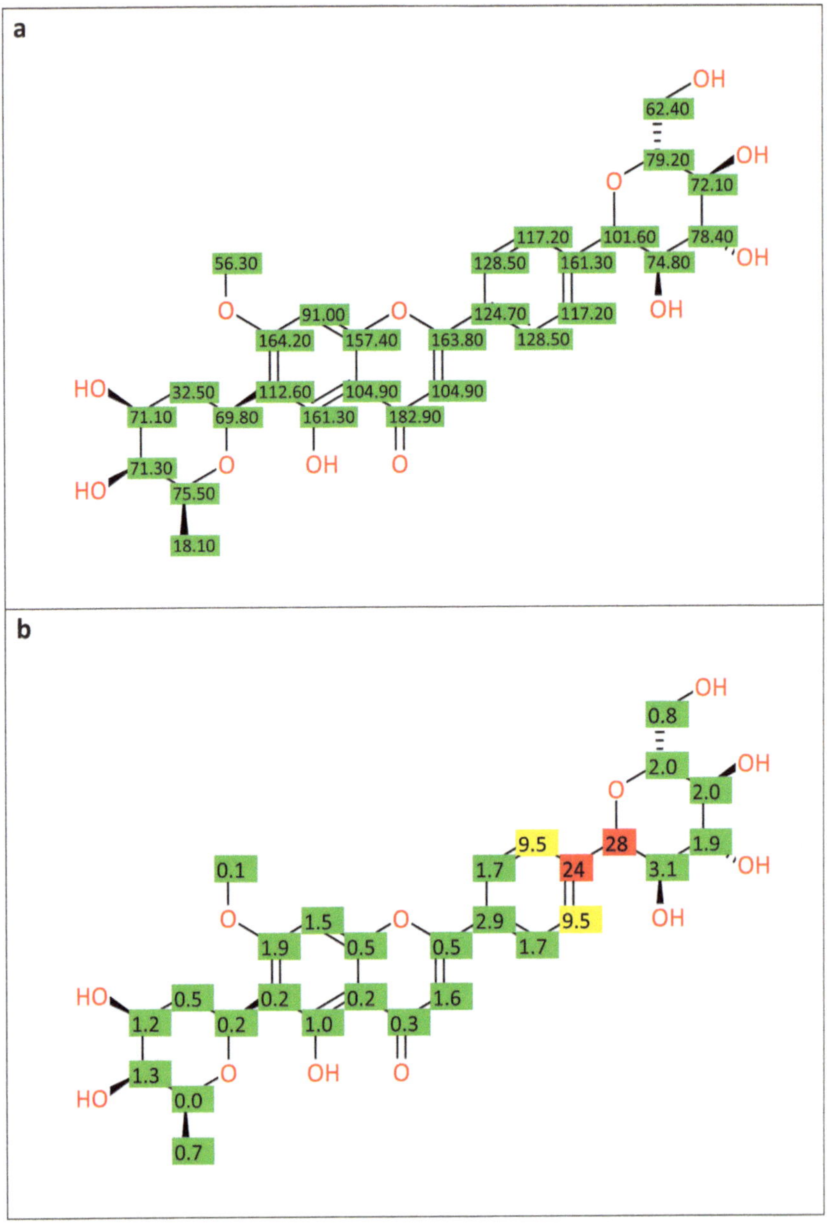

Fig. 9 The structure of compound 2 from [102] having a molecular formula of $C_{28}H_{32}O_{12}$ derived from the structural diagram. (**a**) Measured ^{13}C NMR chemical shift values are shown. (**b**) Differences between the experimental and the predicted chemical shift values. (**c**) Carbons involved in the search for alternative structure proposals (red: large deviation between experiment and prediction and blue: neighbors of carbon atoms having large deviations). (**d**) Best "real-world" alternative structure having a different molecular formula of $C_{28}H_{32}O_{13}$ together with the differences between experimental and predicted chemical shift values that fit very well

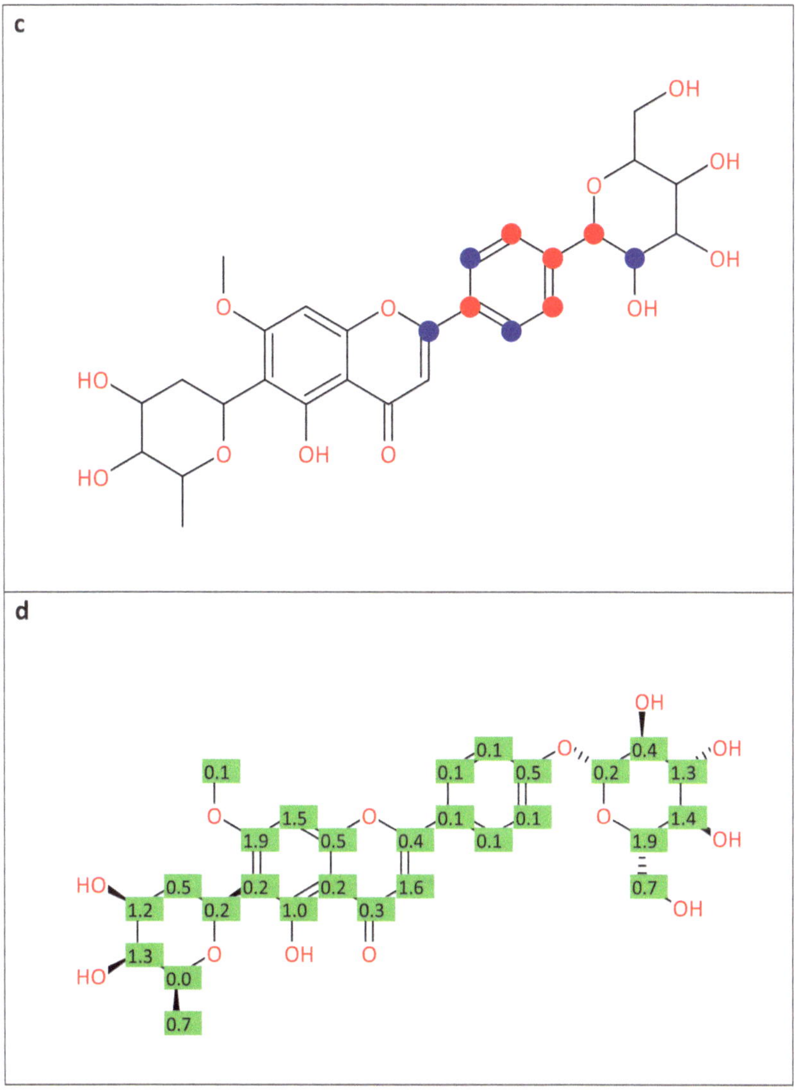

Fig. 9 (continued)

was searched, the PubChem compounds "CID: 21580252" and "CID: 73804419", which differ only in terms of stereochemical information, were ranked first. These two structures are identical to the proposal derived from structure generation with respect to their two-dimensional topology proven by their common 2D-InChIKey information [38]. When reading the experimental part in more detail, it can be seen from the MS data that the assumption of a molecular formula having O_{13} instead of O_{12} is justified, therefore a simple drawing error in this paper with respect to compound 2 was assumed. Using the corrected structure and the same peak list for an evaluation, this led to a result of "Minor Revision", based mainly on a small number of inconsistent assignments in the underlying "CSEARCH" database used for spectrum prediction. It should be noted that even a small inconsistency in the data queried and/or in the underlying data used as reference information will lead to "Minor Revision" for security reasons in order to alert the user to the need to check the available data.

3.7 Structural Revision of Drymaritin

Drymaritin was published in [102] with an assignment error combined with a wrong structure proposal despite the structure has been investigated using various NMR spectroscopic methods, including ^{1}H, ^{13}C, DEPT, HMQC, and ^{1}H-^{13}C HMBC at 400 MHz and ^{1}H-^{15}N HMBC at 600 MHz. The incorrect structure proposal, namely, that of a 5-methoxycanthin-4-one, was arrived at by misinterpretation of the ^{1}H-^{13}C HMBC spectrum. The central problem was the assumption of a $^{3}J_{CH}$ correlation instead of a $^{4}J_{CH}$ correlation between C-15 and H-6, which is a quite common error during the structure elucidation process. The structure revision was performed by Wetzel, Allmendinger, and Bracher [105] by total synthesis, leading to reassignment as a canthin-6-one rather than a canthin-4-one derivative. This structure revision has been used as to test an isomer generator program [7] based on 2D NMR correlations, achieving the same result as demonstrated by chemical synthesis. The "CSEARCH-Robot-Referee" used only the given structure proposal and the 1D ^{13}C peak list as given in [102], with no correlation peaks considered necessary from 2D NMR techniques. The data inputted are summarized in Fig. 10, giving a rating of "Reject" as result of the evaluation process, based mainly on the large differences between the experimentally obtained and predicted chemical shift values for five carbons, as shown in Fig. 10b. The subsequent structure scanning was restricted to the highlighted positions (Fig. 10c), giving 147 alternative proposals. After sorting the list of chemical structures obtained according to the similarity between the experimental and predicted ^{13}C NMR spectra, the correct solution [105] was at position 5 with an average deviation of 2.47 ppm, whereas the originally proposed structure can be found at position 62 with an average deviation of 5.57 ppm.

When applying an identical workflow approach to the correct structure proposal for drymaritin [105] together with the correct assignment of the resonance signals, the evaluation ended with a rating of "Accept". The corrected structure and the improved assignment together with the deviations between the experimental and the predicted spectrum are shown in Fig. 11a, b.

The example of drymaritin, because its structural revision is well-established by total synthesis [105], has been used as test case to show the robustness of the applied NMR database evaluation scheme. The result of a prediction of ^{13}C NMR chemical shift values for a given chemical structure is dependent on the content of the underlying database and the algorithms used. The HOSE (Hierarchical

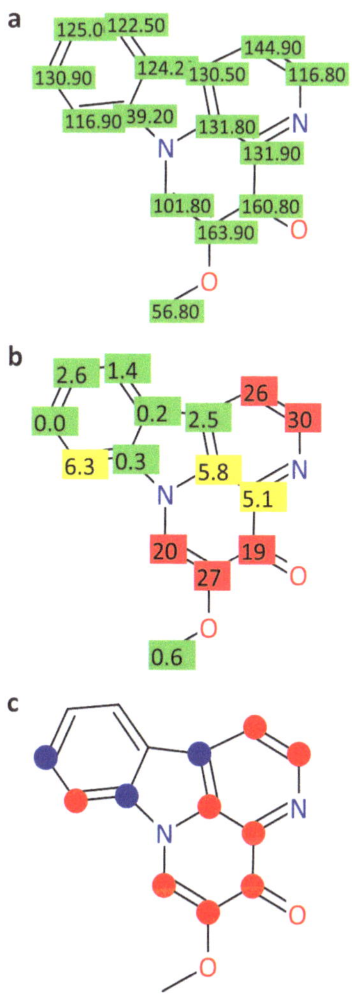

Fig. 10 (**a**) Original data for drymaritin as given in [102]. (**b**) Differences between the experimentally produced and predicted chemical shift values. The incorrect signal assignment in the pyridine ring is evident as well as the wrong enone system. (**c**) Carbon positions that are systematically modified during scanning for alternative structures (for color coding, see Fig. 9)

Organization of Spherical Environments)-code technology [36] reproduces exactly the content of the database and is therefore very sensitive to assignment errors, but gives an excellent performance when very similar and well-assigned reference information with respect to the structure being queried is available. Neural networks are more error-tolerant, however. Within the "CSEARCH-Robot-Referee" system, both techniques are available, and, depending on statistical parameters as well as on the similarity between the structure queried and the reference structures,

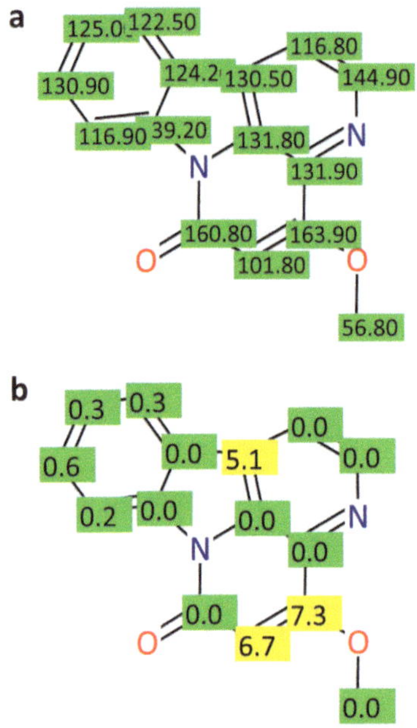

Fig. 11 (**a**) Correct structure for the given data obtained by the "Structure Generator" program and proven by total synthesis [105]. (**b**) Differences between experimentally obtained and predicted ^{13}C NMR chemical shift values, showing a much better similarity compared to Fig. 10b

a weighted mean value of the predicted chemical shift values obtained by both methods may be taken for comparison against the experimental chemical shift values. This would explain the robustness of the evaluation procedure. It should be pointed out that a final rating is not based solely on the difference between experimentally obtained and predicted chemical shift values, but also on the consistency of the reference data for each carbon taken into account. A large inconsistency in the reference data and/or the availability of only a few similar structures increases the accepted deviation for a specific carbon prediction, making the system even more robust.

4 Analysis of Selected Chemical Literature Focused on Natural Products in the Public Domain

In this section, an analysis of selected publications found in the chemical literature dealing mostly with the chemical characterization of natural products and their precursors, their biosynthesis pathways, and their total synthesis is discussed. The data used consisted only of the given structure and the assigned 1D ^{13}C NMR spectroscopic data as summarized in the publication. Each analysis was performed using the "CSEARCH-Robot-Referee", as available on the Internet, without any user interactions [13]. The intention was to answer two basic questions:

- Is it possible to verify or to reject the proposed structure on the sole basis of the published 1D ^{13}C NMR data by an automatic evaluation protocol?
- Is it possible to generate a set of reasonable alternative structural proposals, either using the "Structure Generator" or the "Spectroscopic Similarity Search" features?

There was no systematic attempt to check for known structure revisions and nor was it intended to use data for those compounds where a structure revision had been proposed. The selection of the examples featured was based on the large difference between experimentally detected ^{13}C NMR chemical shift values and those predicted. The following discussion is focused on the four key journals, namely, "Chemical and Pharmaceutical Bulletin", "Fitoterapia", the "Journal of Natural Products", and "Phytochemistry". From each journal, up to five publications were selected describing compounds for which problems were identified in their structure elucidation process. The selection of these four journals results from their importance in the field of natural product chemistry as well as the availability of the relevant ^{13}C NMR data in the "CSEARCH" database. Any suggestion that other related specialist journals contain a lesser number of incorrect assignments leading to the need for compound structural revisions, is not intended. The author posits that about 20% of the data out of some 700,000 ^{13}C NMR spectra evaluated would be classified either as "Major Revision" or "Reject". Furthermore, there is evidence that high-quality spectroscopic data do not necessarily come from more highly ranked journals [15]. All results presented here are based on the published 1D ^{13}C NMR data and disregard any information using 2D NMR spectroscopic techniques.

4.1 *Isoflavonoids from Erythrina variegata*

In [106], the structures of four new isoflavonoids isolated from the stem bark of *Erythrina variegata* are presented together with their ^{1}H and ^{13}C NMR data, supported with COSY, DEPT, and HMQC experiments. The evaluation for these four compounds yielded a rating of "Minor Revision", in each case. For compound 1, a few inconsistencies in the underlying data were perceived, with the average deviation per carbon ($\Delta\delta_C$) being 1.38 ppm. No carbons were involved in the structure generation process that showed the given data to be consistent with the proposed structure. The same situation is apparent for compounds 3 and 4, which gave a $\Delta\delta_C$ of 1.19 and 1.33 ppm.

Compound 2, containing an oxirane fragment, had published ^{13}C NMR chemical shift values of 68.8 and 79.8 ppm for the oxirane carbon atoms [106], which is outside the expected range. From these data, both the oxirane carbon atoms as well as the two attached methyl groups are involved in the overall process of structure generation, leading to 786 alternative possibilities. The structure of compound 2 as given in [106] is ranked at position 36 among these possible structures, in having a $\Delta\delta_C$ value of 3.19 ppm, whereas the best proposal ($\Delta\delta_C = 1.88$ ppm) is a known structure from the PubChem collection. The result of this analysis is summarized in Fig. 12.

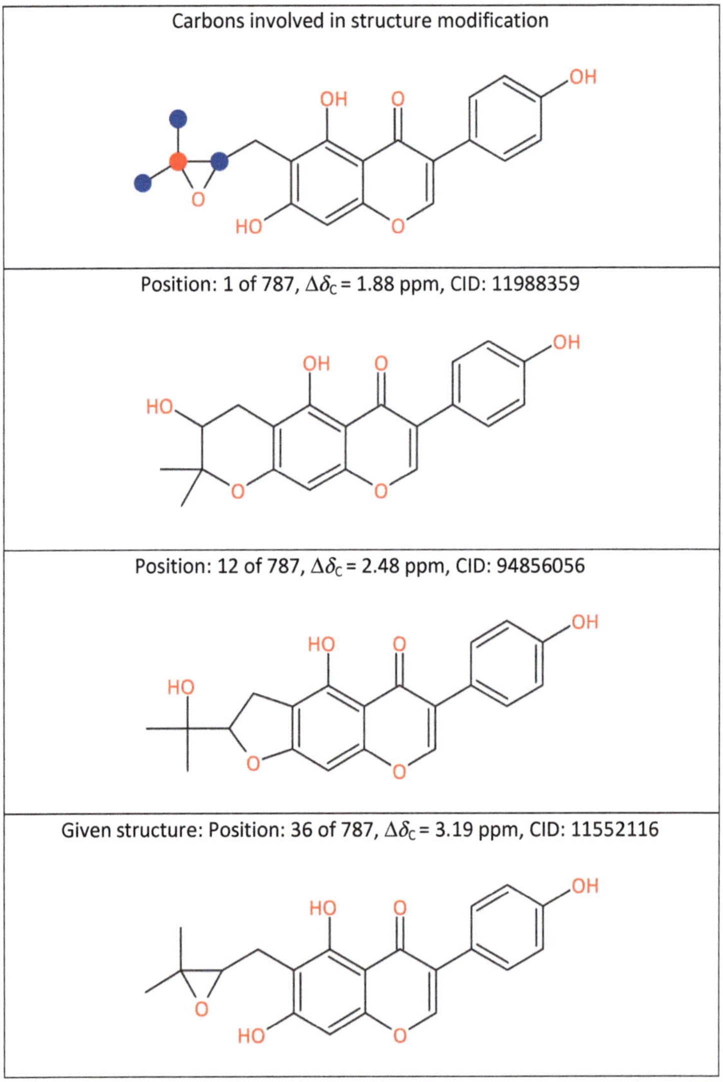

Fig. 12 Possible alternative structures when using the data of compound 2 from [106] for evaluation and subsequent structure generation

4.2 Miscellaneous Constituents of *Aphanamixis grandifolia*

Reference [107] describes the ^{13}C NMR data of four new chemical constituents from the aerial parts of the arbor tree, *Aphanamixis grandifolia*, collected in southern mainland China. Two new compounds (1 and 2) were identified as cycloartane derivatives. Compound 1 showed an excellent correlation between the predicted and measured ^{13}C NMR chemical shift values. The evaluation gave a rating of "Minor Revision" based on the fact that the underlying data used for the prediction had some inconsistencies therein. For this compound, the structure generation and the subsequent ranking process placed its identity at position 1 of structural possibilities, making the structure elucidation extremely reliable. Furthermore, the structure of compound 1 was verified by X-ray structural analysis [107]. However, compound 2 showed deviations between the measured and predicted chemical shift values in both the D ring and the side chain, therefore this structure proposal was ranked at position 651 of 2908 structural proposals having $\Delta\delta_C = 1.65$ ppm. This value is unusually large for steroidal compounds, because they are very well represented in terms of knowledge of their spectroscopic parameters. All alternative structures generated were unknown with respect to the PubChem collection. The result from this analysis is that the side chain of compound 2 as published needs further detailed investigation and it appears that the structure proposed in [107] might be in error (Fig. 13).

Compound 6 in [107] was elucidated as a cyclopentenone derivative with a chemical shift value of 198.1 ppm for the oxo group carbon, which is outside the typical range for this functionality. Carbonyl carbon atoms usually resonate around 209 ppm in cyclopentenones, whereas a value of around 199 ppm is more characteristic for cyclohexenones. A detailed analysis revealed very large deviations between the predicted and measured ^{13}C NMR chemical shift values, as shown in Fig. 14, leading to a classification of "Major Revision". The "Structure Generator" created 972 topologies, of which two were contained in the PubChem collection. In Fig. 15, the two most likely cyclohexenone derivatives are shown, but because of minor differences in their similarity measures, no final decision could be made.

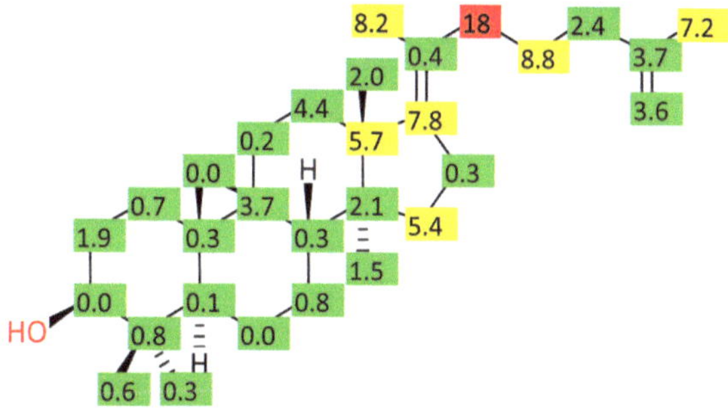

Fig. 13 Compound 2 from [107] labeled with the difference between the experimentally obtained and predicted chemical shift values leading to the conclusion that the side chain might be in error

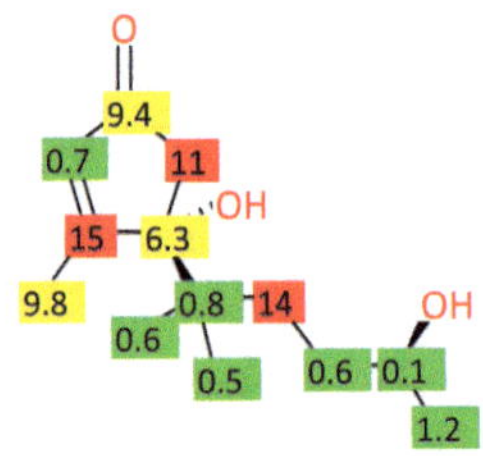

Fig. 14 Compound 6 from [107]: Deviations between experimentally obtained and predicted chemical shift values

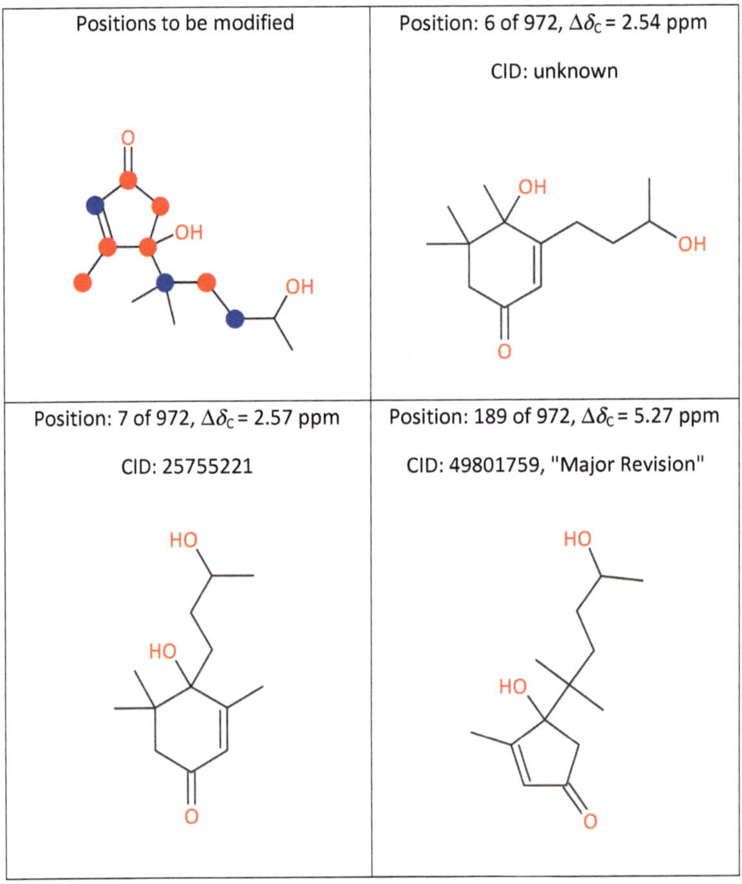

Fig. 15 Compound 6 from [107]), in which the given cyclopentenone derivative is ranked at position 189, whereas two cyclohexenone derivatives can be found at positions 6 and 7

When a spectroscopic similarity search was conducted using the peak list published, this also led to the same alternative structures for compound 6 as obtained by application of the "Structure Generator".

Compound 7 in [107] is a 1,4-naphthoquinone derivative with an average deviation between the measured and predicted ^{13}C NMR chemical shift values of 2.86 ppm, and, after ranking, this structure was found at position 525 of 5838 possibilities during structure generation. Exchanging the hydroxy and the methyl group in the naphthoquinone part of the molecule improves the ranking from position 525 ($\Delta\delta_C = 2.86$ ppm) to position 5 having a $\Delta\delta_C = 1.96$ ppm. From this evaluation it cannot be claimed that this is the correct solution, but it is a reasonable structural alternative and there is a high probability that the published structure is in error. Additional experimental work to resolve this ambiguity would need to be undertaken.

4.3 Lignans from *Pseuderanthemum carruthersii* var. *atropurpureum*

Pseuderesinol, compound 1 in [108], was isolated from the roots of *Pseuderanthemum carruthersii* var. *atropurpureum*, and the compound was purported to be a new lignan according to the abstract of the paper. When analyzing the ^{13}C NMR data of the 3′,5′-dioxygenated phenyl ring, it became evident the chemical shift values at 145.5 and 146.7 ppm are in contradiction with the proposed substitution pattern. The computerized evaluation classified this example as "Minor Revision", with a subsequent structure generation process producing 4352 alternative proposals. The original structure has a $\Delta\delta_C = 4.13$ ppm and was ranked at position 982 of these possibilities. The 3′-methoxy-4′-hydroxy derivative was ranked at position 1 having $\Delta\delta_C = 1.64$ ppm, whereas the 3′-hydroxy-4′-methoxy analog was at position 2 having $\Delta\delta_C = 1.77$ ppm; Since the first of these alternative structural proposals is available in the PubChem collection (CID: 21632950), the "new" lignan described in [108] thus seems to be an already well known compound. The proton NMR data of pseuderesinol were given in [108], but decisive information about $^3J_{HH}$ and $^4J_{HH}$ couplings was missing.

4.4 Kiusianins A-D from *Tilia kiusiana*

Four compounds named kiusianins A-D were isolated and characterized structurally from the leaves of the Japanese endemic plant, *Tilia kiusiana* [109], using their ^{1}H and ^{13}C NMR signals assigned with COSY, NOESY, DEPT, HMQC, and HMBC experiments. Kiusianins B, C, and D were proposed as cholestane-type sterols with a C-3 carbonyl group, conjugated with a C=C double bond between positions C-4 and C-5 in kiusianin B, with kiusianins C and D having a double bond between positions C-7 and C-8. For all three examples, the chemical shift value of the oxo group carbon atom was given between 199.5 and 199.7 ppm, fitting quite well to a conjugated carbonyl group. The absence of conjugation in kiusianins C and D would increase the chemical shift value to approximately 210 ppm in each case. Furthermore, kiusianin A, assigned as a lanostane-type triterpenoid, in having a 7-ene functionality, showed a chemical shift value of 145.7 ppm for C-8, whereas kiusianins C and D gave values around 164 ppm. Since this increase by approximately 20 ppm cannot be attributed to the influence of a 6β-OH group, a reasonable explanation is an enone system that would deshield the β-position. From this point of view, either a 6-hydroxy-3-on-4-ene or 3-hydroxy-6-on-7-ene functionality could be expected.

The automated computer evaluation of kiusianin A gave a rating of "Minor Revision", based mainly on the fact that there were some inconsistencies in the assignment of the reference chemical shift data used for the prediction and incorrect assignments of C-13 and C-14. A subsequent structure generation program created 1626 alternative proposals, showing the original structure at position 50 with a small average deviation of 1.36 ppm per carbon. This result is not considered a structural verification for kiusianin A, but only indicates that the structure proposed and the measured ^{13}C NMR data are compatible.

Kiusianin B showed very small deviations between the measured and predicted chemical shift values ($\Delta\delta_C = 0.3$ ppm), leading to an evaluation of "Accept". In this case, there is a 3-on-4-ene moiety, which is fully compatible with a chemical shift of 199.5 ppm for the C=O group in conjugation to a C=C bond. Carbon atom C-5 was found to resonate at 171.4 ppm, which is also in excellent agreement with the literature data, and underwent a strong deshielding effect in the β-position with respect to the C=O functionality.

On performing the same type of computerized analysis for kiusianin C, this led to a rating of "Minor Revision", based mainly on the poor prediction for carbon atoms C-2, C-3, C-4, C-5, and C-8 in the A and B rings of the steroidal skeleton. The ^{13}C NMR data measured, the differences with respect to the predicted values, and the positions varied during the structure generation process, are shown in Fig. 16.

A total of 3189 alternative structure proposals for kiusianin C was created, with four out of these included in PubChem and having deviations of 0.79, 2.09, 3.93, and 4.27 ppm. These can be found at positions 1, 57, 561, and 802 in the sorted "hit list" of alternatives, showing that the original structure has a low probability of being the correct solution to this structural problem, whereas a 3-hydroxy-6-on-7-ene derivative seems to be the correct solution. In Fig. 17, the structures of previously known compounds from the PubChem collection are together with their similarity measures are shown.

Kiusianin D also gave a result of "Minor Revision" when the same type of evaluation was applied as before, with the largest deviations between the predicted and measured ^{13}C NMR chemical shift values evident for carbon atoms located in rings A and B. The proposed structure in [109] was found at position 307 of 5729 alternative structural possibilities, with an average deviation of 2.92 ppm. In turn, an analog with a 3-hydroxy-6-on-7-ene system could be found having an average deviation of 1.46 ppm at position 7, and a 3-on-4-ene-6-hydroxy unit-containing

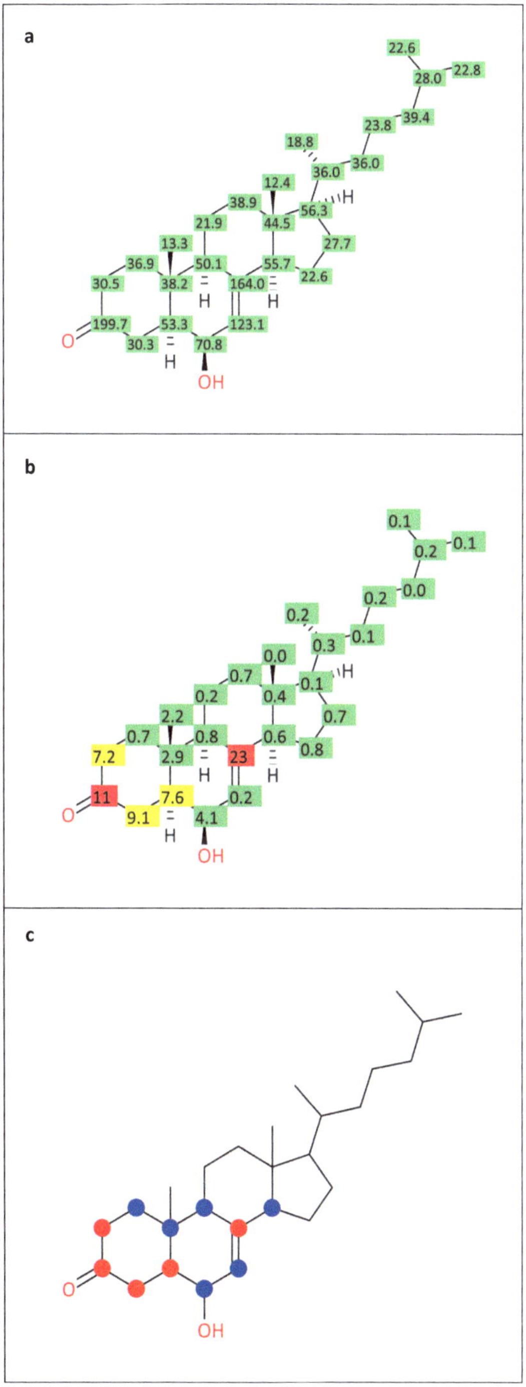

Fig. 16 (**a**) ^{13}C NMR data of kiusianin C as given in [109]. (**b**) Differences between predicted and measured chemical shift values. (**c**) Variable positions involved in the process of structure generation

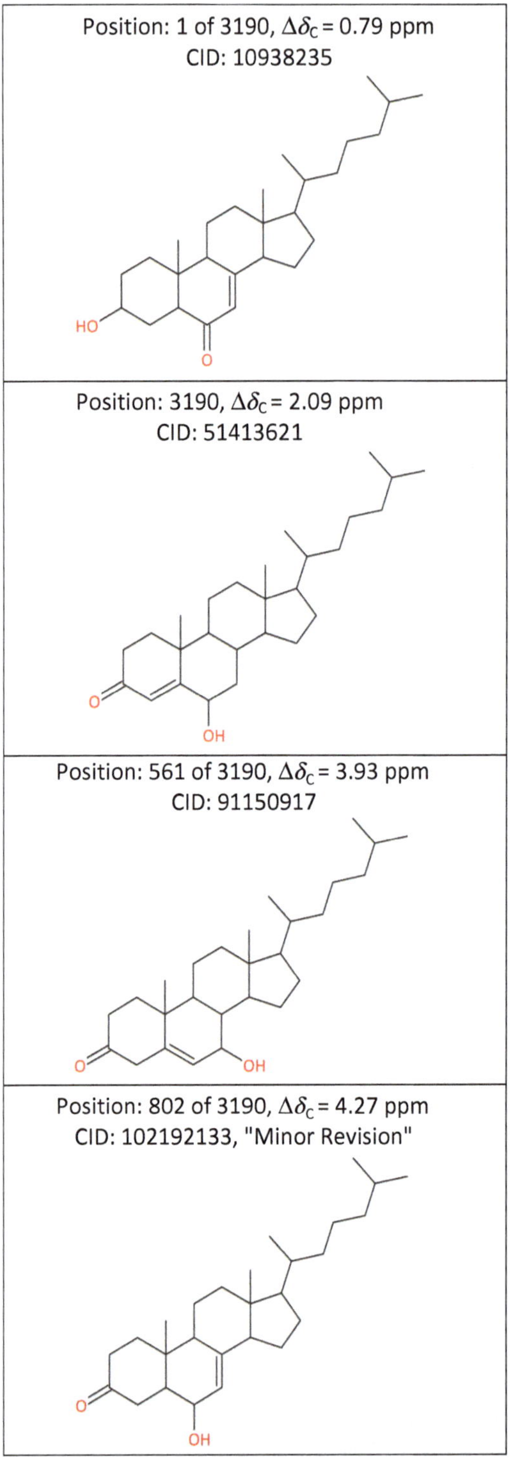

Fig. 17 Alternative structures in the ranked list generated for kiusianin C that are available in the PubChem collection together with their positions in this list and their similarity measures. The proposed structure for kiusianin C from [109] was found at position 802 of 3190 possibilities

derivative was located at position 82 with a deviation of 2.43 ppm. The original structure as published for kiusianin D can be found in the PubChem collection (CID: 102368177), but all the other structure proposals are not included in this collection.

Accordingly, of the four compounds described in [108], it appears that two of these are correct, while the other two structures seem to be wrong. However, it should be kept in mind that the central role of the "CSEARCH-Robot-Referee" is only in verifying the compatibility of a given structure with the measured 1D ^{13}C NMR data. Generating alternative proposals, sorting them by similarity, and comparing them against the PubChem collection is a valuable additional feature, but the technology used aims to generate a set of similar structures and is therefore not an isomer generator program creating all possible structures compatible with all given constraints. When using an isomer generator program, the user is required to enter a large number of correlation signals from the 2D NMR data on hand in order to reduce the possible structural space to a restricted number of isomers that can be handled within a reasonable period of time. It is well established that very large data input (as would be necessary even for a medium-sized molecule) is prohibitive with respect to general use. However, if a MOLfile and the peak list are copied and pasted into a graphical user interface this seems to be acceptable.

4.5 Arteminin from Artemisia apiacea

In [110], a compound named arteminin (numbered as "compound 5"), and reported in 2002 from the oriental herb, *Artemisia apiacea*, was elucidated from its ^{1}H, ^{13}C, and HMBC NMR data as 5-hydroxy-6,8-dimethoxycoumarin (CAS-RN: 466639-11-2). Methoxy groups located at a phenyl ring are known to resonate around 56 ppm when at least one *ortho*-position is unsubstituted, whereas the ^{13}C NMR chemical shift value is increased to approximately 61 ppm when both *ortho*-positions are substituted. Furthermore, 1,2,4,5-tetraoxygenated benzene derivatives usually have chemical shift values below 150 ppm at the 4 *ipso*-positions. When examining the data reported for compound 5, one methoxy group was found to resonate at 61.8 ppm, showing that both *ortho*-positions should be substituted. However, there were two signals above 150 ppm that are incompatible with a 1,2,4,5-tetraoxygenated system. The "CSEARCH-Robot-Referee" led to a classification of these data as shown in Fig. 18a as "Major Revision", and subsequently, the "Structure Generator" and the "Spectroscopic Similarity Search" were started. In Fig. 18b, the large differences between the experimental and predicted chemical shift values are presented. The used algorithm decided which prediction technology (HOSE-code or Neural Network, see Sect. 2.5) should be preferred for a specific carbon resulting in a superior robustness of the prediction, even when the wrong dataset is contained in the underlying database.

The process of structure generation created 1561 structural proposals, and seven coumarins were ranked better ($\Delta\delta_C = 1.89$ to 3.18 ppm) than the structure proposed in [110] (position 78 out of the alternative structure proposals; $\Delta\delta_C = 3.55$ ppm) and are included in either the PubChem or the "CSEARCH" collections. Six of the

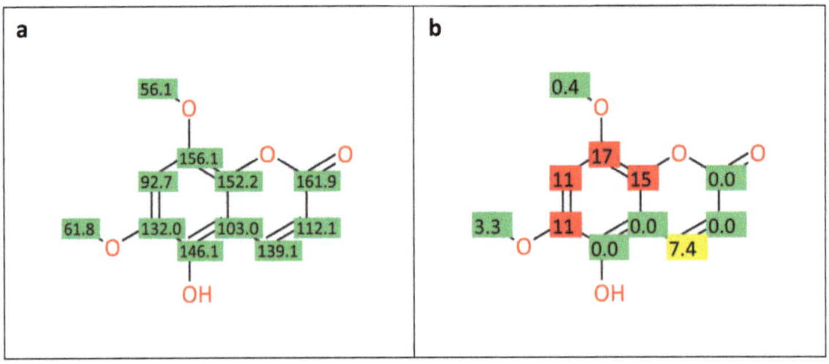

Fig. 18 (**a**) ^{13}C NMR data for arteminin from [110]. (**b**) Differences between the experimentally produced and predicted chemical shift values

above-mentioned seven structural possibilities have a methoxy group with both *ortho*-positions substituted, as shown in Fig. 19. The "Spectroscopic Similarity Search" using the PubChem collection generated a quite similar result, but again permitted no clear decision between the various reasonable alternatives because of very minor differences in their similarity measures. However, from this evaluation, it follows that the structure proposed for arteminin is incompatible with the published ^{13}C NMR data.

The underlying database used by the "CSEARCH-Robot-Referee" for this particular evaluation consisted of approximately 321,000 spectra. The structure queried and the peak list published were used as starting points for either an "Identical Structure Search", based on the two-dimensional topology of the given molecule, or an "Identical Spectrum Search", using an allowed deviation of 0.4 ppm between corresponding lines, but retrieving only compounds having a different topology with respect to the queried structure. This procedure detects spectra that have been published twice with different structural proposals, in order to detect known or even unknown revisions. Use of the "Identical Structure Search" revealed that this entry as given in literature was contained once in the dataset and use of the "Identical Spectrum Search" did not give an alternative structure proposal. When increasing the limit to 0.6 ppm and using the full "CSEARCH" database holding some 700,000 entries, a closely comparable ^{13}C NMR spectrum belonging to compound III in [111] was found, namely, 5-hydroxy-6,7-dimethoxycoumarin (tomentin). This publication from 2008 did not cite the paper on arteminin from 2002. The authors of [111] have in essence performed a structure revision for arteminin when changing this from 5-hydroxy-6,8-dimethoxycoumarin to 5-hydroxy-6,7-dimethoxycoumarin, based on the nearly identical ^{13}C NMR data obtained. A change of structure from arteminin to tomentin has been performed automatically by the "CSEARCH-Robot-Referee". Furthermore, almost the same spectroscopic data were published for two different compounds, which could have been avoided if the purported new structure of arteminin [110] had been supported by X-ray crystallography, as is done for many new natural product structures. The necessity to deposit NMR data, as has been done with X-ray data for many decades, is also understandable from this example [112].

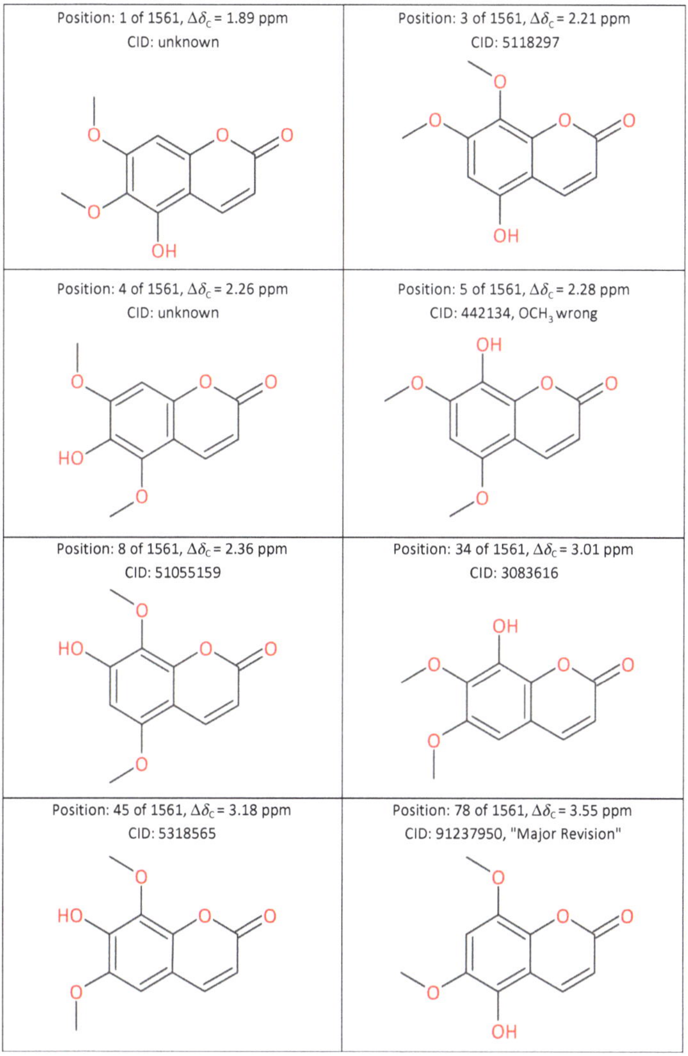

Fig. 19 Seven structure proposals for arteminin present in either the "CSEARCH" or the PubChem collections having a smaller average deviation than the proposed structure in [110]; the position within the ranked priority list of alternative structures together with the average deviation and the "Compound-Identifier" from PubChem is given. The structure published in [110] for "arteminin" is located at position 78, whereas the revised structure from [111] is located at position 1

4.6 *Fuscacarpans A-C from Erythrina fusca*

Epoxides are common structural fragments frequently found in natural products. The published structure elucidation of the pterocarpan fuscacarpan A, compound 1 in [113] isolated from the stems of *Erythrina fusca*, revealed a side chain bearing

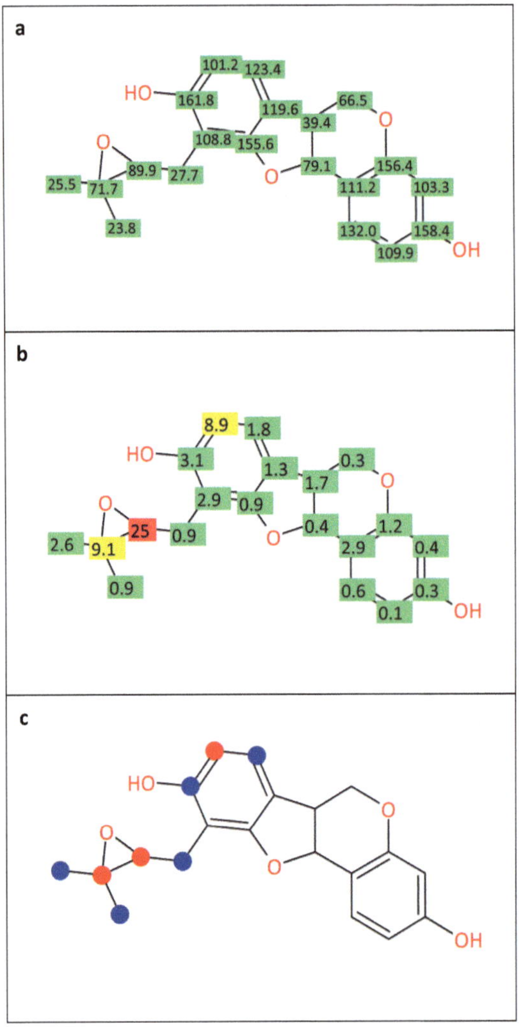

Fig. 20 (**a**) Fuscacarpan A together with its ^{13}C NMR data from [113]. (**b**) Differences between predicted and experimentally obtained chemical shift values. (**c**) Carbon positions varied during the "Structure Generator" process

an oxirane moiety attached to a tetracyclic system. This structure is shown in Fig. 20a, together with the measured ^{13}C NMR chemical shift values. The signals at 89.9 and 71.7 ppm were assigned to the carbon atoms within the oxirane ring, which are far away from the expected range of between 55 and 65 ppm. The differences between the predicted and measured ^{13}C NMR chemical shift values are presented in Fig. 20b, showing large differences in the three-membered ring as well as near a hydroxy group, which might be involved in a cyclization reaction. In

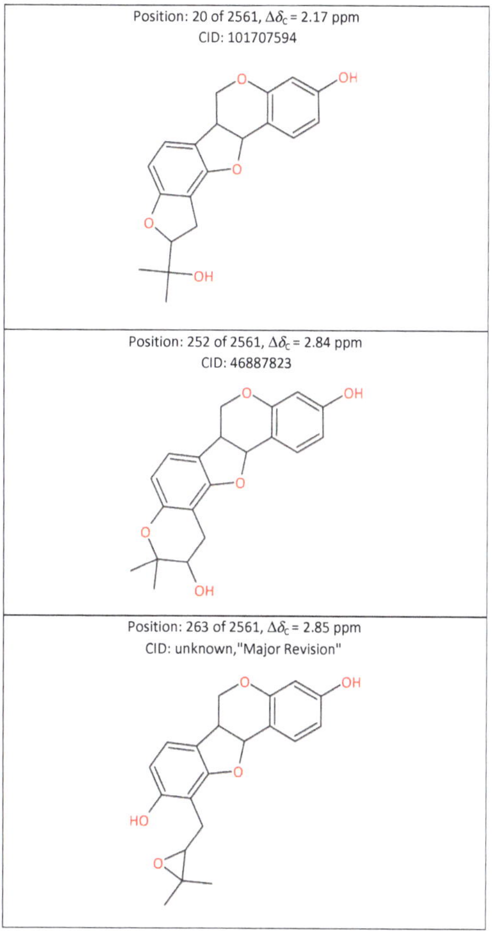

Fig. 21 Two better fitting structural proposals for the data of fuscacarpan A [113], with both known in PubChem. The evaluation of fuscacarpan A gave a result of "Major Revision"

Fig. 20c, carbon atoms with unusual chemical shift values are highlighted with red dots, with neighboring carbons highlighted in blue. Both groups of marked carbon atoms were allowed to be varied during the subsequent computerized structure generation process, since this combination of structure and ^{13}C NMR data was classified as "Major Revision" during the evaluation step.

The structure generation process created 2561 structure proposals, and, after this ranking procedure, the original structure as elucidated in [113] was found at position 263 with an average deviation of 2.85 ppm, whereas two possible cyclization products were at positions 20 and 252, having deviations of 2.17 and 2.84 ppm, as shown in Fig. 21.

From these results, the structure ranked at position 20 is a reliable alternative structure for fuscacarpan A. A “Spectroscopic Similarity Search” using the 74 million predicted spectra for the PubChem compounds ranked also a very similar pentacyclic ring-containing compound in first place (PubChem CID: 11551842), with all structural features as the proposed structure obtained by the process of structure generation.

The other two structures characterized as new in [113] (fuscacarpans B and C) were evaluated as “Minor Revision”, on the basis of minor inconsistencies in the underlying data. The “Structure Generator” system failed to propose better fitting topologies for these data, so, therefore, these structures as published in [113] seem to be correct.

4.7 *Flavans from Livistona chinensis*

In [114], the structures of three new flavans (compounds 1–3) were proposed. Of these, compound 1 has according to its structural diagram a 3,5-dihydroxyphenyl ring. However, this was not consistent with the ^{13}C NMR data, since a 3,5-dihydroxyphenyl ring should display resonance lines around 156 ppm, whereas values at 144.4 ppm are tabulated for C-3′ and C-5′. The use of this structural proposal and its ^{13}C NMR data for an evaluation as shown in Fig. 22a gave a result of “Minor Revision”, based mainly on the finding that there was some deviation between the experimental and predicted chemical shift values. The deviation was focused on one carbon atom because the incorrect data were again contained in the database, leading to a strongly biased result of the prediction.

The subsequent structure generation process yielded 1590 structures, and, after the alternative structures were ranked, the isomer having a 3,4-dihydroxyphenyl ring was found at position 1, with an average deviation of 1.31 ppm compared to 4.43 ppm for the original structure, which was found at position 351 within the sorted list of possibilities. Further evidence was given by the fact that a simple “Identical Spectrum Search” over the 321,000 spectra in the underlying “CSEARCH” database retrieved four structures, which were different from the molecule being queried, but having identical reported chemical shift values [115–118].

Use of the two independent technologies, “Structure Generation” with subsequent ranking, and “Identical Spectrum Search”, led to the same alternative structure and a list of references was retrieved wherein this compound has been reported.

Compounds 2 and 3 described in [114] both contain highly oxygenated phenyl moieties that have chemical shift values outside the typical expected ranges [119–121]. The computerized evaluation recommended therefore “Major Revision” for

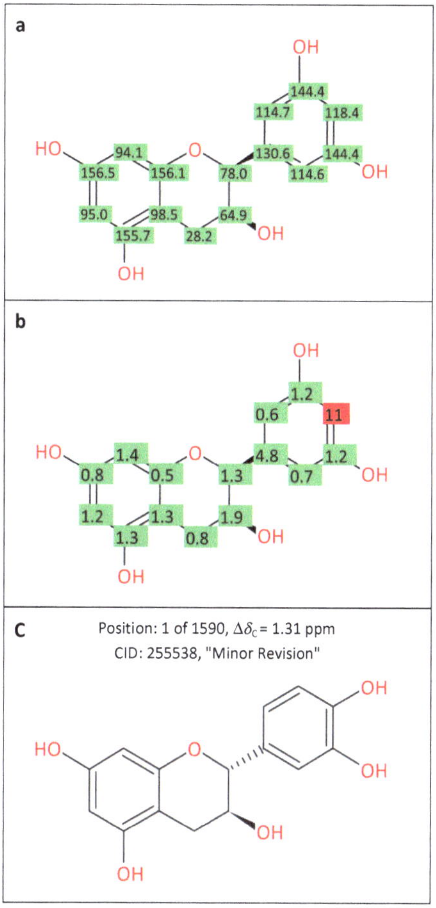

Fig. 22 (**a**) ^{13}C NMR data as given in [114]. (**b**) Differences between the experimentally obtained and the predicted chemical shift values. (**c**) The best alternative structure proposal having an average deviation $\Delta\delta_C = 1.31$ ppm that was ranked at position 1 of 1590 possibilities

both of these compounds. The structure generation procedure failed to provide reasonable alternatives, and also the "Spectroscopic Similarity Search" did not provide any useful information, because of the large number of hydroxylated flavans having different substitution patterns being found. The B ring of compound 3 is shown having C-3′,C-5′-dihydroxylation, but the same arguments as discussed for compound 1 were applied that led again to a C-3′,C-4′ hydroxylation pattern. The resultant evaluation recommended "Major Revision". However, the two above-mentioned additional processes did not clarify the situation, because the inconsistency in the A ring was predominant. Even if the generation of alternative structure proposals was not successful, the classification has shown clearly a need to perform further experimental work in order to elucidate the correct structures.

4.8 Polyketides from the Endolichenic Fungus *Ulocladium* sp.

In [122], two small-molecule polyketides were reported as new in 2012 from the endolichenic fungus, *Ulocladium* sp., obtained from a lichen (*Eveniastrum* sp.) collected in the southern region of the People's Republic of China, namely, compounds 1 and 2. The experimental ^{13}C NMR data of two compounds isolated from this work, numbered as compounds 1 and 4 are given. Compound 4 was identified as a previously known structural analog of compound 1. The detailed analysis of compound 1 is shown in Fig. 23a, showing both the proposed structure

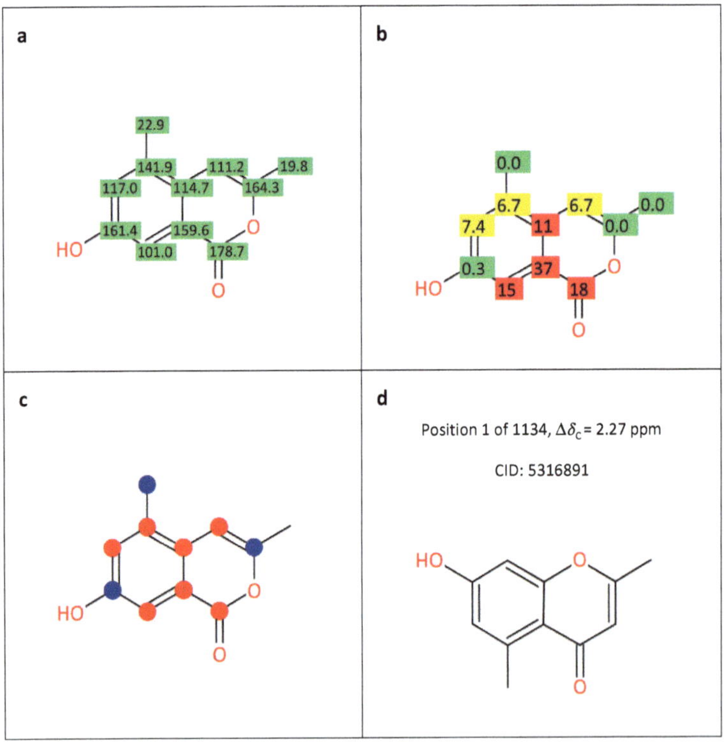

Fig. 23 (**a**) Structure of compound 1 and ^{13}C NMR chemical shift values as presented in [122]. (**b**) Deviation between experimentally obtained and predicted chemical shift values, giving an evaluation of "Reject". (**c**) Carbon positions subject to variation during the "Structure Generator" process. (**d**) Best alternative structure obtained

together with the measured chemical shift values attached to each carbon atom. In Fig. 23b, differences between the experimental and the predicted ^{13}C NMR spectroscopic values are presented. The largest deviation was found in the vicinity of the carboxyl group, leading to an impression that the structure of this proposed isochromene derivative might be in doubt. The outcome of the automatic computerized evaluation process was "Reject", pointing to the incompatibility between the structural proposal given and the associated NMR data. The two subsequent processes were then applied, namely, "Structure Generator" and "Spectroscopic Similarity Search", using a database of 74 million predicted ^{13}C NMR spectra based on the PubChem compounds. In Fig. 23c, the carbon atoms involved in the subsequent structure generation process are marked. However, despite nearly all positions of this molecule being available for variation, a total of 1134 structures (dramatically less than 0.01% of the possible number of isomers for a molecule having a formula of $C_{11}H_{10}O_3$) were created within 80 s and sorted by the coincidence of the experimental and predicted spectroscopic data. The original structural proposal in [122] can be found at position 213 with an average deviation of 5.32 ppm per carbon, whereas the best alternative proposal has an average deviation of 2.27 ppm. This alternative automatically created structure is shown in Fig. 23d, and corresponds to a previously known compound in the PubChem collection. The "Spectroscopic Similarity Search" also found this same compound in PubChem ranked at position 2, although having 11 carbon atoms in the molecular formula, among 2578 possible structures having similar spectra.

From all these completely automatic evaluations, it could be concluded that the proposed structure for compound 1 in [122] is incompatible with the given 1D ^{13}C NMR data. Furthermore, two different strategies led to the same alternative structural proposal, a compound for which the ^{13}C NMR data had been available in the literature since 1990 [123, 124].

When applying the same computerized automatic workflow to compound 4 of [122], the result of the evaluation was again "Reject", with the two analogous processes for searching for alternative structure proposals then being activated. In a similar manner to the previous example, the structure generation process proposed instead of the published isochromene derivative for compound 4, the analogous chromene derivative. The original structure in [122] could be found at position 566 of the structural alternatives, having an average deviation of 5.55 ppm, whereas the proposed revised structure is found at position 1 with an average deviation of 1.92 ppm. The similarity search ranked the same structure at position 2 within a list of 3806 structures having similar spectra; the first entry differed by its molecular weight and therefore was readily excluded.

The new compound 2 in [122] was evaluated as "Minor Revision", and the subsequent isomer generation process led to 2126 alternative structures, but the deviations of these alternatives were equivalent or only slightly better than structure actually proposed. The same arguments can be used for compound 3, wherein the structure generation process produced 1451 alternatives. Of these, the published structure was ranked at position 3 with a very small deviation of 1.47 ppm per carbon.

In [122], thirteen compounds were reported, with two of these postulated as being new (1 and 2). Detailed characterization by 1D and 2D NMR techniques for four compounds (1–4) led to two structures with major incompatibilities (compounds 1 and 4) between their proposed constitution and the given spectroscopic data as determined automatically by the "CSEARCH-Robot-Referee". The structure elucidation of compound 1 was based mainly on HMBC correlations, with the actual chemical shift assignments neglected. A simple comparison of the chemical shift value of 178.7 ppm for C-1 in compound 1 with the known data of benzoic acid would have revealed that this value to be quite distant from the expected range for a carbon having this environment. The same argument could have been used for carbon C-8a, where a chemical shift value of 159.6 ppm together with a value of 161.4 ppm for C-7 would point towards the occurrence of a *meta*-substituted aromatic dihydroxy derivative.

4.9 Neolignans from *Leontopodium leontopodioides*

In [125], two new compounds were reported from the aerial parts of the Chinese plant, *Leontopodium leontopodioides*, where compound 1 was a neolignan and compound 2 a benzofuran derivative. The automated computerized evaluation process gave for both compounds "Minor Revision". The ^{13}C NMR data for compound 1 showed two peaks at 146.3 (C-5) and 147.5 (C-3) ppm, which were in contradiction with a *meta*-dioxygenated aromatic fragment but fit quite well with an *ortho*-dioxygenated phenyl group. The data as given in [125] are shown in Fig. 24a. These obviously incorrect values are contained in the database used for prediction based on the HOSE-code technology, therefore leading to predicted values that are only slightly different from the experimentally determined ones, as shown in Fig. 24b. When comparing the HOSE-code derived values with the NN-prediction, large differences could be observed, as shown in Fig. 24c. This is because the HOSE-code technology tends to exactly reproduce the content of the database, whereas the NN-method is more tolerant and aims to generalize increments from a large amount of data. As a result, all carbons in the phenyl ring as well as two carbon atoms in the bicyclic system were involved in the structure generation process, as represented in Fig. 24d.

The structure generation process created 3838 structure proposals within 280 s, of which three out of these 3838 were present either in the "CSEARCH" or the PubChem collections. Of these alternative structures, the subsequent ranking placed the structural proposal for compound 1 in [125] at position 678 with an average deviation of 3.58 ppm, whereas the most reasonable alternatives were found at positions 2 and 3 having an average deviation of 1.37 and 1.40 ppm. From this evaluation, it was observed that the structure for compound 1 [125] is erroneous. Two alternative structures having nearly identical similarity values could be proposed, but this allowed no final decision to be made based only on these calculations. However, this result provided valuable information for selecting appropriate

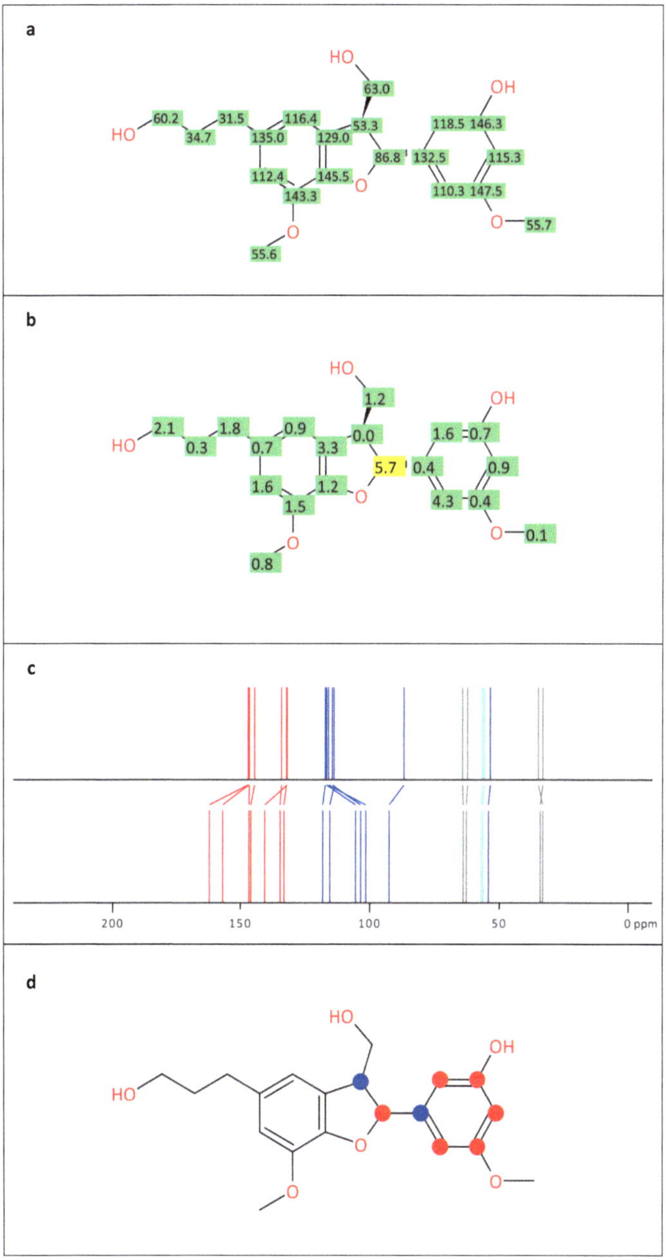

Fig. 24 (**a**) Structure of compound 1 and its ^{13}C NMR chemical shift values, as given in [125]. (**b**) Deviation between experimentally obtained and predicted chemical shift values. (**c**) NN-values (bottom) showing large differences when compared against the HOSE-code based values (top), because the identical (incorrect) structure was contained in the database. (**d**) Carbon positions that were involved in the process of structure generation

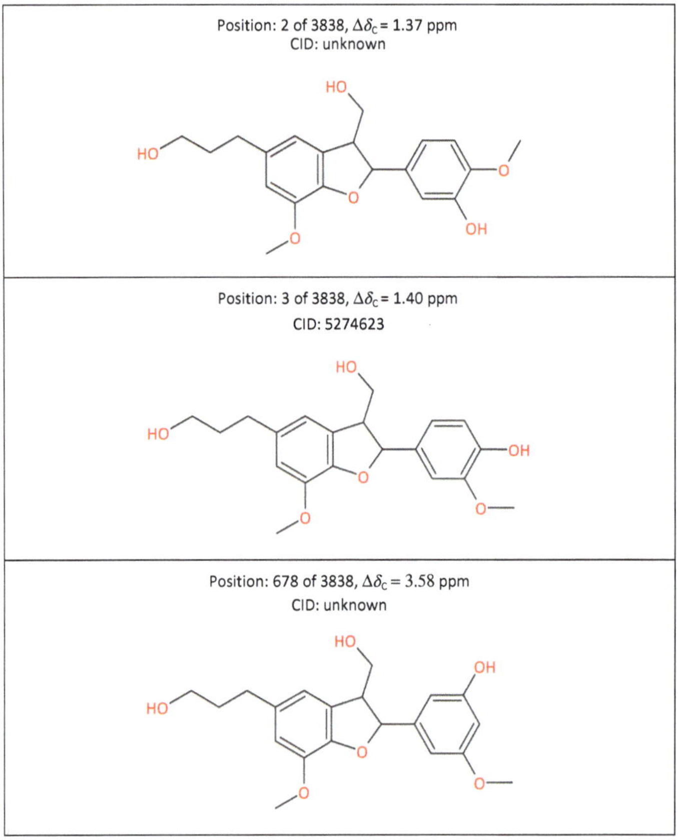

Fig. 25 The original structure proposed in [125] showed an average deviation of 3.58 ppm and a ranking at position 678 in the list of alternative structures; both reasonable alternative structure proposals can be found at positions 2 and 3 having an average deviation of 1.37 and 1.40 ppm. The substitution pattern thus changes from *meta*- to *ortho*-dihydroxylation

additional experimental approaches in order to distinguish between these two isomers (Fig. 25).

4.10 Vismiaphenones D-G from Vismia cayennensis

The isolation and structure elucidation of four compounds named vismiaphenones D to G is described in [126]. The evaluation of vismiaphenones D, E, and F gave in all three cases a result of "Minor Revision" showing that the differences between the measured and predicted chemical shift values were very small, making these structures reliable proposals for the given ^{13}C NMR data. However, the data for vismiaphenone G showed inconsistencies, with the largest differences evident at the carbon atoms of the oxirane ring system, suggesting strongly that the structure proposed is not consistent with the published data. The subsequent structure generation produced a series of pyran derivatives having similarity measures within a very small range, but allowing no final decision to be made between these alternative proposals without further experiments. For all four compounds, all the isomers generated are unknown in the PubChem database. From a chemical point of view, ring opening of the oxirane ring and addition of the solvent (CD_3OD) cannot be excluded without inspection of the original data. Since 1999, when this paper was published, it has become usual to publish experimental NMR spectra in the supporting information of natural product isolation chemistry papers as PDF files. However, if the raw NMR data had been included also in the supporting information of [120], this would have permitted further investigation of the structures proposed (Fig. 26).

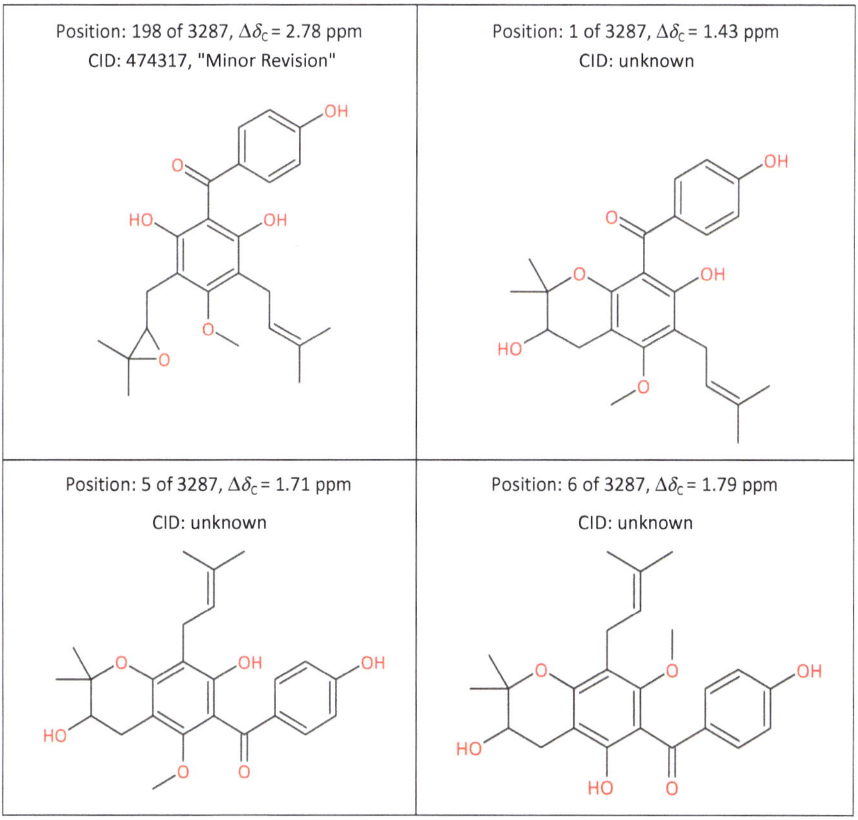

Fig. 26 The original structure proposal and its ^{13}C NMR data of [126] are ranked at position 198 of 3287 created structural proposals. Three possible alternative proposals are presented having much smaller average deviations

4.11 *Drymaritin from Drymaria diandra*

The structure elucidation of drymaritin [102] using 400 and 600 MHz NMR spectra including ^{15}N-^{1}H HMBC measurements showed clearly the limitations of using a manual approach to the interpretation of correlation spectra that disregarded alternative structure proposals. This example of an incorrect structural proposal, because it was verified by total synthesis, was used extensively as a test set of data during the development of the software toolkit described in this chapter. A more detailed description was provided in Sect. 3.7.

4.12 *Bellalosides A and B from Belamcanda chinensis*

In [127], three new compounds, bellalosides A–C (compounds 1–3) were reported from the rhizomes of the Thai medicinal plant, *Belamcanda chinensis*. The ^{13}C NMR data of four compounds are given in this paper, compounds 1–3 and the already known substance, 4. Using the previously mentioned automated computer evaluation, this resulted in ratings of "Major Revision" for compounds 1 and 2, "Accept" for compound 3, and "Minor Revision" for compound 4. A detailed inspection of the protocol for compound 4 showed an exact coincidence between the measured and predicted chemical shift values, but the resulting need for "Minor Revision" was based solely on the fact that the underlying data used for prediction exhibited some inconsistencies. The compound identified is included 16 times in the database with slightly different assignments, which were detected automatically and influenced the classification of the resultant dataset.

Bellalosides A and B (compounds 1 and 2 in [127]), according to their structures, contain a 1,2-dioxygenated phenyl ring, in agreement with the ^{13}C NMR data given. However, obviously incorrectly assigned chemical shift values at 148.6 (C-3) and 150.2 (C-4) ppm were in evidence for both compounds. Subsequent use of the "Structure Generator" showed a high probability of these structures being correct, in spite of having erroneous ^{13}C NMR signal assignments. In the case of bellalosides A and B, their published structures were ranked at positions 74 and 92 of the potential alternative structures, having $\Delta\delta_C$ values of 1.33 and 1.49 ppm. The best reasonable alternatives were, in turn, at positions 188 and 1032, with similarity measures of 1.51 and 2.55 ppm, among 5286 generated possible proposals for bellaloside A and 5156 for bellaloside B. A visualization of these findings is given in Fig. 27 for bellaloside A; the situation with bellaloside B was quite similar.

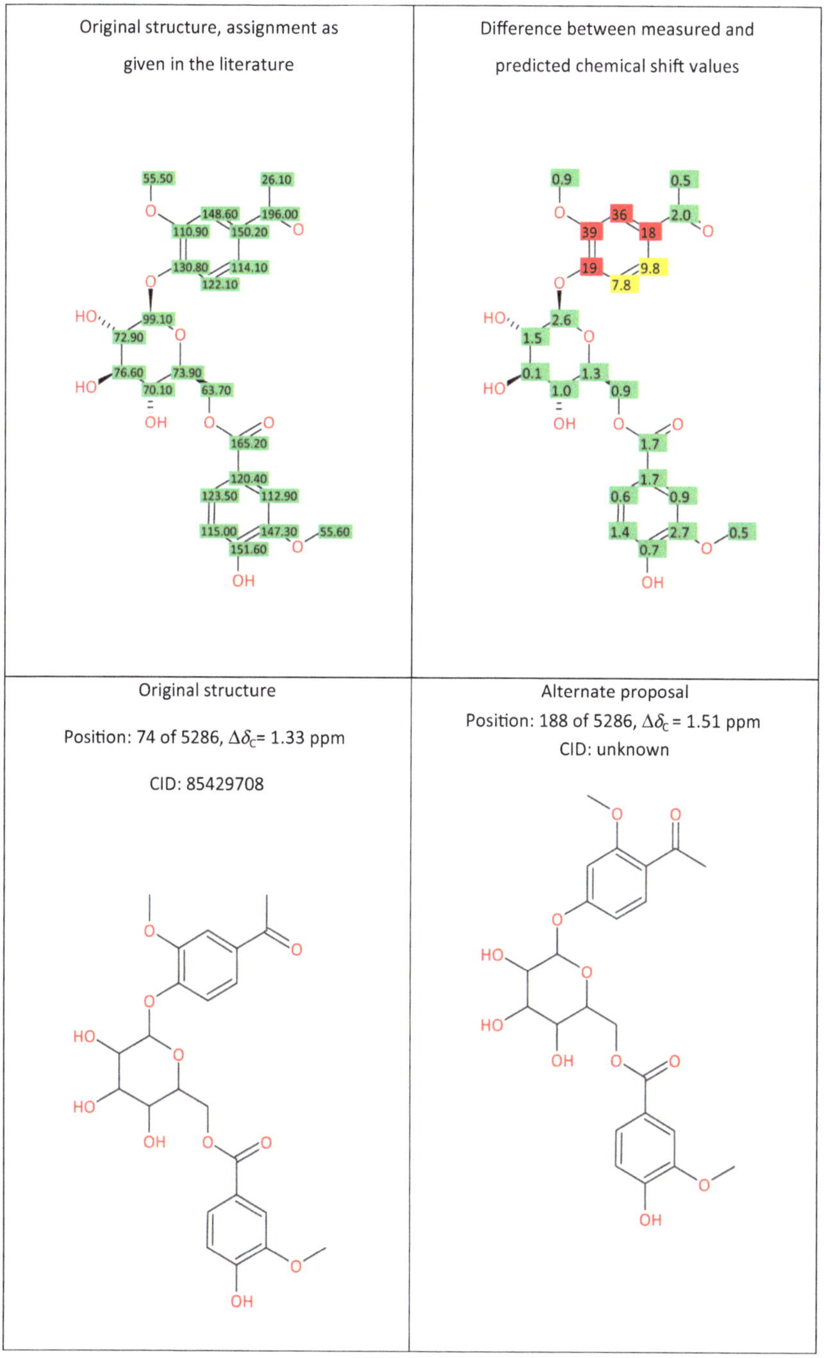

Fig. 27 ^{13}C NMR data for bellaloside A of [127] showing assignment errors in one phenyl ring. The “Structure Generator” process created 5286 proposals, with the original structure being ranked at position 74, having a quite small average deviation of 1.33 ppm

4.13 Brominated Metabolites from *Suberea mollis*

In [128], four brominated metabolites were reported as new compounds from the Red Sea sponge *Suberea mollis*, with ^{1}H, ^{13}C, and HMBC NMR spectroscopy being among the methods used for their structural determination. On application of the automated computerized procedure, the new bromotyrosine-derived alkaloids subereamollines A and B were classified as "Accept", while the new brominated phenolic compounds, subereaphenols B and C (compounds 7 and 9 in [128]), were rated as "Minor Revision". The ^{13}C NMR signals assigned to subereaphenol B showed deviations at all positions within the aromatic ring system, with the chemical shift values of 149.7 (C-1) and 143.7 (C-3) ppm being inconsistent with a 1,3-hydroxybenzene derivative, as proposed [128]. The "Structure Generator" process led to 1256 alternative structural proposals, in which those having a 1,4-dihydroxy pattern were ranked first, followed by compounds having a 1,2-dihydroxylation pattern. The originally proposed structure occurred at position 209 having a $\Delta\delta_C$ of 4.65 ppm, showing this structure was incompatible with the ^{13}C NMR data assigned. More than ten regioisomers were found with a smaller deviation than 4.65 ppm making the selection of the correct structure quite difficult. Three compounds available within the PubChem collection are given in Fig. 28, together with the structure published for subereaphenol B. However, any discrimination between the best proposals based solely on this evaluation cannot be recommended because of the small differences in their similarity measures.

4.14 Asperjinone from the Thermophilic Fungus, *Aspergillus terreus*

Asperjinone, isolated from a thermophilic specimen of *Aspergillus terreus*, was reported as new, with its structure elucidation process involving ^{1}H, ^{13}C, and HMBC NMR measurements [130]. Asperjinone was projected as a *nor*-neolignan having an oxirane moiety. The resonance lines at 68.8 (C-8″) and 77.0 (C-9″) ppm were assigned to the oxirane carbons, which were far outside the expected range. Evaluation by the "CSEARCH-Robot-Referee" recommended "Minor Revision", and the use of the "Structure Generator" procedure created 2511 alternative chemical structures. After being ranked, the structure proposed in [130] was found at position 96 with a $\Delta\delta_C$ value of 2.89 ppm, whereas a corrected structure [131] was found at position 1 ($\Delta\delta_C = 1.64$ ppm). The published structural reassignment for asperjinone utilized the assigned ^{13}C NMR chemical shift values and the HMBC correlation signals [131] (Fig. 29).

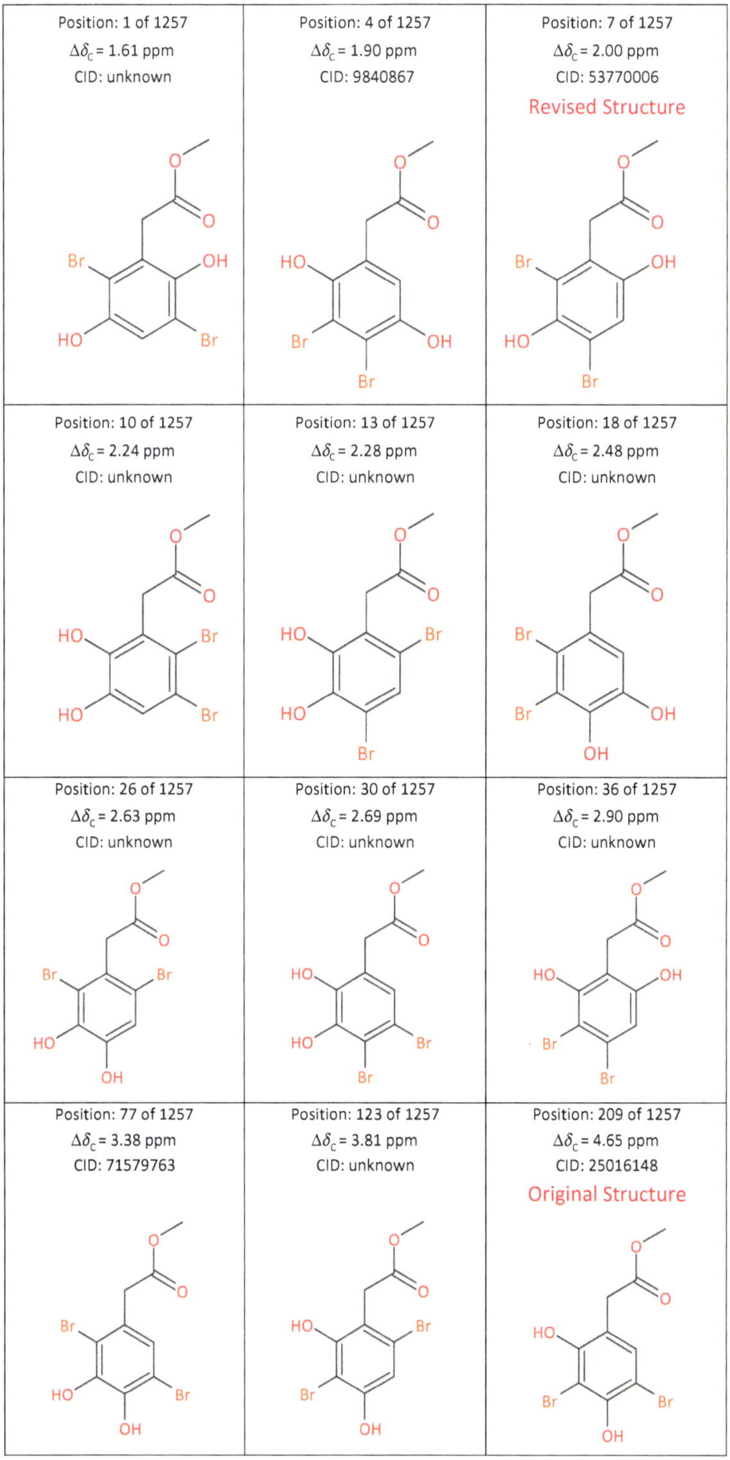

Fig. 28 Evaluation based on the original data of subereaphenol B of [128]. A total of 1257 structure proposals was created, with the original structure is ranked at position 209. In total, 11 reasonable alternatives are shown, and the revised structure published in [129] can be found at position 7

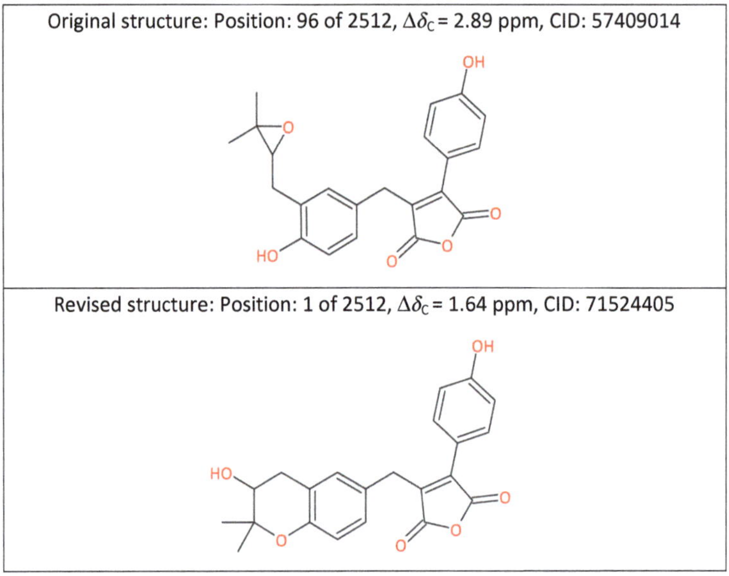

Fig. 29 Ranking of the original structure asperjinone in [130] at position 96 of the structural alternatives, and the revised structure for this compound as in [131] at position 1

4.15 Triterpenoid Constituents of Adiantum venustum

Reference [132], published in 2000, described three triterpenoids (compounds 1–3) from the Himalayan fern, *Adiantum venustum*. The ^{13}C NMR spectroscopic data of these three triterpenoids and the acetylated derivative (4) of compound 3 were measured at 600 MHz using 1D and 2D NMR techniques for their individual structure elucidation and signal assignments. Compound 1 was named adiantulupanone and elucidated as 30-normethyllupane-20-one. On application of the "CSEARCH-Robot-Referee" system, a few deviations were evident between the measured and predicted chemical shift values for this compound, and the resulting evaluation was "Minor Revision". A structure search conducted using SciFinder® enabled the retrieval of eight publications showing this structure, with the first published in 1973 [133]. Within the "CSEARCH" databases the identical structure could be found taken from [134]. The "Structure Generator" was used to modify the original topology, which was ranked at position 268 of the possible 2276 alternative structures generated, having a $\Delta\delta_C$ of 1.01 ppm, to 30-*nor*-21β-hopan-22-one. This modified structure was ranked at position 2 with a $\Delta\delta_C$ of 0.73 ppm. It is notable that when inverting the search strategy and using the spectroscopic data instead of the chemical structure as a query, this resulted in the finding of the known compound adiantone [135].

Evaluation of the ^{13}C NMR data of compound 2 (Fig. 30) from [132] resulted in a classification of "Major Revision". This compound was given the name

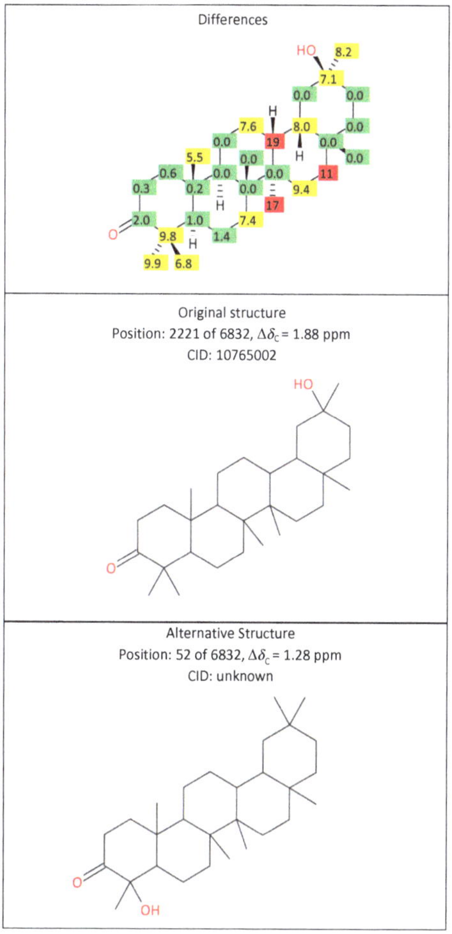

Fig. 30 Evaluation of compound 2 from [132], in which the large differences between the measured and predicted ^{13}C NMR chemical shift values resulted in a classification of "Major Revision". The "Structure Generator" created 6832 structures, with the original structure is ranked at position 2221, and a reasonable alternative structure found at position 52

adiantuoleanthone and assigned with the structure 30-*nor*-methyl-olean-3-one-30β-ol. The "Structure Generator" process as well as the "Spectroscopic Similarity Search" using the data of this compound pointed to alternative structures having oxo and hydroxy groups at neighboring positions. The alternative that was ranked first in the "Spectroscopic Similarity Search" was 30-*nor*-gammaceran-21-one-22-ol (CAS-RN: 18004-20-1), whereas the best answer obtained from the "Structure Generator" procedure is shown in Fig. 30. To avoid any misunderstanding, it is not claimed that one of these two most probable structures represents the correct solution of this structure elucidation problem, but it is only indicated that the structure proposed for compound 2 in [132] is highly unlikely and the two reasonable alternatives were presented, in order to demonstrate the questionable structure elucidation procedure for this compound.

The evaluation of compounds 3 and 4 (the acetylated derivative of compound 3) in [132] resulted in a classification of "Minor Revision". Both of these lanosterol derivatives show large differences between the measured and predicted ^{13}C NMR chemical shift values at positions C-3, C-8, C-13, and C-14 in the ring skeleton and at positions C-24 and C-28 in the side chains (Fig. 31).

In both cases, the "Structure Generator" did not create an alternative that could be selected readily; the published structures occupied, in turn, position 2784 of 8448 options ($\Delta\delta_C = 2.29$ ppm) and position 1584 of 5802 options ($\Delta\delta_C = 2.15$ ppm). From a more detailed search with the complete "CSEARCH" database, a considerable spectroscopic similarity of compound 3 to filicenol B [136] was found.

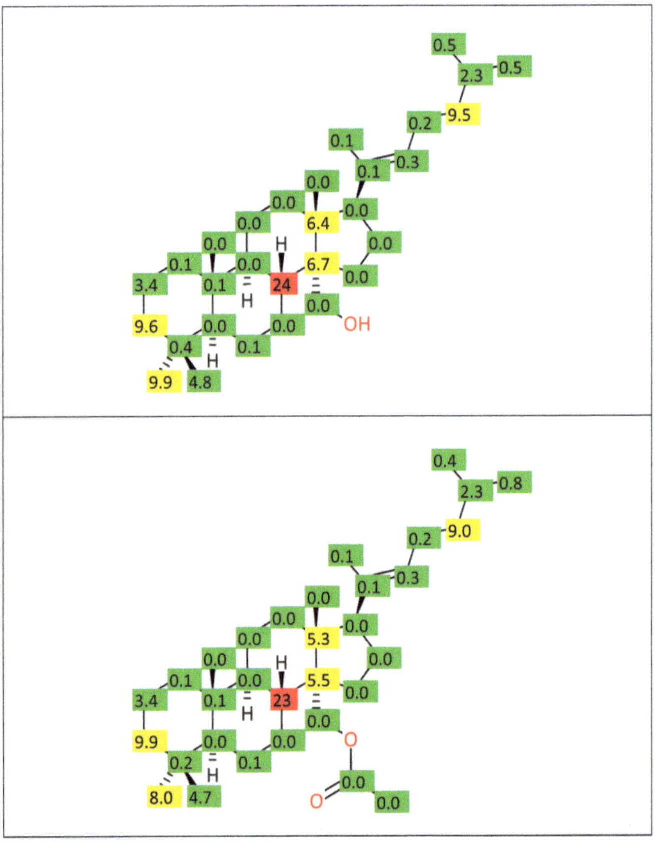

Fig. 31 Compounds 3 and 4 from [132], showing large differences between their measured and predicted ^{13}C NMR chemical shift values

4.16 *Pentaoxygenated Xanthones from Bredemeyra brevifolia*

In [137], two pentaoxygenated xanthones put forward as being new, and several known substances were reported from the roots of the Brazilian plant, *Bredemeyra brevifolia*. The authors gave ^{13}C NMR data for a total number of five compounds, with two (compounds 2 and 3) isolated from the plant investigated, whereas compounds 7–9 were used as model compounds from the literature. An analysis using the "CSEARCH-Robot-Referee" gave evaluations of "Accept" for compound 2 and "Minor Revision" for compound 3 (Fig. 32). The result for compound 3 was based on the fact that there were some inconsistencies in the underlying database used for spectrum prediction as well as an assignment problem for two methoxy groups present in the molecule.

Application of the same automated computerized workflow to the reference compounds 7–9 gave "Minor Revision" for compound 9, "Reject" for 7, and "Major Revision" for 8. In all three cases, the subsequent structure generation process led to better-fitting alternative structures. For compound 9, the improvement was very small, with $\Delta\delta_C = 2.11$ ppm for the original structure and

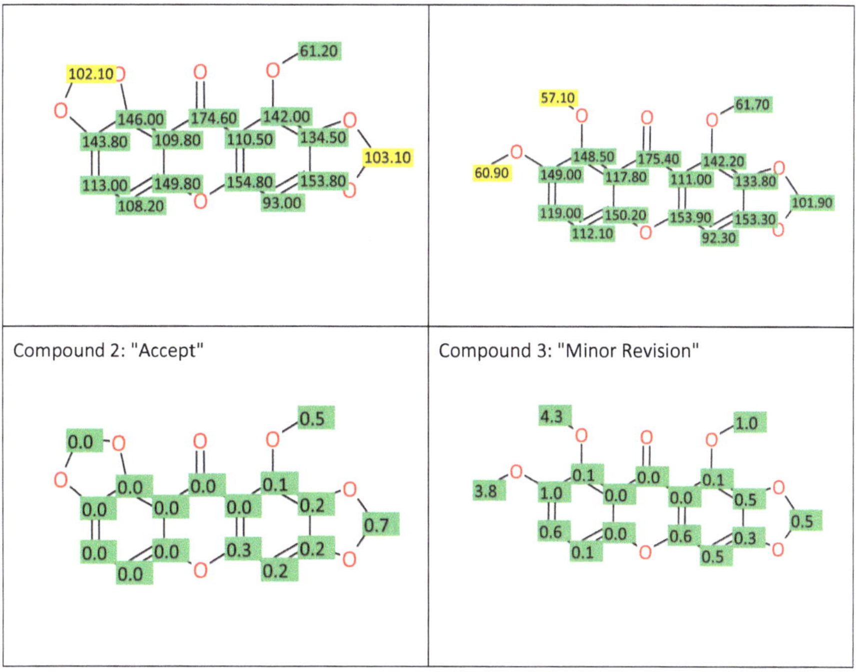

Fig. 32 Compounds 2 and 3 from [137], for which the evaluation of the data of compound 2 resulted in a rating of "Accept", while the "Minor Revision" assessment for compound 3 was based partly on an assignment problem at two methoxy groups. The top row shows the experimentally obtained data and the bottom row the differences between the measured and predicted chemical shift values

Compound 7 as given in [137] Position: 374 of 2883, $\Delta\delta_C = 3.28$ ppm "Reject", CID: unknown 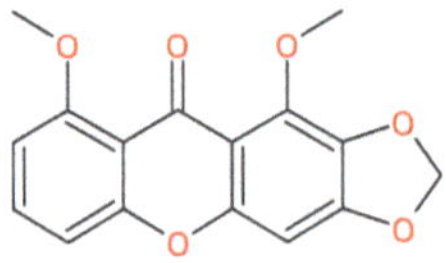	Alternative structure proposal = Compound 8 [137] Position: 1 of 2883, $\Delta\delta_C = 1.22$ ppm CID: 85670503 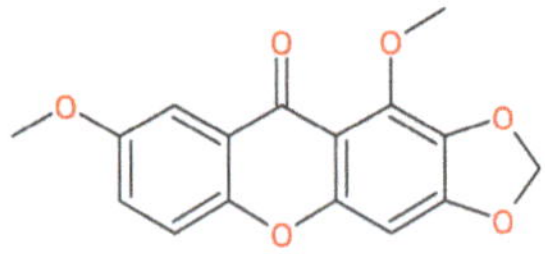
Compound 8 as given in [137] Position: 149 of 3025, $\Delta\delta_C = 3.00$ ppm "Major Revision", CID: 85670503 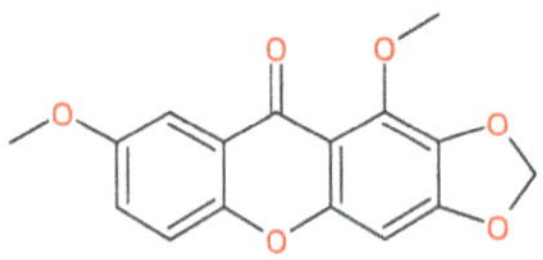	Alternative structure proposal = Compound 7 [137] Position: 1 of 3025, $\Delta\delta_C = 1.57$ ppm CID: unknown 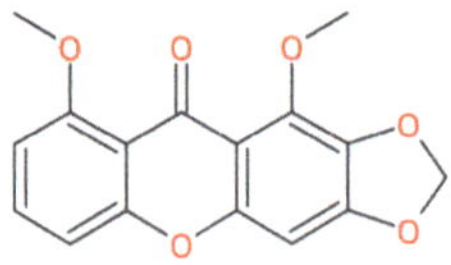

Fig. 33 A combination of structural data and given chemical shift values for compounds 7 and 8 as given in [137], which led in each case to the other isomer. When starting with the data of compound 7, the best ranked alternative structure proposal is compound 8, and vice versa. Compounds 7 and 8 were used as model compounds from earlier publications and their structure drawings were transposed in [137]

$\Delta\delta_C = 2.00$ ppm for the alternative. This alternative could be excluded readily using the chemical shift value for the methoxy groups, since two of these had chemical shift values of 61.5 and 61.6 ppm. This requires two di-*ortho*-positioned methoxy groups, as found in compound 9, but not in the proposed alternative structure.

Detailed analysis of the data given for reference compounds 7 and 8 as tabulated in [137] led to the finding that the structural drawings of compounds 7 and 8 were interchanged. For the conclusions drawn in [137] from these data this error is irrelevant, but when using these interchanged data as reference information in the future a proliferation of this error might well occur (Fig. 33).

4.17 Dulcinone from Garcinia dulcis

In [138], four new compounds called dulcisxanthones C–F were reported from the flowers of *Garcinia dulcis* collected in Thailand. Two comprehensive tables showing HMBC correlation signals were provided. An evaluation using "CSEARCH-Robot-Referee" of dulcinone, a fifth new xanthone proposed [138], resulted in a "Major Revision" rating, with major incompatibilities evident in the phenyl ring. The chemical shift values of 160.9 and 154.6 ppm are consistent with a 5,7-hydroxylation profile instead of the postulated 6,8-hydroxy group pattern

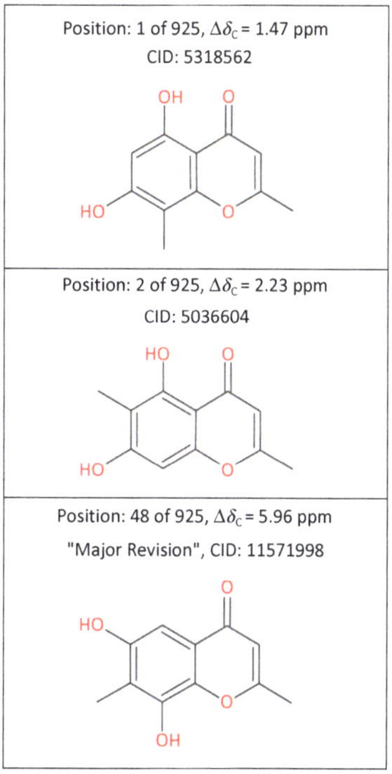

Fig. 34 Structure generation starting from the data given for dulcinone of [138], leading to two reasonable alternative structural proposals ranked at positions 1 and 2. These gave improved similarity measures of 1.47 ppm and 2.23 ppm, respectively, instead of 5.96 ppm as obtained for the original structure proposed

proposed. The "Structure Generator" created 925 structures, and, among these, two isomers having the expected 5,7-dihydroxy pattern could be located at positions 1 and 2 within the ranked list of alternative structures, having significantly smaller differences in their similarity measures, as shown in Fig. 34.

4.18 Moracins Q-U from Morus mesozygia

The structure elucidation of five benzofuran derivatives named moracins Q–U is described in [139]. These compounds were isolated from the trunk bark of *Morus mesozygia* collected in Cameroon. The result of the evaluation of these compounds using the automated computerized system resulted in a "Minor Revision" evaluation for each of these substances. When analyzing the data in more detail, the average deviation between the measured and predicted ^{13}C NMR chemical shift values was in the range 1.38 to 1.91 ppm for moracins R to U, with no better known structural alternatives found. However, moracin Q showed a much higher deviation ($\Delta\delta_C = 2.58$ ppm) and was ranked at position 84 within the list of structural alternatives, having an already known structure ranked at position 1 with a $\Delta\delta_C$ of 1.09 ppm. The pyran-derived structure as given in the literature should be modified to the appropriate furan analogue and this alternative ranked first by its similarity measure among 5031 created structures.

These two possible structural motifs can be distinguished readily by their ^{13}C NMR data spectroscopic pattern. The furan-derived structures have a quaternary carbon around 71 ppm and a CH group around 91 ppm, whereas the pyran-derived analogues have the quaternary carbon around 76 ppm and the CH group around 72 ppm.

A very interesting finding showing the questionable style of structure storing and publication can be demonstrated with this example. Thus, the ^{13}C NMR chemical shift values are taken from the above-mentioned paper (Table 1 in [139]), this structure can be found in the PubChem collection (CID: 42605182). Taking the InChi (IUPAC International Chemical Identifier), which can be found on the PubChem webpage for this compound, and pasting this information into the "Structure Editor Function" used in SciFinder® has created a pyran-derived derivative identical with the structure shown in Fig. 35. Using this structure as a query for a search for identical molecular topologies retrieved no entry within the Chemical Abstracts Service REGISTRY℠ file (performed May 27, 2016). When starting from [139] and retrieving the structures published therein, it was not possible to find this compound. On performing the same procedure using the furan-derived structure (CID: 53306268), this retrieved two compounds (Chemical Abstracts Service (CAS) Registry Numbers®: 1310328-29-0 and 1418764-58-5). It was found that these two compounds are contained in three publications, including one paper published by the same group [140] who were responsible for

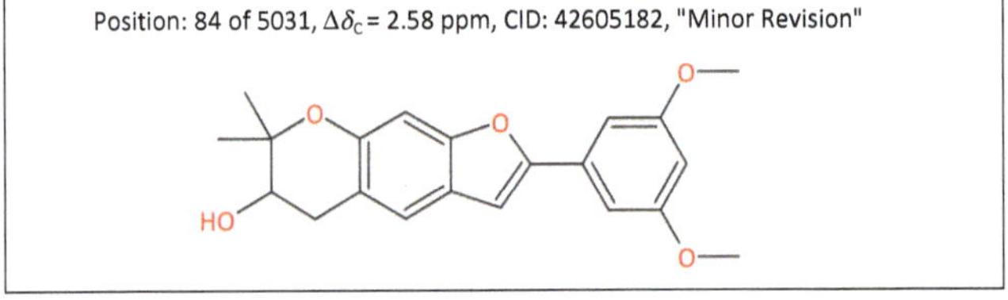

Fig. 35 Evaluation of the data given for moracin Q in [139]. The original structure was ranked at position 84 of alternative structures, whereas the corrected structure published in [140] is found at position 1

[139]. Reference [140] cites [139] without even mentioning a structural revision. As such, [140] is written so as to constitute a completely independent structure elucidation instead of a compound structural revision. Furthermore, the incorrect structure proposed in [139] is no longer included in the CAS REGISTRY^{SM}, which is worthy of mention.

Accordingly, starting from a simple analysis of the ^{13}C NMR data, together with the associated structure proposals as published in [139], this has shown (a) that an incorrect but published structure cannot be found in the CAS REGISTRY^{SM} file, and (b) a structure revision has been performed [140] without either pointing to the original structure and or clearly stating that an incorrect structure has been proposed in a previously published paper [139].

5 Summary

The "CSEARCH-Robot-Referee" database described in this chapter allows for the determination of the compatibility between a chemical structure proposed in the literature and its ^{13}C NMR chemical shift values, and could be an adjunct to the peer review process of technical journals. A number of errors in the chemical literature have been pointed out, and these represent either or both spectroscopic data assignment problems or inaccurate structural assignments. Perhaps these are just the tip of the iceberg of such problems. A steadily increasing number of review articles have appeared dealing with the need for structure revisions of natural products, despite the tremendous progress in the utility of analytical techniques that can be applied to organic structure elucidation problems. Rather than detecting and correcting incorrect structures, it would be much better to avoid such errors from appearing at all during the structure elucidation process. The "CSEARCH-Robot-Referee" system is able to point to regions of a given proposed molecule that look dubious and require further investigation by additional experimental techniques. In addition, when the user enters assignments of the ^{13}C NMR chemical shift values for distinct carbon positions in the molecule, such information can be used in subsequent predictions. It is estimated by the author that at least 80% of the incorrect structures and/or assignments as published in the chemical literature could be detected by systematic application of this type of approach. Furthermore, in many cases reasonable alternative proposals can be generated by application of the "Structure Generator" and "Spectroscopic Similarity Search" systems. Required as the necessary input data are only the structure proposed and the peak list of a ^{13}C NMR measurement as obtained from the peak-picking regimen. Access is already supported by commonly used NMR software suites and could be applied routinely for every new or known compound identified on a routine basis. However, for complex cases like the inversion of one chiral center among many others present, these would structure revision to be conducted only after total synthesis.

References

1. Nicolaou KC, Montagnon T (2008) Molecules that changed the world. Wiley-VCH, Weinheim
2. Nicolaou KC, Snyder SA (2005) Chasing molecules that were never there: misassigned natural products and the role of chemical synthesis in modern structure elucidation. Angew Chem Int Ed 44:1012
3. Suyama TL, Gerwick WH, McPhail KL (2011) Survey of marine natural product structure revisions: a synergy of spectroscopy and chemical synthesis. Bioorg Med Chem 19:6675
4. Maier ME (2009) Structural revisions of natural products by total synthesis. Nat Prod Rep 26:1105
5. Garcia-Rubino ME, Mahfoudh N, Campos JM (2014) Structural elucidation errors in organic chemistry. Curr Org Chem 18:1513
6. Yoo H-D, Nam S-J, Chin Y-W, Kim M-S (2016) Misassigned natural products and their revised structures. Arch Pharm Res 39:143
7. Elyashberg M, Williams AJ, Blinov K (2010) Structural revisions of natural products by Computer-Assisted Structure Elucidation (CASE) systems. Nat Prod Rep 27:1296
8. Elyashberg M (2015) Identification and structure elucidation by NMR spectroscopy. Trends Anal Chem 69:88
9. https://scifinder.cas.org
10. Reynolds WF, Mazzola EP (2015) Nuclear magnetic resonance in the structural elucidation of natural products. Prog Chem Org Nat Prod 100:223
11. Schütz V, Purtuc V, Felsinger S, Robien W (1997) CSEARCH-Stereo: a new generation of NMR database systems allowing three-dimensional spectrum prediction. Fresenius J Anal Chem 359:33
12. Kalchhauser H, Robien W (1985) CSEARCH: a computer program for identification of organic compounds and fully automated assignment of carbon-13 nuclear magnetic resonance spectra. J Chem Inf Comput Sci 25:103
13. http://nmrpredict.orc.univie.ac.at/c13robot/robot.php
14. http://nmrpredict.orc.univie.ac.at/csearchlite/update.htm
15. Robien W (2009) Do high-quality ^{13}C-NMR spectroscopic data really come from journals with high Impact Factors? Trends Anal Chem 28:914
16. Li M-M, Wang K, He J, Peng L-Y, Chen X-Q, Cheng X, Zhao Q-S (2013) Four new labdane-type diterpenoid glycosides from *Diplopterygium laevissimum*. Nat Prod Bioprospect 3:38
17. Hemalatha K, Hareeka N, Sunitha D (2012) Chemical constituents isolated from leaves of *Barleria christata* Linn. Int J Pharm Bio Sci 3:609
18. Hashimoto K, Katsuhara T, Niitsu K, Ikeya Y, Okada M, Mitsuhashi H (1992) Two glycosides from roots of *Asiasarum sieboldi*. Phytochemistry 31:2477
19. Fan H, Yang G-Z, Zheng T, Mei Z-N, Liu X-M, Chen Y, Chen S (2010) Chemical constituents with free-radical-scavenging activities from the stem of *Microcos paniculata*. Molecules 15:5547
20. Tsuge N, Mori T, Hamano T, Tanaka H, Shin-Ya K, Seto H (1999) Cinnatriacetins A and B, new antibacterial triacetylene derivatives from the fruiting bodies of *Fistulina hepatica*. J Antibiot 52:578
21. Mariani C, Braca A, Vitalini S, De Tommasi N, Visioli F, Fico G (2008) Flavonoid characterization and in vitro antioxidant activity of *Aconitum anthora* L. (Ranunculaceae). Phytochemistry 69:1220
22. Lu Y, Sun Y, Foo LY, McNabb WC, Molan AL (2000) Phenolic glycosides of forage legume *Onobrychis viciifolia*. Phytochemistry 55:67
23. Demirkiran O, Mesaik MA, Beynek H, Abbaskhan A, Choudhary MI (2013) Immunosupressive phenolic constituents from *Hypericum montbretii* Spach. Rec Nat Prod 7:210

24. Panthama N, Kanokmedhakul S, Kanokmedhakul K (2009) Galloyl and hexahydroxydiphenoyl esters of phenylpropanoid glucosides, phenylpropanoids and phenylpropanoid glucosides from rhizome of *Balanophora fungosa*. Chem Pharm Bull 57:1352
25. Sichaem J, Kaennakam S, Siripong P, Tip-pyang S (2012) Tabebuialdehydes A-C, cyclopentene dialdehyde derivatives from the roots of *Tabebuia rosea*. Fitoterapia 83:1456
26. Es-Safi N-E, Khlifi S, Kerhoas L, Kollmann A, Abbouyi AE, Ducrot P-H (2005) Antioxidant constituents of the aerial parts of *Globularia alypum* Growing in Morocco. J Nat Prod 68:1293
27. Kanokmedhakul S, Kanokmedhakul K, Kanarsa T, Buayairaksa M (2005) New bioactive clerodane diterpenoids from the bark of *Casearia grewiifolia*. J Nat Prod 68:183
28. Jin D-Z, Min Z-D, Chiou GCY, Iinuma M, Tanaka T (1996) Two *p*-coumaroyl glycerides from *Juncus effusus*. Phytochemistry 41:545
29. Cortez DAG, Young MCM, Marston A, Wolfender J-L, Hostettmann K (1998) Xanthones, triterpenes and a biphenyl from *Kielmeyera coriacea*. Phytochemistry 47:1367
30. Goetz G, Fkyerat A, Métais N, Kunz M, Tabacchi R, Pezet R, Pont V (1999) Resistance factors to grey mould in grape berries: identification of some phenolics inhibitors of *Botrytis cinerea* stilbene oxidase. Phytochemistry 52:759
31. Lee JP, Min BS, An RB, Na MK, Lee SM, Lee HK, Kim JG, Bae KH, Kang SS (2003) Stilbenes from the roots of *Pleuropterus ciliinervis* and their antioxidant activities. Phytochemistry 64:759
32. http://mestrelab.com/software/mnova/verify/
33. ftp://ftp.ncbi.nih.gov/pubchem/Compound/CURRENT-Full/SDF/
34. https://www.bruker.com/products/mr/nmr/nmr-software/software/topspin/overview.html
35. http://mestrelab.com/software/mnova/nmr/
36. Bremser W (1978) HOSE—a novel substructure code. Anal Chim Acta 103:355
37. http://www.doi.org
38. http://www.inchi-trust.org/downloads/
39. https://www.emolecules.com
40. Ross H (2015) EurJOC has come a long way. Eur J Org Chem 2015:4
41. Fuwa H, Ebine M, Bourdelais AJ, Baden DG, Sasaki M (2006) Total synthesis, structure revision, and absolute configuration of (−)-brevenal. J Am Chem Soc 128:16989
42. Pietruszka J, Rieche ACM (2008) Total synthesis of marine oxylipins solandelactones A–H. Adv Synthesis Catal 350:1407
43. Bell RA, Dickson KC, Valliant JF (1999) The total synthesis of a technetium chelate—tamoxifen complex. Can J Chem 77:146
44. Gustafsson T, Saxin M, Kihlberg J (2003) Synthesis of a *C*-glycoside analogue of *β*-D-galactosylthreonine. J Org Chem 68:2506
45. Morales A, Ochoa E, Suárez M, Verdecia Y, González L, Martín N, Quinteiro M, Seoane C, Soto JL (1996) Novel hexahydrofuro[3,4-*b*]-2(1*H*)-pyridones from 4-aryl substituted 5-alkoxycarbonyl-6-methyl-3,4-dihydropyridones. J Heterocycl Chem 33:103
46. Rodriguez H, Martin O, Suárez M, Martín N, Albericio F (2011) Eco-friendly methodology to prepare N-heterocycles related to dihydropyridines: microwave-assisted synthesis of alkyl 4-arylsubstituted-6-chloro-5-formyl-2-methyl-1,4-dihydropyridine-3-carboxylate and 4-aryl substituted-4,7-dihydrofuro[3,4-*b*]pyridine-2,5(1*H*,3*H*)-dione. Molecules 16:9620
47. https://scifinder.cas.org; Predicted NMR data calculated using Advanced Chemistry Development, Inc. (ACD/Labs) Software V11.01
48. Suárez M, Martín N, Martínez R, Verdecia Y, Molero D, Alba L, Seoane C, Ochoa E (2002) ^{1}H and ^{13}C spectral assignment of *o*-chloroformyl substituted 1,4-dihydropyridine derivatives. Magn Reson Chem 40:303
49. Elnagdi NMH, Al-Hokbany NS (2012) Organocatalysis in synthesis: L-proline as an enantioselective catalyst in the synthesis of pyrans and thiopyrans. Molecules 17:4300

50. Suárez M, Molero D, Salfran E, Martín N, Verdecia Y, Martinez R, Ochoa E, Alba L, Quinteiro M, Seoane C (2001) ^{1}H and ^{13}C spectral assignment of 1,4,5,6,7,8-hexahydroquinolines and their oxo-analogues 5,6,7,8-tetrahydro-4*H*-chromenes. Magn Reson Chem 39:105
51. Suárez M, Salfrán E, Verdecia Y, Ochoa E, Alba L, Martín N, Martínez R, Quinteiro M, Seoane C, Novoa H, Blaton N, Peeters OM, De Ranter C (2002) X-ray and theoretical structural study of novel 5,6,7,8-tetrahydrobenzo-4*H*-pyrans. Tetrahedron 58:953
52. Salfrán E, Suárez M, Verdecia Y, Alvarez A, Ochoa E, Martínez-Alvarez R, Seoane C, Martín N (2004) One-step synthesis of aminopyrimidines from 5-oxo-4*H*-benzopyrans. J Heterocycl Chem 41:509
53. Suárez M, Verdecia Y, Ochoa E, Martín N, Martínez R, Quinteiro M, Seoane C, Soto JL, Novoa H, Blaton N, Peeters OM, De Ranter C (2000) Synthesis and structural study of novel 1,4,5,6,7,8-hexahydroquinolines. J Heterocycl Chem 37:735
54. Kaiser CR, Basso EA, Rittner R (2001) Substituent-induced ^{13}C chemical shifts of 3-substituted camphors. Magn Reson Chem 39:643
55. Sivakumar B, Murugan R, Baskaran A, Khadangale BP, Murugan S, Senthilkumar UP (2013) Identification and characterization of process-related impurities of *trans*-resveratrol. Sci Pharm 81:683
56. Pretsch E, Bühlmann P, Badertscher M (2009) Structure determination of organic compounds: tables of spectral data. Springer, Berlin Heidelberg
57. Ewing DF (1979) ^{13}C substituent effects in monosubstituted benzenes. Org Magn Reson 12:499
58. Liang S, Tian J-M, Feng Y, Liu X-H, Xiong Z, Zhang W-D (2011) Flavonoids from *Daphne aurantiaca* and their inhibitory activities against nitric oxide production. Chem Pharm Bull 59:653
59. Dinda B, Debnath S, Banik R (2011) Naturally occurring iridoids and secoiridoids. An updated review, Part 4. Chem Pharm Bull 59:803
60. Coy Barrera CA, Coy Barrera ED, Granados Falla DS, Delgado Murcia G, Cuca Suarez LE (2011) *seco*-Limonoids and quinoline alkaloids from *Raputia heptaphylla* and their antileishmanial activity. Chem Pharm Bull 59:855
61. Silva CMBL, Garcia FP, Rodrigues JHDS, Nakamura CV, Ueda-Nakamura T, Meyer E, Ruiz ALTG, Foglio MA, Carvalho JED, Costa WFD, Sarragiotto MH (2012) Synthesis, antitumor, antitrypanosomal and antileishmanial activities of benzo[4,5]canthin-6-ones bearing the *N*′-(substituted benzylidene)-carbohydrazide and *N*-alkylcarboxamide groups at C-2. Chem Pharm Bull 60:1372
62. Tsujimoto M, Lowtangkitcharoen W, Mori N, Pangkruang W, Putongking P, Suwanborirux K, Saito N (2013) Chemistry of ecteinascidins Part 4: preparation of 2′-*N*-acyl ecteinascidin 770 derivatives with improved cytotoxicity profiles. Chem Pharm Bull 61:1052
63. Ukida K, Doi T, Sugimoto S, Matsunami K, Otsuka H, Takeda Y (2013) Schoepfiajasmins A-H: *C*-glycosyl dihydrochalcones, dihydrochalcone glycoside, *C*-glucosyl flavanones, flavanone glycoside and flavone glycoside from the branches of *Schoepfia jasminodora*. Chem Pharm Bull 61:1136
64. Zhang B-B, Shi K, Liao Z-X, Dai Y, Zou Z-H (2011) Phenylpropanoid glycosides and triterpenoid of *Pedicularis kansuensis* Maxim. Fitoterapia 82:854
65. Li C, Fu J, Yang J, Zhang D, Yuan Y, Chen N (2012) Three triterpenoid saponins from the roots of *Polygala japonica* Houtt. Fitoterapia 83:1184
66. Ding L, Jiang Z, Liu Y, Chen L, Zhao Q, Yao X, Zhao F, Qiu F (2012) Monoterpenoid inhibitors of NO production from *Paeonia suffruticosa*. Fitoterapia 83:1598
67. Ning J, Di Y-T, Fang X, He H-P, Wang Y-Y, Li Y, S-L LI, Hao X-J (2010) Limonoids from the leaves of *Cipadessa baccifera*. J Nat Prod 73:1327

68. Fang L, Du D, Ding G-Z, Si Y-K, Yu S-S, Liu Y, Wang W-J, Ma S-G, Xu S, Qu J, Wang J-M, Liu Y-X (2010) Neolignans and glycosides from the stem bark of *Illicium difengpi*. J Nat Prod 73:818
69. Cai S, Sun S, Zhou H, Kong X, Zhu T, Li D, Gu Q (2011) Prenylated polyhydroxy-*p*-terphenyls from *Aspergillus taichungensis* ZHN-7-07. J Nat Prod 74:1106
70. Yang M-L, Kuo P-C, Hwang T-L, Wu T-S (2011) Anti-inflammatory principles from *Cordyceps sinensis*. J Nat Prod 74:1996
71. Murillo E, McLean R, Britton G, Agócs A, Nagy V, Deli J (2011) Sapotexanthin, an A-provitamin carotenoid from red Mamey (*Pouteria sapota*). J Nat Prod 74:283
72. Kavala M, Mathia F, Kozísek J, Szolcsányi P (2011) Efficient total synthesis of (+)-dihydropinidine, (−)-epidihydropinidine, and (−)-pinidinone. J Nat Prod 74:803
73. Barber JM, Quek NCH, Leahy DC, Miller JH, Bellows DS, Northcote PT (2011) Lehualides E−K, cytotoxic metabolites from the Tongan marine sponge *Plakortis* sp. J Nat Prod 74:809
74. Nelson KM, Salomon CE, Aldrich CC (2012) Total synthesis and biological evaluation of transvalencin Z. J Nat Prod 75:1037
75. Xiong L, Zhu M, Zhu C, Lin S, Yang Y, Shi J (2012) Structure and bioassay of triterpenoids and steroids isolated from *Sinocalamus affinis*. J Nat Prod 75:1160
76. Jain SK, Pathania AS, Meena S, Sharma R, Sharma A, Singh B, Gupta BD, Bhushan S, Bharate SB, Vishwakarma RA (2013) Semisynthesis of mallotus B from rottlerin: evaluation of cytotoxicity and apoptosis-inducing activity. J Nat Prod 76:1724
77. Chen D, Chen W, Liu D, van Ofwegen L, Proksch P, Lin W (2013) Asteriscane-type sesquiterpenoids from the soft coral *Sinularia capillosa*. J Nat Prod 76:1753
78. Hu Q-F, Zhou B, Ye Y-Q, Jiang Z-Y, Huang X-Z, Li Y-K, Du G, Yang G-Y, Gao X-M (2013) Cytotoxic deoxybenzoins and diphenylethylenes from *Arundina graminifolia*. J Nat Prod 76:1854
79. Huang A-C, Wilde A, Ebmeyer J, Skouroumounis GK, Taylor DK (2013) Examination of the phenolic profile and antioxidant activity of the leaves of the Australian native plant *Smilax glyciphylla*. J Nat Prod 76:1930
80. Liu X, Gan M, Dong B, Zhang T, Li Y, Zhang Y, Fan X, Wu Y, Bai S, Chen M, Yu L, Tao P, Jiang W, Si S (2013) 4862F, a new inhibitor of HIV-1 protease, from the culture of *Streptomyces* I03A-04862. Molecules 18:236
81. Starha P, Popa I, Trávnícek Z, Vanco J (2013) N6-Benzyladenosine derivatives as novel N-donor ligands of platinum(II) dichlorido complexes. Molecules 18:6990
82. Wang F, Han S, Hu S, Xue Y, Wang J, Xu H, Chen L, Zhang G, Zhang Y (2014) Two new secondary metabolites from *Xylaria* sp. cfcc 87468. Molecules 19:1250
83. Rapado LN, Freitas GC, Polpo A, Rojas-Cardozo M, Rincón JV, Scotti MT, Kato MJ, Nakano E, Yamaguchi LF (2014) A benzoic acid derivative and flavokawains from *Piper* species as schistosomiasis vector controls. Molecules 19:5205
84. Eerdunbayaer OMAA, Aoyama H, Kuroda T, Hatano T (2014) Structures of two new flavonoids and effects of licorice phenolics on vancomycin-resistant *Enterococcus* species. Molecules 19:3883
85. Sun J, He X-M, Zhao M-M, Li L, Li C-B, Dong Y (2014) Antioxidant and nitrite-scavenging capacities of phenolic compounds from sugarcane (*Saccharum officinarum* L.) tops. Molecules 19:13147
86. De Sousa Luis JA, Filho JMB, Lira BF, Medeiros IA, de Morais LCSL, dos Santos AF, de Oliveira CS, de Athayde-Filho PF (2010) Synthesis of new imidazolidin-2,4-dione and 2-thioxo-imidazolidin-4-ones via *C*-phenylglycine derivatives. Molecules 15:128
87. Yasar S, Özcan EÖ, Gürbüz N, Cetinkaya B, Özdemir Í (2010) Palladium-catalyzed Heck coupling reaction of aryl bromides in aqueous media using tetrahydropyrimidinium salts as carbene ligands. Molecules 15:649
88. Wang CF, Yang K, Zhang HM, Cao J, Fang R, Liu ZL, Du SS, Wang YY, Deng ZW, Zhou L (2011) Components and insecticidal activity against the maize weevils of *Zanthoxylum schinifolium* fruits and leaves. Molecules 16:3077

89. Jurcek O, Ikonen S, Buricová L, Wimmerová M, Wimmer Z, Drasar P, Horníček J, Galandáková A, Ulrichová J, Kolehmainen ET (2011) Succinobucol's new coat—conjugation with steroids to alter its drug effect and bioavailability. Molecules 16:9404
90. Pokhrel M, Ma E (2011) Synthesis and screening of aromatase inhibitory activity of substituted C_{19} steroidal 17-oxime analogs. Molecules 16:9868
91. Chaturvedula VSP, Upreti M, Prakash I (2011) Diterpene glycosides from *Stevia rebaudiana*. Molecules 16:3552
92. Han L, Ji L, Boakye-Yiadom M, Li W, Song X, Gao X (2012) Preparative isolation and purification of four compounds from *Cistanches deserticola* Y.C. Ma by high-speed counter-current chromatography. Molecules 17:8276
93. Wen Q, Lin X, Liu Y, Xu X, Liang T, Zheng N, Kintoko K, Huang R (2012) Phenolic and lignan glycosides from the butanol extract of *Averrhoa carambola* L. root. Molecules 17:12330
94. Erenler R, Yilmaz S, Aksit H, Sen O, Genc N, Elmastas M, Demirtas I (2014) Antioxidant activities of chemical constituents isolated from *Echinops orientalis* Trauv. Rec Nat Prod 8:32
95. Sarikaya BB, Zencir S, Somer NU, Kaya GI, Onur MA, Bastida J, Berenyi A, Zupko I, Topcu Z (2012) The effects of arolycoricidine and narciprimine on tumor cell killing and topoisomerase activity. Rec Nat Prod 6:381
96. Huang X, Mu B, Lin W, Qiu Y (2012) Pterocarpin and isoflavan derivatives from *Canavalia maritima* (Aubl.) Thou. Rec Nat Prod 6:166
97. Hwang D, Hyun J, Jo G, Koh D, Lim Y (2011) Synthesis and complete assignment of NMR data of 20 chalcones. Magn Reson Chem 49:41
98. Mattiza JT, Meyer VJ, Duddeck H (2010) Experimental verification of diverging mechanisms in the binding of ether, thioether, and sulfone ligands to a dirhodium tetracarboxylate. Magn Reson Chem 48:192
99. Devi P, Wahidullah S, Rodrigues C, Souza LD (2010) The sponge-associated bacterium *Bacillus licheniformis* SAB1: A source of antimicrobial compounds. Mar Drugs 8:1203
100. Siless GE, Knott ME, Derita MG, Zacchino SA, Puricelli L, Palermo JA (2012) Synthesis of steroidal quinones and hydroquinones from bile acids by Barton radical decarboxylation and benzoquinone addition. Studies on their cytotoxic and antifungal activities. Steroids 77:45
101. Krstic NM, Bjelakovic MS, Pavlovic VD, Robeyns K, Juranic ZD, Matic I, Novakovic I, Sladic DM (2012) New androst-4-en-17-spiro-1,3,2-oxathiaphospholanes. Synthesis, assignment of absolute configuration and in vitro cytotoxic and antimicrobial activities. Steroids 77:558
102. Hsieh P-W, Chang F-R, Lee K-H, Hwang T-L, Chang S-M, Wu Y-C (2004) A new anti-HIV alkaloid, drymaritin, and a new *C*-glycoside flavonoid, diandraflavone, from *Drymaria diandra*. J Nat Prod 67:1175
103. Bremser W, Wagner H, Franke B (1981) Fast searching for identical ^{13}C NMR spectra via inverted files. Org Magn Reson 15:178
104. http://nmrpredict.orc.univie.ac.at/similar/eval.php
105. Wetzel I, Allmendinger L, Bracher F (2009) Revised structure of the alkaloid drymaritin. J Nat Prod 72:1908
106. Li X, Wang N, Sau WM, Chen ASC, Yao X (2006) Four new isoflavonoids from the stem bark of *Erythrina variegata*. Chem Pharm Bull 54:570
107. Liu Q, Chen C-J, Shi X, Zhang L, Chen H-J, Gao K (2010) Chemical constituents from *Aphanamixis grandifolia*. Chem Pharm Bull 58:1431
108. Vo TN, Nguyen PL, Tuong LT, Pratt LM, Vo PN, Nguyen KPP, Nguyen NS (2012) Lignans and triterpenes from the root of *Pseuderanthemum carruthersii* var. *atropurpureum*. Chem Pharm Bull 60:1125
109. Shimada M, Ozawa M, Iwamoto K, Fukuyama Y, Kishida A, Ohsaki A (2014) A lanostane triterpenoid and three cholestane sterols from *Tilia kiusiana*. Chem Pharm Bull 62:937

110. Kim KS, Lee S, Shin JS, Shim SH, Kim B-K (2002) Arteminin, a new coumarin from *Artemisia apiacea*. Fitoterapia 73:266
111. Hammoda HM, Aboul Ela MA, El-Lakany AM, El-Hanbali O, Zaki CS, Ghazy NM (2008) New constituents of *Artemisia monosperma* Del. Pharmazie 63:611
112. https://www.ccdc.cam.ac.uk/deposit
113. Innok P, Rukachaisirikul T, Phongpaichit S, Suksamrarn A (2010) Fuscacarpans A–C, new pterocarpans from the stems of *Erythrina fusca*. Fitoterapia 81:518
114. Zeng X, Qiu Q, Jiang C, Jing Y, Qiu G, He X (2011) Antioxidant flavanes from *Livistona chinensis*. Fitoterapia 82:609
115. Shen C-C, Chang Y-S, Ho L-K (1993) Nuclear magnetic resonance studies of 5,7-dihydroxyflavonoids. Phytochemistry 34:843
116. Osakabe N, Yamagishi M, Sanbongi C, Natsume M, Takizawa T, Osawa T (1998) The antioxidative substances in cacao liquor. J Nutr Sci Vitaminol 44:313
117. Zhang M, Jagdmann GE Jr, Van Zandt M, Sheeler R, Beckett P, Schroeter H (2013) Chemical synthesis and characterization of epicatechin glucuronides and sulfates: bioanalytical standards for epicatechin metabolite identification. J Nat Prod 76:157
118. Liu K-Q, Cheng X, Mi Z, Peng L, Li B-C (2014) Chemical constituents of the aerial parts of *Cynanchum chinense* R. Br. J Chem Pharm Res 6:990
119. Moyo F, Gashe BA, Majinda RRT (1999) A new flavan from *Elephantorrhiza goetzei*. Fitoterapia 70:412
120. Venkataraman R, Gopalakrishnan S (2002) A lignan from the root of *Ecbolium linneanum* Kurz. Phytochemistry 61:963
121. Deachathai S, Mahabusarakam W, Phongpaichit S, Taylor WC (2005) Phenolic compounds from the fruit of *Garcinia dulcis*. Phytochemistry 66:2368
122. Wang Q-X, Bao L, Yang X-L, Guo H, Yang R-N, Ren B, Zhang L-X, Dai H-Q, Guo L-D, Liu H-W (2012) Polyketides with antimicrobial activity from the solid culture of an endolichenic *Ulocladium* sp. Fitoterapia 83:209
123. Kashiwada Y, Nonaka G-I, Nishioka I (1990) Chromone glucosides from rhubarb. Phytochemistry 29:1007
124. Ayer WA, Racok JS (1990) The metabolites of *Talaromyces flavus*: Part 1. Metabolites of the organic extracts. Can J Chem 68:2085
125. Li X, Luo J-G, Wang X-B, Luo J, Wang J-S, Kong L-Y (2012) Phenolics from *Leontopodium leontopodioides* inhibiting nitric oxide production. Fitoterapia 83:883
126. Fuller RW, Westergaard CK, Collins JW, Cardellina JH II, Boyd MR (1999) Vismiaphenones D–G, new prenylated benzophenones from *Vismia cayennensis*. J Nat Prod 62:67
127. Monthakantirat O, De-Eknamkul W, Umehara K, Yoshinaga Y, Miyase T, Warashina T, Noguchi H (2005) Phenolic constituents of the rhizomes of the Thai medicinal plant *Belamcanda chinensis* with proliferative activity for two breast cancer cell lines. J Nat Prod 68:361
128. Abou-Shoer MI, Shaala LA, Youssef DTA, Badr JM, Habib A-AM (2008) Bioactive brominated metabolites from the Red Sea sponge *Suberea mollis*. J Nat Prod 71:1464
129. Shaker KH, Zinecker H, Ghani MA, Imhoff JF, Schneider B (2010) Bioactive metabolites from the sponge *Suberea* sp. Chem Biodivers 7:2880
130. Liao Y, Shen C-N, Lin L-H, Yang Y-L, Han H-Y, Chen J-W, Kuo S-C, Wu S-H, Liaw C-C (2012) Asperjinone, a *nor*-neolignan, and terrein, a suppressor of ABCG2-expressing breast cancer cells, from thermophilic *Aspergillus terreus*. J Nat Prod 75:630
131. Elyashberg M, Blinov K, Molodtsov S, Williams AJ (2013) Structure revision of asperjinone using computer-assisted structure elucidation methods. J Nat Prod 76:113
132. Alam MS, Chopra N, Ali M, Niwa M (2000) Normethyl pentacyclic and lanostane-type triterpenes from *Adiantum venustum*. Phytochemistry 54:215
133. Vystrcil A, Blecha Z (1973) Triterpenes. XXXI. Absolute configuration at $C_{(20)}$ in 30-*nor*-20ξ-lupanol derivatives. Collect Czech Chem Commun 38:3648

134. Melos JLR, Silva LB, Peres MTLP, Mapeli AM, Faccenda O, Anjos HH, Torres TG, Tiviroli SC, Batista AL, Almeida FGN, Flauzino NS, Tibana LA, Hess SC, Honda NK (2007) Constituintes químicos e avaliacao do potencial alelopatico de *Adiantum tetraphyllum* Humb. & Bonpl. ex Willd. (Pteridaceae) Quim Nova 30:292
135. Grammes C, Burkhardt G, Becker H (1994) Triterpenes from *Fossombronia* liverworts. Phytochemistry 35:1293
136. Shiojima K, Arai Y, Kasama T, Ageta H (1993) Fern constituents: triterpenoids isolated from the leaves of *Adiantum monochlamys*. Filicenol A, filicenol B, isoadiantol B, hakonanediol, and epihakonanediol. Chem Pharm Bull 41:262
137. de Oliveira MDCF, Silveira ER (2000) Pentaoxygenated xanthones and fatty acids from *Bredemeyera brevifolia*. Phytochemistry 55:847
138. Deachathai S, Mahabusarakam W, Phongpaichit S, Taylor WC, Zhang Y-J, Yang C-R (2006) Phenolic compounds from the flowers of *Garcinia dulcis*. Phytochemistry 67:464
139. Kapche GDWF, Fozing CD, Donfack JH, Fotso GW, Amadou D, Tchana AN, Bezabih M, Moundipa PF, Ngadjui BT, Abegaz BM (2009) Prenylated arylbenzofuran derivatives from *Morus mesozygia* with antioxidant activity. Phytochemistry 70:216
140. Fozing CDA, Ali Z, Ngadjui BT, Choudhary MI, Kapche GDWF, Abegaz BM, Khan IA (2012) Phosphodiesterase I-inhibiting Diels-Alder adducts from the leaves of *Morus mesozygia*. Planta Med 78:154

Wolfgang Robien has been Associate Professor at the Institute for Organic Chemistry at the University of Vienna, Austria since 1997. He obtained a Master's degree and a Ph. D. degree in Chemistry from the University of Vienna. He worked as a postdoctoral Research Associate at Stanford University, CA, USA, in the field of peptide and protein NMR spectroscopy (1985). Dr. Robien has co-authored some 100 peer-reviewed publications and several book chapters. His research interests cover the structure elucidation of natural products and is focused presently on algorithm development in the field of computer-assisted structure elucidation, with the technologies developed within his CSEARCH project having been implemented into a large variety of non-commercial and commercial computer programs for processing NMR data.

MIX
Papier aus verantwortungsvollen Quellen
Paper from responsible sources
FSC® C105338

If you have any concerns about our products,
you can contact us on
ProductSafety@springernature.com

In case Publisher is established outside the EU,
the EU authorized representative is:
Springer Nature Customer Service Center GmbH
Europaplatz 3, 69115 Heidelberg, Germany

Printed by Libri Plureos GmbH
in Hamburg, Germany